About IFPRI

The International Food Policy Research Institute (IFPRI), established in 1975, provides research-based policy solutions to sustainably reduce poverty and end hunger and malnutrition. The Institute conducts research, communicates results, optimizes partnerships, and builds capacity to ensure sustainable food production, promote healthy food systems, improve markets and trade, transform agriculture, build resilience, and strengthen institutions and governance. Gender is considered in all of the Institute's work. IFPRI collaborates with partners around the world, including development implementers, public institutions, the private sector, and farmers' organizations.

About IFPRI's Peer Review Process

IFPRI books are policy-relevant publications based on original and innovative research conducted at IFPRI. All manuscripts submitted for publication as IFPRI books undergo an extensive review procedure that is managed by IFPRI's Publications Review Committee (PRC). Upon submission to the PRC, the manuscript is reviewed by a PRC member. Once the manuscript is considered ready for external review, the PRC submits it to at least two external reviewers who are chosen for their familiarity with the subject matter and the country setting. Upon receipt of these blind external peer reviews, the PRC provides the author with an editorial decision and, when necessary, instructions for revision based on the external reviews. The PRC reassesses the revised manuscript and makes a recommendation regarding publication to the director general of IFPRI. With the director general's approval, the manuscript enters the editorial and production phase to become an IFPRI book.

Agricultural Productivity in Africa

Trends, Patterns, and Determinants

Edited by Samuel Benin

A Peer-reviewed Publication

International Food Policy Research Institute
Washington, DC

International Food Policy Research Institute
2033 K Street, NW
Washington, DC 20006-1002, USA
Telephone: 202-862-5600

DOI: http://dx.doi.org/10.2499/9780896298811

Library of Congress
Cataloging in Publication Program
101 Independence Avenue, S.E.
Washington, DC 20540-4283
Library of Congress Cataloging-in-Publication Data
Names: Benin, S. (Samuel), editor.
Title: Agricultural productivity in Africa : trends, patterns, and determinants / edited by Samuel Benin.
Description: Washington, DC : International Food Policy Research Institute, [2016] | Includes bibliographical references and index.
Identifiers: LCCN 2015041365 | ISBN 9780896298811 (pbk. : alk. paper)
Subjects: LCSH: Agricultural productivity--Africa.
Classification: LCC S472.A1 A346 2016 | DDC 338.1/6096--dc23 LC record available at http://lccn.loc.gov/2015041365
Cover Design: Anne C. Kerns, Anne Likes Red
Project Manager: Patricia Fowlkes, IFPRI
Book Layout: David Peattie, BookMatters

Contents

Tables, Figures, and Boxes

Tables

Figures

Boxes

ABBREVIATIONS AND ACRONYMS

AEZ	agroecological zone
AFDB	African Development Bank
AFSI	L'Aquila Food Security Initiative
AHSRP	Animal Health Services Rehabilitation Programme
AI	artificial insemination
APEP	Agriculture Productivity Enhancement Programme
APZs	Agricultural productivity zones
ASARECA	Association for Strengthening Agricultural Research in Eastern and Central Africa
ASDP	Agricultural Sector Development Programme
ASDS	Agricultural Sector Development Strategy
AU	African Union
AU-NEPAD	African Union–New Partnership for Africa's Development
BMGF	Bill & Melinda Gates Foundation
BXW	banana *Xanthomonas* wilt
C3P	Crop Crisis Control Project
CA	conservation agriculture
CAADP	Comprehensive Africa Agriculture Development Programme
CAP1	Conservation Agriculture Project 1
CBOs	community-based organizations
CDC	Commonwealth Development Corporation

CEDP	Cassava Enterprise Development Project
CEN-SAD	Community of Sahel–Saharan States
CFU	Conservation Farming Unit
CMD	cassava mosaic disease
COMESA	Common Market for Eastern and Southern Africa
CPA	stock of land suitable for crop production
CRS	constant returns to scale
CSPR	Civil Society for Poverty Reduction
DANIDA	Danish International Development Agency
DEA	data envelopment analysis
DVS	Department of Veterinary Services
EAC	East African Community
EADD	East Africa Dairy Development Project
ECA	eastern and central Africa
ECCAS	Economic Community of Central African States
ECOWAS	Economic Community of West African States
EIA	environmental impact assessment
EPRC	Economic Policy Research Centre
ERPs	economic recovery programs
FADGIP	FARM Africa Dairy Goat Improvement Project
FAO	Food and Agriculture Organization of the United Nations
FFs	farmer field schools
FISBP	Farm Input Subsidy Program
FISPP	Farmer Input Support Program
FPIS	Fuve Panganai Irrigation Scheme
FTSP	Fodder Trees and Shrubs Project
G1–G4	Groups 1 through 4
G8	Group of Eight
G20	Group of Twenty
GAFSP	Global Agriculture and Food Security Program
GDP	gross domestic product

GIS	geographic information system
GRUMP	Global Rural Urban Mapping Project
ha	hectare
I$	international dollars
ICIPE	International Centre of Insect Physiology and Ecology
ICRAF	World Agroforestry Centre
IDA	International Development Association
IFAD	International Fund for Agricultural Development
IFPRI	International Food Policy Research Institute
IGAD	Intergovernmental Authority on Development
IITA	International Institute of Tropical Agriculture
ILO	International Labor Organization
ILRI	International Livestock Research Institute
IMF	International Monetary Fund
IPCC	Intergovernmental Panel on Climate Change
ISO	International Organization for Standardization
ISPs	input subsidy programs
IWMI	International Water Management Institute
JICA	Japan International Cooperation Agency
KARI	Kenya Agricultural Research Institute
KASCOL	Kaleya Smallholders Company Limited
KASFA	Kaleya Smallholder Farmers' Association
KDDP	Kenya Dairy Development Programme
kg	kilogram
kg/ha	kilogram per hectare
KIP	Kaleya Irrigation Project
KIPPRA	Kenya Institute for Public Policy Research and Analysis
km	kilometer
km^2	square kilometer
km/hr	kilometers per hour
LP	linear programming

m	meter
m^2	square meters
M&E	monitoring and evaluation
MAFC	Ministry of Agriculture, Food Security and Cooperatives
MAUP	modifiable areal unit problem
MCE	multicriteria evaluation
MDTF	Multi-Donor Trust Fund
Mha	million hectares
MI	middle income
MM	metafrontier Malmquist index
MSE	mean square error
NAADS	National Agricultural Advisory Services
NAEIP	National Agricultural Extension Intervention Program
NAFSIPs	National Agricultural and Food Security Investment Plans
NAIPs	national agricultural investment plans
NAIVS	National Input Voucher System
NARS	National Agricultural Research System
NDVI	normalized difference vegetation index
NEPAD	New Partnership for Africa's Development
NERICA	New Rice for Africa upland rice
NFSD	Novartis Foundation for Sustainable Development
NGO	nongovernment organization
O&M	operation and maintenance
OECD	Organisation for Economic Co-operation and Development
OMO	Operation *Mwolyo* Out
OPEC	Organization of Petroleum Exporting Countries
PADETES	Participatory Demonstration and Training System
PAPSTA	Support Project for the Strategic Plan for the Transformation of Agriculture
PCU	Projects Coordinating Unit
PFP	partial factor productivity

PIDP	Participatory Irrigation Development Programme
PMSU	Project Management Support Unit
PPS	production possibility set
PPT	Push–Pull Technology
PRSPs	Poverty Reduction Strategy Papers
R&D	research and development
RECs	Regional Economic Communities
RELMA	Regional Land Management Unit
RESAKSS	Regional Strategic Analysis and Knowledge Support System
SADC	Southern African Development Community
SAP	structural adjustment program
SAPRIN	Structural Adjustment Participatory Review International Network
SCP	Specialty Coffee Program
SG 2000-AP	Sasakawa Global 2000 Agricultural Program
SOAS	School of Oriental and African Studies
SOFA	The State of Food and Agriculture
SPAM	Spatial Production Allocation Model
SRI	System of Rice Intensification
SSA	Africa south of the Sahara
TFP	total factor productivity
TGR	technology gap ratio
TI	tropicality index
TPA	total potential agricultural area
UMA	Union du Maghreb Arabe
UNDP	United Nations Development Programme
UNFFE	Uganda National Farmers Federation
URT	United Republic of Tanzania
USAID	United States Agency for International Development
VRS	variable returns to scale
WARDA	West Africa Rice Development Association

WHO	World Health Organization
WSS	within sum of squares
WUA	water use association
WWIDP	Wei Wei Integrated Development Project
ZSC	Zambia Sugar Company

FOREWORD

Agricultural Productivity in Africa: Trends, Patterns, and Determinants presents updated and new analyses of land, labor, and total productivity trends in African agriculture. It brings together analyses of a unique mix of data sources and evaluations of public policies and development projects to recommend ways to increase agricultural productivity in Africa. This book is timely in light of the recent and ongoing growth recovery across the continent.

The good news is that agricultural productivity in Africa increased at a moderate rate between 1961 and 2012, although there are variations in the rate of growth in land, labor, and total factor productivities depending on country and region. Differences in input use and capital intensities in agricultural production in the various farming systems and agricultural productivity zones also affect advancements in technology. One conclusion based on the book's research findings derives from the substantial spatial variation in agricultural productivity. For areas with similar agricultural productivity growth trends and factors, what works well in one area can be used as the basis for formulating best-fit, location-specific agricultural policies, investments, and interventions in similar areas. This finding along with others will be of particular interest to policy- and decisionmakers.

By asking and answering pointed questions, *Agricultural Productivity in Africa* offers succinct recommendations for specific situations as well as broad development objectives. How can Africa further raise labor productivity to reduce mass poverty? Can increasing land productivity (yields) make a difference in averting future food crisis? How does Africa effectively take full advantage of regional and subregional alliances that promote and disseminate appropriate technologies capable of reversing the declining growth in land productivity, sustain or strengthen the recent rapid growth in labor productivity, and expand the technological frontier on an ongoing basis?

The authors have successfully framed their research questions, mapping out the body of evidence they present which has resulted in an informative book that will assist researchers in understanding the various ways agricultural productivity can increase and help policymakers and those in decision-making positions determine what options are best for their country, subregion, and region.

Shenggen Fan
Director General

ACKNOWLEDGMENTS

Funding was provided by the United States Agency for International Development and the Bill & Melinda Gates Foundation, through their support to the Regional Strategic Analysis and Knowledge Support System program in Africa. Several people have contributed to the production of this book, including Greg Traxler and Ousmane Badiane, whose assistance during conceptualization of the agricultural productivity study was invaluable. Melanie Bacou, Angga Pradesha, Linden McBride, and Heather Wyllie provided data and analytical support. Additional assistance was provided by the International Food Policy and Research Institute's (IFPRI's) Publications Unit of the Communications and Knowledge Management Division, especially Patricia Fowlkes and Andrea Pedolsky. We would also like to thank participants of IFPRI's conference on "Increasing Agricultural Productivity & Enhancing Food Security in Africa: New Challenges and Opportunities," held on November 1–3, 2011, in Addis Ababa, Ethiopia, for their feedback on a presentation of this study. We are especially indebted to the anonymous reviewers for providing insightful comments and directions for additional work.

Chapter 1

INTRODUCTION

Samuel Benin, Stanley Wood, and Alejandro Nin-Pratt

Fostering higher agricultural productivity and accelerating agricultural growth in Africa are commonly seen as core strategies for overall development in the continent (Lewis 1954; Fei and Ranis 1961; Hayami and Ruttan 1985; Hazell and Haggblade 1991; Binswanger and Townsend 2000; World Bank 2007).[1] Because the majority of Africa's poor and malnourished population depends largely on farming, these strategies can be particularly effective in reducing poverty and hunger. Yet, agricultural growth in Africa lags behind overall economic growth, and the continent's agricultural performance has fallen further behind that of other developing regions of the world.

The development literature offers many hypotheses to help explain the chronic underperformance of Africa's agriculture sector. One particularly fundamental perception by those making critical policy and investment decisions is the ambiguity of agriculture's role in development. Additionally, the quality and relevance of data and analysis provided to those individuals to allow them to measure potential costs and benefits, consider trade-offs, and make informed decisions are questioned. The recent high global food prices of 2007–2008 and later periods, which gave rise to food crises in many African countries and drew varied and often productivity-reducing responses from several governments across the continent (Benson, Mugarura, and Wanda 2008; Wodon and Zaman 2010; Headey et al. 2012; Benson et al. 2013), have renewed concern about knowledge gaps surrounding appropriate strategies for raising and maintaining higher levels of agricultural productivity.

1 Diao et al. (2007) provides a solid review of the literature on the role of agriculture in development, spanning the classical thinking of a passive role where agriculture serves as a reserve of labor and capital, to one where agriculture plays an active role through production and consumption linkages, including its role in rural, as opposed to national, development because of spatially differentiated constraints in production and market linkages. They also review more recent discourse about the agriculture–nutrition nexus, agriculture's role in stabilizing food prices and ensuring food security, and the unique decisionmaking processes associated with managing the sector.

This book raises explicit questions for policy analysts in African countries and development agencies who advise policymakers on strategies to accelerate productivity growth and presents new and updated analyses of agricultural productivity trends for African countries and subregions. These analyses offer greater economic and spatially disaggregated insights than is typical for studies encompassing all of Africa, and suggest some critical conclusions for the viability of a rapid acceleration of agricultural productivity and value addition in Africa.

To fully contextualize the book, the remainder of this introductory chapter examines the competing hypotheses for Africa's poor agricultural performance from a historical perspective, beginning in the colonial era, through the structural adjustment periods, to the current crop of agricultural development strategies guiding the continent under the auspices of the Comprehensive Africa Agriculture Development Programme (CAADP). Following an assessment of the challenges faced in implementing these strategies, the chapter concludes with a summary of the organization of the remainder of the book.

History of African Agriculture and Hypotheses Regarding Its Poor Performance

A starting point for the contextualization of this book begins with a historical overview that, by necessity, offers stylized facts about Africa's development. These facts also bring into sharp focus the sweeping generalizations made about Africa that effectively led to some simplistic approaches to agricultural development that lacked an understanding of the continent's diversity and variation, presaging the critical necessity of higher-resolution data and analysis that are the later focus of this book.

During the colonial period in Africa, agriculture was the most important economic activity. Farmers were required or incentivized by many colonial administrations to grow cash crops for export, primarily to provide raw materials for industrial production in the metropolitan countries (Anthony et al. 1979). The dominant cash crops for export included cocoa, coffee, tea, palm oil, and rubber in the rainforest areas of central and West Africa; groundnuts and cotton in the Sahel belt of West Africa; sisal, tea, and coffee in East Africa; and sisal, sugarcane, and tobacco in southern Africa. In general, food crops were not promoted, and farmers grew them for subsistence only. Colonial administrations invested heavily in transportation systems to facilitate the movement of cash crops from the interior to the coastal ports, as well as the flow of manufactured goods imported from the metropolitan countries into the interior. To bolster their aims, administrations also invested in farm support, research, extension, and marketing infrastructure directed to those commodities.

Also during the colonial period, Africa was developed essentially as an agricultural-exporting economy. This goal was achieved with some success, as evidenced by the number of African countries being top global producers of tropical cash crops.[2] This orientation of agricultural production toward exports of primary products persisted during the 1960s, the era of Africa's independence from colonial rule, except now the export revenues and development assistance in many countries were concentrated on financing ambitious domestic manufacturing activities under import substitution industrialization strategies and on developing the urban sector (Lawrence 2005). This was consistent with the "dual-economy" models of development, which viewed agriculture as a low-productivity supplier of food, raw materials, and surplus labor to a modern and more urbanized industrialization process (Adelman 2001). As such, there was underinvestment in agriculture and in the rural sector (Fan 2008). Investments in agriculture were concentrated on input subsidies; government-provided services (marketing, infrastructure, extension, research); and the establishment of input and commodity marketing parastatals to promote the export crops of the colonial era, which now provided African governments with their major source of foreign exchange (along with minerals in some countries).

To ensure low food prices in the urban areas, food price controls and government-run estate farms and food marketing and distribution cooperatives (which consumed the bulk of subsidies on farm inputs and machinery) were established. However, the import substitution manufacturing strategy was unsustainable for a variety of reasons. Protectionist policies employed by countries within and outside the continent constrained demand for manufactured goods to the size of the domestic market, which is small for many African countries. Groups of countries tried to overcome this constraint through customs unions. The East African Community, for example, had agreements concerning the location of specific manufacturing plants, so that

2 In the 1960s, the earliest periods when data were available, the highest-ranked African country and the total number of African countries in the top 20 agricultural producers in the world were listed as follows: cocoa beans (Ghana was ranked number 1 in the world, with a total of 10 African countries in the top 20); green coffee (Côte d'Ivoire 3, total 8); unshelled groundnuts (Nigeria 3, total 12); palm oil (Nigeria 1, total 12); rubber (Liberia 6, total 6); sisal (Tanzania 2, total 10); tea (Kenya 7, total 6); tobacco (Zimbabwe 20, total 1); and cassava (Democratic Republic of the Congo 3, total 12) (FAO 2014).

production was not duplicated across the community; however, these agreements were not always adhered to. Furthermore, the factories were highly dependent on expensive imported capital and expatriate labor for producing mostly basic consumption goods (such as food processing, textiles and clothing, and shoes) and processing primary products for exports, except in a few cases where intermediate goods were produced, such as fertilizer in Tanzania (Lawrence 2005). Neglect of smallholder farmers who produced the bulk of the food crops resulted in diminishing food production and rising food prices.

These developments—in addition to leadership problems, economic mismanagement, and corruption on the one hand and political turmoil and internal conflicts on the other—which many African countries experienced in the 1970s and 1980s, characterized the complex development issues in the continent at the time. The oil and drought shocks of the 1970s complicated the issues further. In general, the 1970s and 1980s are often associated with the beginning of the chronically poor performance of African agriculture. Between 1971 and 1980, for example, agricultural output in Africa south of the Sahara grew by only 1 percent per year on average, compared with 3 percent in Asia and other developing regions of the world, and land productivity (output per unit area) was about two to three times lower (Table 1.1; Fuglie and Nin-Pratt 2013).

The 1980s and 1990s ushered in the structural adjustment programs (SAPs) and the economic recovery programs (ERPs) of the International Monetary Fund (IMF) and the World Bank. The programs constituted conditions for receiving new loans or international development assistance, and involved cutting government expenditures, dismantling the parastatals, ending commodity and input subsidies, removing price controls, devaluing currencies, and stimulating private-sector investments to occupy the spaces left by the government-run agencies. While the overall impacts of the SAPs and ERPs are still debated, the prevailing view is negative, especially with regard to their impact on poverty (for example, Killick 1995; SAPRIN 2004; Easterly 2005).[3] Because SAPs promoted economic output based on direct export and resource extraction, they also exacerbated the lack of attention on the rural sector, smallholder farmers, and food crops. For example, because devaluation makes local goods cheaper for foreigners to buy and foreign goods more expensive to

3 There is vast scholarly literature on the SAPs and ERPs. Killick (1995) provides a good review of the literature and highlights the difficulties in making generalizations about the effects of SAPs and ERPs because of data and methodological problems in dealing with the complex and varied instruments employed in the SAPs and ERPs on the one hand, and the many different, but connected, outcomes on the other.

TABLE 1.1 Annual average agricultural growth, productivity, and public spending in Africa and other selected developing regions of the world, 1970–2010

Indicator and region	Years and values			
Agricultural output growth rate (%)[1]	1971–1980	1981–1990	1991–2000	2001–2010
Africa south of the Sahara	1.0	2.7	3.1	2.6
Asia[a]	3.0	4.1	4.0	3.5
Latin America and the Caribbean	2.9	2.4	3.1	3.2
Agricultural output per hectare of land (constant 2004–2006 US$)[1]	1980	1990	2000	2009
Africa south of the Sahara	163	182	192	219
Asia[a]	494	607	704	773
Latin America and the Caribbean	326	368	394	424
Government agriculture expenditure (% of total expenditure)[2]		1981–1990	1991–2000	2001–2010
Africa south of the Sahara		7.1	3.3	3.1
Asia[b]		7.2	5.0	5.5
Latin America and the Caribbean		3.6	3.2	2.0
Government agriculture expenditure (% of agriculture value-added)[2]		1981–1990	1991–2000	2001–2010
Africa south of the Sahara		4.9	3.0	3.9
Asia[b]		3.7	3.0	4.6
Latin America and the Caribbean		7.2	7.5	7.4
Agriculture R&D in Africa south of the Sahara[3]	1971–1980	1981–1990	1991–2000	2001–2008
Growth rate in expenditure (%)	1.7	0.6	1.0	2.4
Growth rate in full-time-equivalent staff (%)	5.4	3.8	1.3	2.8

Source: Authors' calculations based on [1] Fuglie and Nin-Pratt (2013), [2] IFPRI (2014a), and [3] Beintema and Stads (2011).

Notes: [a] Made up of Northeast, South, and Southeast Asia. [b] South Asia. R&D = research and development.

import, it provides incentives for SAP-implementing countries to export more and import less in the long run. However, by simultaneously devaluing the currency and removing subsidies, the immediate effect of structural adjustment was to raise the prices of agricultural inputs, especially those of yield-enhancing technologies, such as fertilizers, pesticides, and machinery, which are typically imported. The consequences were higher farm production costs, low adoption of high-yielding technologies, low agricultural productivity, and low incomes to smallholder farmers. Furthermore, private-sector investments did not materialize as expected, and new problems related to market failures surfaced (Dorward, Kidd, and Poulton 1998; Kherallah et al. 2002).

The austerity measures imposed by the SAPs led to a drastic reduction in government spending on agriculture in general (IFPRI 2014a), and an erosion of critical agricultural investments in national research and extension systems in particular (Beintema and Stads 2011). For example, in Africa south of the Sahara, the share of government agriculture expenditure declined from an average of 7.4 percent per year of the total budget in the 1980s to 3.3 percent in the 1990s, whereas the growth rate in the amount spent on agriculture research and development (R&D) declined from an annual average of 1.7 percent in the 1970s to 0.6 percent in the 1980s and 1.0 percent in the 1990s (Table 1.1). Therefore, although growth in African agriculture was higher in the 1980s and 1990s than in the 1970s—thanks largely to area expansion, rather than to the adoption of yield-enhancing technologies—agricultural productivity remained very low compared with levels achieved in other developing regions of the world, especially in Asia, where the Green Revolution was taking root. In 1980 and 1990, for example, agricultural output per hectare of land in Africa south of the Sahara was $163 and $180, respectively—about one-third of the values achieved in Asia (Fuglie and Nin-Pratt 2013; Table 1.1).[4]

It is important to remember that the Green Revolution in Asia occurred before the SAPs were established in Africa. After starting in Mexico, the Green Revolution quickly spread to Asia, where it is widely acknowledged to have doubled both output and yields of key food staples—rice and wheat—in just 20 years. These successes helped promote a broader reassessment of agriculture's role in Africa's development, which we will return to shortly.

The start of the new millennium introduced a greater emphasis on a more comprehensive approach to poverty reduction, in which agriculture was called on to play a more significant role. National strategies were formalized into Poverty Reduction Strategy Papers (PRSPs), required by the IMF and the World Bank for countries requiring debt relief and seeking new development assistance. While PRSPs have been described by some as simply an extension of SAPs (e.g., SAPRIN 2004), they are based—in theory if not always in practice—on a more broadly based articulation of development, including the need for poverty-focused growth, participatory processes in strategic planning, public–private partnerships, and other principles that are expected to ensure that the benefits of growth are distributed to all members of society.

Although the impact of the PRSPs in Africa is still being debated, their primary focus on poverty—a particularly prevalent phenomenon in rural

4 All currency is in US dollars, unless specifically noted as "international dollars."

areas—suggests that proper implementation of such plans should increasingly favor agricultural and rural development. For example, the use of agricultural input and farm support subsidies, which was discouraged under the SAPs, has returned strongly, particularly following the recent high food and input prices crisis. This is consistent with several studies prior to the start of the PRSPs, which recommended that the World Bank and IMF revisit their position on input subsidies by considering their merits in the broader context of agricultural intensification, in addition to their macroeconomic feasibility (for example, Lele, Christiansen, and Kadiresan 1989; Reardon et al. 1999; World Bank 1994).

The successes of the Green Revolution in Asia also helped promote this movement, although this is not apparent in the PRSPs. For example, whereas most of the PRSPs state in various ways that raising agricultural output and productivity will be accomplished by promoting and supporting the use of yield-enhancing technologies and modern management practices, as done during the Green Revolution in Asia, only a few country PRSPs made direct reference to employing lessons or technologies from India (for example, Ghana 2003; IMF 2006), while Madagascar's PRSP made explicit reference to creating a Green Revolution there (Madagascar 2007). These PRSPs have promoted greater adoption of yield-enhancing technologies and modern management practices and helped return agricultural productivity to levels achieved prior to the decline in the 1970s—although still much lower than levels achieved in other developing regions of the world (Table 1.1).

Against this historical narrative, the literature examining the poor performance of African agriculture has largely formulated hypotheses based on partial analyses, local contexts, and particular points in time. Associated findings and recommendations, therefore, often fall short of addressing the fundamental issues in their entirety. For example, it is reasonable to assume that Africa needs a movement similar to Asia's Green Revolution. What we now know about that brief period in history is that it involved more than just high-yielding, semi-dwarf rice and wheat varieties. It also included investments in irrigation infrastructure, modernization of farm management techniques, supportive public policies, a strong geopolitical undercurrent, and a clear smallholder focus tied to its geopolitical motivation (Djurdfeldt et al. 2005), and had significant environmental consequences (for example, Shiva 1991). Nevertheless, the Green Revolution offers one—and only one, possibly irreplicable—model for the intensification and modernization of agriculture in Africa.

Why may Asia's Green Revolution not be replicable in Africa? This question derives from some of the arguments that have been advanced for the poor performance of African agriculture, which also are consistent with different parts of the historical narrative or with different geographical contexts of the continent, including agroecological complexities and heterogeneity that make it difficult to exploit intercontinental technology spillovers (for example, Pardey et al. 2007); poor economic policies and, in particular, lack of openness to international markets or access to ports (for example, Sachs and Warner 1997); and the low productivity and high cost of labor (Karshenas 2001; Collier and Dercon 2009; Woodhouse 2009).

Regarding the agroecological complexities and technology spillover constraints, for example, many countries in Africa have small economies and limited capacities and resources for adopting or adapting technologies that fit their own national interests and needs. Thus, although regional agricultural R&D systems can help fill these gaps and facilitate economies of scale,[5] high transaction costs associated with political, institutional, and administrative barriers can rapidly erode the potential gains—gains that can differ substantially by commodity and by the degree of agroecological similarity between technology source and the target areas for technology application. The agroecological complexities mitigating the replicability of Asia's Green Revolution in Africa are further complicated by climate change and global warming, a topic that, because of its recent emergence and its potential substantial effects on potential growth and development pathways, is logically absent in the historical narrative of the performance of African agriculture. We will address the implications of this topic for future development strategies after we have examined the current African agricultural development strategy.

The labor constraint arguments for the poor performance of African agriculture are representative of the issues related to the supply of key factors of production, including capital. All of these factors require an economywide, rather than a sectoral, approach to development because of the strong forward and backward linkages between the agriculture and nonagriculture sectors (Diao et al. 2012). For labor specifically, the possibility of increasing labor productivity depends on the availability of appropriate labor-saving technologies in agriculture, having profitable exit options out of agriculture into other sectors of the economy, and having high-quality labor to be able to earn a higher wage or return to labor in the other sectors.

---

5 See Omamo et al. (2006), Nin-Pratt et al. (2011), and Johnson et al. (2014) on the potential gains from implementing such regional agricultural R&D strategies in different subregions of the continent.

Current African Agricultural Development Strategy

In July 2003, African heads of state at the Second Ordinary Session of the Assembly of the African Union launched CAADP in Maputo, Mozambique. This agriculture-led integrated framework of development priorities in Africa is aimed at reducing poverty and increasing food security in the continent (AU-NEPAD 2003). The program shares many of the principles articulated in the PRSPs, including poverty-focused growth, participatory processes in strategic planning and implementation, country ownership, public–private partnership, and mutual accountability, to ensure that the benefits of growth are equitably distributed to all members of society.

The main difference between CAADP and preceding development strategies in Africa is that it emphasizes the role of agriculture as the engine of economic growth and development in its compact-signing countries. Furthermore, CAADP deliberately categorizes investment into four mutually reinforcing pillars (land and water management, market access, food security, and agricultural R&D) and cross-cutting enabling factors, including institutional capacity strengthening. It also prescribes specific policies and programs to be implemented, in addition to specific targets to be achieved. CAADP has two overarching targets: (1) achieving an annual average agricultural growth rate of 6 percent, and (2) spending 10 percent of the national budget on agriculture—popularly known as the Maputo Declaration (AU 2003). Various processes at national, regional, and continental levels have been put in place to ensure evidence-based planning, to facilitate implementation of CAADP according to the declared principles, to monitor and evaluate progress, and to promote mutual learning (AU-NEPAD 2014c).

The impact of CAADP on agricultural and economic growth, poverty, and food and nutrition security is yet to be assessed. However, a little more than a decade since its launch in 2003, CAADP can point to several achievements. For example, CAADP has significantly raised the political profile of agriculture; has contributed to more specific, purposeful, and incentive-orientated agricultural policies; and has promoted greater participation of multiple state and nonstate actors in agricultural policy dialogue and strategy development (AU-NEPAD 2010). Some of the specific tools, mechanisms, and processes that have contributed to these achievements include the annual CAADP Partnership Platform and Business meetings since 2006 that bring the different stakeholders at different levels together to review progress and make plans for the future (AU-NEPAD 2014a); preparation of the four pillar framework documents to guide adaptation of the CAADP principles and targets into national and regional policymaking (AU-NEPAD 2010); establishment of the knowledge systems to provide analyses that track progress, document success, and derive lessons for the implementation of the CAADP agenda (IFPRI 2014b); development of a monitoring and evaluation (M&E) framework (Benin, Johnson, and Omilola 2010) and a mutual accountability framework (Oruko et al. 2011); and establishment of the CAADP Multi-Donor Trust Fund (MDTF) to finance the CAADP processes at all levels (AU-NEPAD 2010). By the end of 2014, 40 African countries had signed their CAADP compacts with their main stakeholder groups, and many of them had developed detailed country investment plans (or National Agricultural Investment Plans [NAIPs] or National Agricultural and Food Security Investment Plans [NAFSIPs]). Furthermore, a majority of the strategies and plans are based on economywide analysis in order to identify coherent growth options and quantify the aggregate public agricultural resources required to support different growth paths (for example, Diao et al. 2012).

Despite these and other achievements that can be attributed to CAADP, several challenges have arisen. First is assessing the impact of CAADP, where the major issue involves attributing change in the outcome indicators to CAADP. This assessment is difficult because many governments and countries were already engaged in policy reforms in harmony with the CAADP principles, and much of the CAADP framework was derived from earlier strategies and successful agricultural reforms in those African countries. The issue, therefore, will be how to isolate CAADP's specific contributions.

A second challenge is the delayed response in adapting a continental-level agenda and commitments to fit regional- and national-level priorities or vice versa. For example, when CAADP was launched in 2003, the heads of state set a five-year timeline for implementation (see AU 2003, Declaration 7(II).2). By 2008, however, only Rwanda had a signed CAADP Compact to demonstrate a concrete implementation progress (AU-NEPAD 2014b). These delays reflected inherent political, institutional, and administrative barriers across national boundaries. Therefore, the heads of state renewed their commitment through a resolution at the 13th Ordinary Session of the Assembly of the African Union (AU) in Sirte, Libya, in July 2009 by requesting

> the AU Commission, the NEPAD [New Partnership for Africa's Development] Secretariat and the RECs [Regional Economic Communities] to continue to mobilize the necessary technical expertise and financial resources to support capacity development and

> related policy reforms to accelerate CAADP implementation in all Member States, including the signing of country CAADP Compacts indicating the policy measures, investment programs, and required funding to achieve the six percent (6 percent) growth and ten percent (10 percent) budget share targets for the agricultural sector by 2015. (AU 2009, Declaration 2(XIII).5)

Consequently, 12 more countries and the Economic Community of West African States signed their compacts in 2009, an acceleration spurred by the establishment of the MDTF to finance CAADP processes in 2008 and the establishment of the Global Agriculture and Food Security Program (GAFSP) in 2009 to assist in securing Group of Twenty (G20) pledges that would support the financing of NAIPs.

Other challenges faced by CAADP related to achieving the 10 percent budget allocation and the 6 percent growth rate targets and to completing development of the NAIPs. With regard to progress made toward the 10 percent budget allocation target for agriculture, Table 1.1 shows that Africa south of the Sahara managed to reach only 3.1 percent on average between 2001 and 2010 (IFPRI 2014a). Since 2003, only 13 countries in all of Africa have managed to surpass the target in any year (Benin and Yu 2013). NEPAD also has set a national agricultural R&D investment target of at least 1 percent of agricultural value-added, which only a few countries have been able to achieve so far (Figure 1.1)—especially Botswana, Mauritius, Namibia, and South Africa, all of which have relatively well-established and well-funded agricultural research systems and relatively small contributions of agriculture to gross domestic product (Beintema and Stads 2011). Regarding progress toward achieving the 6 percent agricultural growth target, Table 1.1 shows that Africa south of the Sahara managed to reach only 2.6 percent on average between 2001 and 2010 (Fuglie and Nin-Pratt 2013). Between 2003 and 2009, for example, only six countries—Angola, Ethiopia, Guinea, Mozambique, Nigeria, and Rwanda—met or surpassed the target (Benin et al. 2011).

Looking now at the plans for the future, results of the economic modeling used in CAADP planning indicate that although it is possible for many African countries to reach the 6 percent annual average agricultural growth rate target, it will require substantial additional growth across different key subsectors and commodities. This in turn will require substantial additional investments to stimulate the necessary acceleration in growth in the key subsectors (for example, Diao et al. 2012). In many cases, the additional

FIGURE 1.1 Public expenditure on agricultural research and development in selected African countries, 1996–2008 (annual average % of agricultural value-added)

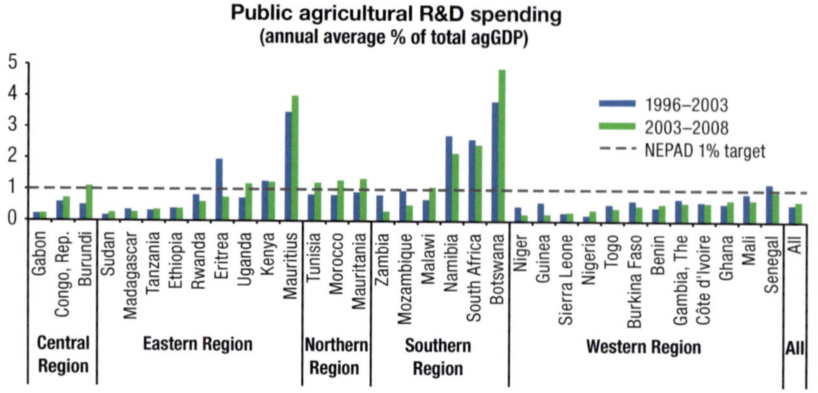

Source: Author's calculations based on IFPRI (2013).

investments required are in excess of the 10 percent of total expenditures commitment agreed upon under the Maputo Declaration. Such large demands on fiscal resources are necessary because of inadequate or no technical change in the sector (for example, Irz and Thirtle 2004; Nin-Pratt and Yu 2008). As countries enter the operational stage of CAADP investment program design and execution, a fundamental question and technical and institutional challenge is how to achieve and sustain significantly higher levels of agricultural productivity across different parts of Africa. Given the limits to boosting productivity that is achievable through area expansion, as has been experienced in many parts of Africa for long periods of time, agricultural productivity gains in the future must rely heavily on technological change.

A review of the NAIPs shows that individual countries have formulated different strategic responses to these common policy, technology, and institutional challenges. Such variation is expected, for example, since climate and natural resource endowments that condition strategic agricultural development options differ considerably among countries. Table 1.2 and Figure 1.2 show some of the clear differences in investment and development approaches among NAIPs in terms of the proportion of the total agriculture budget that is allocated to different priority areas and investments. With regard to the general approach to agricultural development, for example, Table 1.2 records budget allocations according to the overall agriculture sector goal, the four

CAADP pillars, the CAADP cross-cutting theme, and other areas. The table illustrates that achieving the agriculture sector goal of increasing agricultural productivity, growth, or income represents the dominant strategy in many of the African countries reported. However, in several of the other countries—for example, Ethiopia, The Gambia, Liberia, Malawi, Niger, and Sierra Leone—food and nutrition security (pillar 3) and natural resource management (pillar 1) are given higher priority. Pillars 2 and 4 and the cross-cutting theme were accorded lower priority in terms of the stated budget allocations. Whereas drawing conclusions from these budget shares is difficult, because different countries may invest differently to achieve different goals and objectives, the disparate results are consistent with a fundamental knowledge gap about the drivers of high levels of agricultural productivity growth across the continent.

A similar implication derives from Figure 1.2, which reports allocations to specific subprograms or functions that are known to be critical for overall agricultural productivity growth, including research, extension, irrigation, natural resource management, and farm support subsidies. Although these represent the major functions that were articulated in the NAIPs, the results in Figure 1.2 show that budgets were not necessarily allocated accordingly. The figure also highlights differences across governments and stakeholders in individual countries in terms of making explicit resource allocation commitments to such specific agricultural functions. Clearly, commitments to invest in natural resource management and providing farm support subsidies were favored or seemed easier to make in many countries in terms of attracting large shares of the agriculture budgets. These commitments were followed by investment in irrigation. Although investing in research and extension has been found to have large and long-lasting impacts on agricultural growth and other development outcomes (for example, Fan, Hazell, and Thorat 2000; Fan 2008; Mogues et al. 2012), they were stated priorities in only a handful of countries, including Benin, Burundi, Côte d'Ivoire, and Uganda.

Therefore, although we expect countries to have different strategic responses to achieving the CAADP targets and their own national objectives in ways that reflect their own national contexts that are also shaped by such noneconomic factors as political, cultural, social, historical, and linguistic factors, the contrasting results shown in Figure 1.2 would suggest that there is a knowledge gap about the drivers of high levels of agricultural productivity growth across Africa.

TABLE 1.2 **Stated budget allocation to the top three programs in selected African countries (percentage of total NAIP budget)**

African countries	Sector goal	CAADP pillar/theme					Other
		Pillar 1	Pillar 2	Pillar 3	Pillar 4	Cross-cutting	
Benin, 2010–2015	51.9	2.7	—	44.7	—	—	0.7
Burkina Faso, 2011–2015	67.9	—	17.7	—	—	11.9	2.5
Burundi, 2012–2017	55.9	—	19.0	—	—	20.1	4.9
Côte d'Ivoire, 2010–2015	41.8	—	14.9	—	—	24.3	19.0
Ethiopia, 2010–2020	3.4	57.4	—	17.1	—	—	22.1
The Gambia, 2011–2015	—	27.9	30.3	15.2	—	—	26.6
Ghana, 2011–2015	55.7	—	—	36.9	3.4	—	4.0
Kenya, 2010–2015	36.0	42.0	13.1	—	—	—	8.9
Liberia, 2011–2015	—	—	32.6	39.9	—	14.4	13.0
Malawi, 2011–2014	—	36.6	—	46.9	6.2	—	10.4
Niger, 2010–2012	—	34.4	—	—	—	12.6	53.0
Nigeria, 2011–2014	35.5	40.9	12.7	—	—	—	10.8
Rwanda, 2009–2012	77.7	—	15.1	—	—	4.9	2.3
Senegal, 2011–2015	59.4	31.0	—	—	—	—	9.6
Sierra Leone, 2010–2014	17.3	—	23.6	33.7	—	—	25.4
Tanzania, 2012–2016	71.1	13.7	—	—	—	7.8	7.4
Togo, 2010–2015	66.1	—	—	—	9.0	15.3	9.6
Uganda, 2011–2015	68.6	—	25.0	—	—	4.2	2.2

Key: Pillar 1: Natural resource management (land, water, climate, etc.);
Pillar 2: Competitiveness, market trade, and private-sector development;
Pillar 3: Food and nutrition security and emergency preparedness;
Pillar 4: Science and technology;
Cross Cutting: Enabling environment (Policies, institutions, good governance)

Source: Authors' calculations based on national agricultural investment plans (NAIPs). The plans can be viewed and downloaded at www.resakss.org and http://www.caadp.net/library-country-status-updates.php.

Notes: This table has been prepared to show allocations to the top three programs only. Because the budgets in the different NAIPs were presented in different formats, the six programs identified here try to capture allocations to the overall agriculture sector goal (productivity, growth, income); the four CAADP pillars; and the cross-cutting theme. Furthermore, because not all the NAIPs had budget allocations for these programs, the blank spaces are intentional, so as to not crowd the table. The calculations are based on the stated amounts allocated to different programs, in terms of share of total budget. A blank space means that the calculated share allocated to the related program is not in the top three programs when compared with the shares allocated to different programs. Because different NAIPs have different programs, a blank space may also indicate that the related program is not stated in that country's NAIP. Therefore, the last column, labeled "Other," collects the remaining shares outside of the top three programs, so that the total for the row or country adds up to 100 percent. — = not applicable.

FIGURE 1.2 Stated budget allocation to selected agricultural functions in selected African countries (percentage of total NAIP budget)

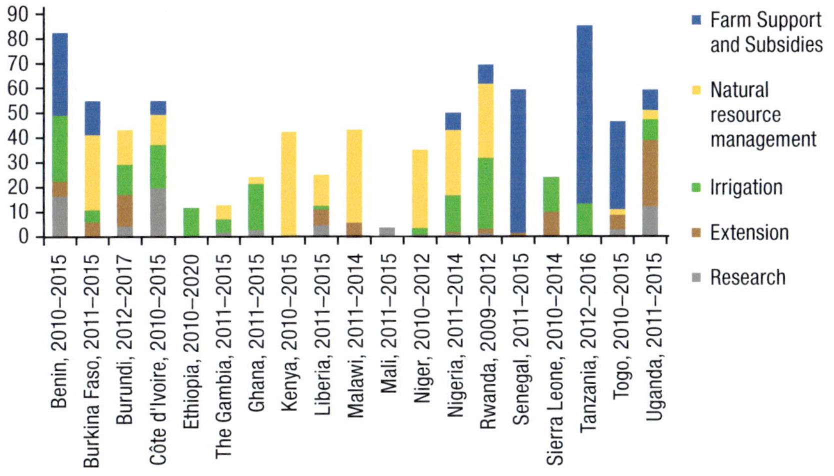

Source: Authors' calculations based on national agricultural investment plans (NAIPs). The plans can be viewed and downloaded at www.resakss.org and http://www.caadp.net/library-country-status-updates.php.

Notes: This figure has been prepared to show the stated budget allocations in the NAIPs for the five agricultural functions only, which represent the major functions articulated in the NAIPs and are consistent with the major tenets of the Green Revolution. Because the budgets in the different NAIPs were presented in different formats, not all the NAIPs had budget allocations for these agricultural functions. However, because all five functions were identified in all of the NAIPs as being important for achieving their respective development objectives, a zero share applied to any of the five functions indicates that there was no information to estimate the share of the budget for that function. As a result, the percentages do not add up to 100, because the total budget was not allocated exhaustively to the five functions.

Another challenge faced by CAADP is following through with the various and increasing number of policies, initiatives, and principles, particularly with those emerging since its initial launch in 2003. For example, the Global Agriculture and Food Security Program (GAFSP 2014), the L'Aquila Food Security Initiative (AFSI, G8 2009), and the New Alliance for Food Security and Nutrition (G8 2012), which have emerged and are expected to complement CAADP, require additional funding-eligibility processes. Therefore, although these new initiatives clearly state that CAADP compliance is an essential prerequisite for securing potential country support, the total amount of additional resources they provide relative to the status quo is often unclear (see, for example, Benin 2014 on the contribution of AFSI). Because the goal of CAADP here is to develop partnerships to meet the necessary policy, budgetary, and development assistance needs of the NAIPs, the cost of involvement in and management of multiple partnerships may not be apparent.

A different kind of challenge that the NAIPs may only now be starting to internalize is global warming and climate change, which could affect agriculture in several ways, including productivity effects in terms of the quantity and quality of outputs; husbandry effects through changes in water availability and use of yield-enhancing technologies; environmental effects, such as soil erosion, water pollution, and reduction of diversity; land use, such as through land valuation and speculation; and adaptation in response to changes in the functional characteristics of organisms and ecological systems. Several studies (Kurukulasuriya et al. 2006; IPCC 2007; Seo et al. 2008; Nelson et al. 2010) provide strong evidence that climate change caused by accumulating greenhouse gases is likely to impose serious costs on agricultural growth. Nelson et al. (2010), for example, show that the negative effect of climate change on crop yields will increase over time, whereas Seo et al. (2008) show that the impacts of climate change will vary across different agroecological zones in Africa—farms in the savanna areas are expected to be the most vulnerable to higher temperature and reduced precipitation, while those in subhumid or humid forests could gain even from severe climate change. Because of the agroecological complexities in Africa, having information on specific local regions will be critical for identifying climate-smart agricultural interventions among the numerous possibilities to increase the resilience of livelihoods and production systems and to maximize the effects of technological changes on growth and development in a sustainable and equitable manner.

Objectives and Organization of This Book

By improving understanding of the spatial and temporal patterns of a range of productivity measures assessed consistently and comparably across Africa, this book is intended to contribute to the knowledge base of how best to achieve and sustain significantly higher levels of agricultural productivity. While individual countries have taken various investment and development approaches in preparing their NAIPs, a critical question remains: Which strategies work best in which contexts, and do so cost-effectively?

In addressing this question, we base our analysis on the now rapidly expanding base of agricultural data in Africa, including geographically specific information on production system heterogeneity, quality of natural resources, population density, infrastructure, and market access. We present analyses and findings aimed at improving our understanding of the status of and trends in African agricultural productivity and its determinants and, on that basis, identifying opportunities for agricultural productivity growth that lead to more effective design and implementation of agricultural policies and strategies in Africa.

The book's unique mix of data sources, detailed in the relevant chapters, includes time-series data on agricultural production from the Food and Agriculture Organization of the United Nations, national accounts from the World Bank, and public expenditure and project M&E data from governments and multilateral agencies. We acknowledge the legitimate concerns about data reliability, as highlighted in Jerven (2013). We address these concerns by triangulating among a range of independent sources and types of data (for example, static cross-sectional and time-series data, spatial and nonspatial data), which we believe has reduced some potential data pitfalls. For example, because the spatial data used are based on observed measures of outcomes, rather than self-reported data, the measurement errors associated with capturing only the formal sector are eliminated, although measurement errors associated with the methodology used to collect or compile the data remain. The specific approaches taken to combine different data components and some of the challenges involved are described in the individual chapters.

This introductory chapter is followed by analysis of intertemporal trends (Chapter 2) and spatial analysis of different indicators and measures (Chapter 3) of agricultural productivity. Taken together, Chapters 2 and 3 provide a broad overview of the contemporary landscape of African agricultural productivity, and highlight the relevance and utility of different measures of agricultural productivity in M&E. Chapter 2 involves defining, calculating, and interpreting trends in partial and total productivity measures of agricultural productivity using time-series data. In contrast, Chapter 3 brings a more spatially explicit perspective on productivity using a harmonized collection of Africa-wide geographic information system data (some 300,000 10 x 10–kilometer grid cells), in order to explore different production systems (including rainfed and irrigated cropping systems and livestock systems) and partial productivity measures.

Building on Chapter 3, Chapter 4 uses statistical and econometric methods, particularly spatial and cluster techniques, to develop a typology of agricultural productivity zones (APZs) according to similarity in their likely pathways of technology adoption and agricultural productivity growth. Chapter 5 zooms in to examine some of the dominant APZs developed in the preceding chapter, and analyzes the status of and recent trends in patterns of intensification, as well as changes in output composition and input use associated with different intensification patterns. In particular, the chapter examines the use of fertilizer and its role in the intensification pathways followed by different subregions in Africa in recent years, and the implications of those patterns for agricultural growth and policymaking.

Chapter 6 examines case studies of agricultural investment programs and value chains in different parts of Africa that were intended for enhancing agricultural productivity. Using a qualitative and narrative approach, the chapter aims to identify what did or did not work well where and why, by distilling lessons on key factors contributing to the effectiveness of the investment programs.

Chapter 7 summarizes and synthesizes the key insights and findings provided in the preceding chapters, focusing on major challenges to and opportunities for raising African agricultural productivity, including investments in agricultural R&D, cross-border technology spillover, and institutional capacity. While the unique mix of methodologies and data used is a major strength in the book, particularly in terms of its policy relevance, it also reflects the difficulty of compiling coherent and comprehensive sets of sufficiently reliable and interoperable data. We close with a call for governments and development agencies to invest more in strengthening national data and statistical systems (especially for compiling national production and public accounts data), as well as the human capacity to use more accessible, timely, and reliable data to support policy and investment decisionmaking at all levels.

References

Adelman, I. 2001. "Fallacies in Development Theory." In *Frontiers of Development Economics*, edited by G. M. Meier, and J. E. Stigliz. Oxford, UK: Oxford University Press.

Anthony, K. R. M., B. F. Johnston, W. O. Jones, and V. C. Uchendu. 1979. *Agricultural Change in Tropical Africa*. Ithaca, NY: Cornell University Press.

AU (African Union). 2003. "Declarations." Assembly of the African Union, Second Ordinary Session, July 10–12, 2003, Maputo, Mozambique. Accessed April 15, 2014. http://www.nepad.org/system/files/Maputo%20Declaration.pdf.

———. 2009. "Decisions and Declarations." Assembly of the African Union, Thirteenth Ordinary Session, July 1–3, 2009, Sirte, Libya. Accessed April 15, 2014. www.au.int/en/sites/default/files/ASSEMBLY_EN_1_3_JULY_2009_AUC_THIRTEENTH_ORDINARY_SESSION_DECISIONS_DECLARATIONS_%20MESSAGE_CONGRATULATIONS_MOTION_0.pdf.

AU-NEPAD (African Union–New Partnership for Africa's Development). 2003. *Comprehensive Africa Agriculture Development Programme*. Midrand, South Africa.

———. 2010. The Comprehensive Africa Agriculture Development Programme (CAADP) in Practice: Highlighting the Success. Midrand, South Africa.

———. 2014a. "CAADP Partnership Platform." Accessed April 2014. http://www.nepad-caadp.net/library-meeting-documents.php.

———. 2014b. "Countries with Compacts/Investment Plans." http://www.nepad-caadp.net/pdf/Table%201%20Countries%20with%20Investment%20Plans%20ver19.pdf.

———. 2014c. "Implementing the CAADP Agenda." Accessed April 2014. http://www.nepad-caadp.net/implementing-caadp-agenda.php.

Beintema, N. M., and G. J. Stads. 2011. *African Agricultural R&D in the New Millennium: Progress for Some, Challenges for Many.* Washington, DC: International Food Policy Research Institute.

Benin, S. 2014. *Aid Effectiveness: How Is the L'Aquila Food Security Initiative (AFSI) Doing?* IFPRI Discussion Paper 1329. Washington, DC: International Food Policy Research Institute.

Benin, S., M. Johnson, and B. Omilola. 2010. *Monitoring and Evaluation (M&E) System for the Comprehensive Africa Agriculture Development Programme (CAADP).* ReSAKSS Working Paper 6. Washington, DC: International Food Policy Research Institute.

Benin, S., A. Nin-Pratt, S. Wood, and G. Guo. 2011. *Trends and Spatial Patterns in Agricultural Productivity in Africa, 1961–2010.* ReSAKSS Annual Trends and Outlook Report 2011. Washington, DC: International Food Policy Research Institute.

Benin, S., and B. Yu. 2013. *Complying with the Maputo Declaration Target: Trends in Public Agricultural Expenditures and Implications for Pursuit of Optimal Allocation of Public Agricultural Spending.* ReSAKSS Annual Trends and Outlook Report 2012. Washington, DC: International Food Policy Research Institute.

Benson, T., N. Minot, J. Pender, M. Robles, and J. von Braun. 2013. "Information to Guide Policy Responses to Higher Global Food Prices: The Data and Analyses Required." *Food Policy* 38: 47–58.

Benson, T., S. Mugarura, and K. Wanda. 2008. "Impacts in Uganda of Rising Global Food Prices: The Role of Diversified Staples and Limited Price Transmission." *Agricultural Economics* 39 (suppl. 1): 513–524.

Binswanger, H. P., and R. F. Townsend. 2000. "The Growth Performance of Agriculture in Sub-Saharan Africa." *American Journal of Agricultural Economics* 82 (5): 1075–1086.

Collier, P., and S. Dercon. 2009. "African Agriculture in 50 Years: Smallholders in a Rapidly Changing World?" Paper presented at the Organization of the United Nations Economic and Social Development Department Expert Meeting "How to Feed the World in 2050," Rome, June 24–26.

Diao, X., P. Hazell, D. Resnick, and J. Thurlow. 2007. *The Role of Agriculture in Development: Implications for Sub-Saharan Africa*. IFPRI Research Report 153. Washington, DC: International Food Policy Research Institute.

Diao, X., J. Thurlow, S. Benin, and S. Fan, eds. 2012. *Strategies and Priorities for African Agriculture: Economywide Perspectives from Country Studies*. Washington, DC: International Food Policy Research Institute.

Djurfeldt, G., H. Holmen, M. Jirstrom, and R. Larsson. 2005. *The African Food Crisis: Lessons from the Asian Green Revolution*. Cambridge, MA: CABI Publishers.

Dorward, A. R., J. Kidd, and C. Poulton, eds. 1998. *Smallholder Cash Crop Production under Market Liberalisation: A New Institutional Economics Perspective*. Wallingford, UK: CABI Publishers.

Easterly, W. 2005. "What Did Structural Adjustment Adjust? The Association of Policies and Growth with Repeated IMF and World Bank Adjustment Loans." *Journal of Development Economics* 76 (1): 1–22.

Fan, S., ed. 2008. *Public Expenditures, Growth, and Poverty: Lessons from Developing Countries*. Baltimore: Johns Hopkins University Press.

Fan, S., P. Hazell, and S. Thorat. 2000. "Government Spending, Agricultural Growth and Poverty in Rural India." *American Journal of Agricultural Economics* 82 (4): 1038–1051.

FAO (Food and Agriculture Organization of the United Nations). 2014. "FAOSTAT: Commodities by Country." Accessed April 15, 2014. http://faostat.fao.org/site/339/default .aspx.

Fei, J. C., and G. Ranis. 1961. "A Theory of Economic Development." *American Economic Review* 51 (4): 533–565.

Fuglie, K., and A. Nin-Pratt. 2013. "Agricultural Productivity: A Changing Global Harvest." In *2012 Global Food Policy Report*. Washington, DC: International Food Policy Research Institute.

G8 (Group of Eight). 2009. *L'Aquila Joint Statement on Global Food Security: L'Aquila Food Security Initiative (AFSI)*. L'Aquila, Italy. Accessed April 15, 2014. http://www.g8italia2009.it/static/G8_Allegato/LAquila_Joint_Statement_on_Global_Food_Security%5B1%5D,0.pdf.

———. 2012. "New Alliance for Food Security & Nutrition." Accessed April 15, 2014. http://www.usaid.gov/unga/new-alliance.

GAFSP (Global Agriculture and Food Security Program). 2014. "Global Agriculture and Food Security Program." Accessed April 15, 2014. http://www.gafspfund.org/content/about-gafsp.

Ghana (Government of Ghana). 2003. "Analysis and Policy Statement." In *Ghana Poverty Reduction Strategy, 2003–2005: An Agenda for Growth and Prosperity*. Vol. I. *Accra*. IMF. https://www.imf.org/external/pubs/ft/scr/2003/cr0356.pdf.

Hayami, Y., and V. Ruttan. 1985. *Agricultural Development: An International Perspective*. Baltimore: Johns Hopkins University Press.

Hazell, P., and S. Haggblade. 1991. "Rural-Urban Growth Linkages in India." *Indian Journal of Agricultural Economics* 46 (4): 512–529.

Headey, D., F. Bachewe Nisrane, I. Worku, M. Dereje, and A. Seyoum Taffesse. 2012. *Urban Wage Behavior and Food Price Inflation: The Case of Ethiopia*. Ethiopia Strategy Support Program II (ESSP II) Working Paper 41. Addis Ababa: International Food Policy Research Institute.

IFPRI (International Food Policy Research Institute). 2013. "Agricultural Science and Technology Indicators (ASTI) Database." Accessed January 2013. www.asti.cgiar.org/data/.

———. 2014a. DATASET. "Public Expenditure Database: Statistics on Public Expenditure for Economic Development (SPEED)." Accessed on March 15, 2014. http://www.ifpri.org/book–39/ourwork/programs/priorities-public-investment/speed-database.

———. 2014b. "Regional Strategic Analysis and Knowledge Support System (ReSAKSS)." Accessed April 15, 2014. http://www.resakss.org/about.

IMF (International Monetary Fund). 2006. *2nd Generation PRSP, 2007–2011*. Mali: Poverty Reduction Strategy Paper. Washington, DC.

IPCC (Intergovernmental Panel on Climate Change). 2007. *State of the Science*. Contribution of Working Group I to the Fourth Assessment Report of the Intergovernmental Panel on Climate Change. Cambridge, UK: Cambridge University Press.

Irz, X., and C. Thirtle. 2004. "Dual Technological Development in Botswana Agriculture: A Stochastic Input Distance Function Approach." *Journal of Agricultural Economics* 55 (3): 455–478.

Jerven, M. 2013. *Poor Numbers: How We Are Misled by African Development Statistics and What to Do about It*. Ithaca, NY: Cornell University Press.

Johnson, M., S. Benin, L. You, X. Diao, P. Chilonda, and A. Kennedy. 2014. *Exploring Strategic Priorities for Regional Agricultural Research and Development Investments in Southern Africa*. IFPRI Discussion Paper 1318. Washington, DC: International Food Policy Research Institute.

Karshenas, M. 2001. "Agriculture and Economic Development in Sub-Saharan Africa and Asia." *Cambridge Journal of Economics* 25 (3): 315–342.

Kherallah, M., C. Delgado, E. Gabre-Madhin, N. Minot, and M. Johnson. 2002. *Reforming Agricultural Markets in Africa*. Baltimore: Johns Hopkins University Press.

Killick, T. 1995. "Structural Adjustment and Poverty Alleviation: An Interpretative Survey." *Development and Change* 26 (2): 305–330.

Kurukulasuriya, P., R. Mendelsohn, R. Hassan, J. Benhin, M. Diop, H. M. Eid, K. Y. Fosu, et al. 2006. "Will African Agriculture Survive Climate Change?" *World Bank Economic Review* 20 (3): 367–388.

Lawrence, P. 2005. "Explaining Sub-Saharan Africa's Manufacturing Performance." *Development and Change* 36 (6): 1121–1141.

Lele, U., R. Christiansen, and K. Kadiresan. 1989. *Issues in Fertilizer Policy in Africa: Lessons from Development Programs and Adjustment Lending, 1970–87.* MADIA Report. Washington, DC: World Bank.

Lewis, W. A. 1954. "Economic Development with Unlimited Supplies of Labor." *The Manchester School of Economics and Social Studies* 22 (2): 139–191.

Madagascar (Republic of Madagascar). 2007. "Madagascar Action Plan 2007–2012: A Bold and Exciting Plan for Rapid Development." http://www.un.org/esa/coordination/Alliance/MADAGASCAR%20ACTION%20PLAN.htm.

Mogues, T., B. Yu, S. Fan, and L. McBride. 2012. *The Impacts of Public Investment in and for Agriculture.* IFPRI Discussion Paper 01217. Washington, DC: International Food Policy Research Institute.

Nelson, G. C., M. W. Rosegrant, A. Palazzo, I. Gray, C. Ingersoll, R. Robertson, S. Tokgoz, et al. 2010. *Food Security, Farming, and Climate Change to 2050: Scenarios, Results, Policy Options.* Washington, DC: International Food Policy Research Institute.

Nin-Pratt, A., M. Johnson, E. Magalhaes, L. You, X. Diao, and J. Chamberlin. 2011. *Yield Gaps and Potential Agricultural Growth in West and Central Africa*. Washington, DC: International Food Policy Research Institute.

Nin-Pratt, A., and B. Yu. 2008. *An Updated Look at the Recovery of Agricultural Productivity in Sub-Saharan Africa*. IFPRI Discussion Paper 787. Washington, DC: International Food Policy Research Institute.

Omamo, S. W., X. Diao, S. Wood, J. Chamberlain, L. You, S. Benin, U. Wood-Sichra, and A. Tatwangire. 2006. *Strategic Priorities for Agricultural Development in Eastern and Central Africa*. IFPRI Report 150. Washington, DC: International Food Policy Research Institute.

Oruko, L., I. Randall, M. Bwalya, S. Kisira, and M. Wanzala. 2011. "Mutual Accountability Framework." Accessed April 15, 2014. http://www.nepad.org/system/files/CAADP%20MAF%20Final%20Draft%20Report%20%20March%202011.pdf.

Pardey, P. G., J. James, J. Alston, S. Wood, B. Koo, E. Binenbaum, T. Hurley, and P. Glewwe. 2007. *Science, Technology and Skills.* Working Paper 41371. Washington, DC: World Bank. Accessed September 2011. http://siteresources.worldbank.org/INTWDR2008/Resources/2795087–1191427986785/PardeyPEtAl_SciTech&Skills--ALL1.pdf.

Reardon, T., C. Barrett, V. Kelly, and K. Savadogo. 1999. "Policy Reforms and Sustainable Agricultural Intensification in Africa." *Development Policy Review* 17 (4): 375–395.

Sachs, J. D., and A. M. Warner. 1997. "Sources of Slow Growth in African Economies." *Journal of African Economies* 6 (3): 335–376.

SAPRIN (Structural Adjustment Participatory Review International Network). 2004. *The Policy Roots of Economic Crisis and Poverty: A Multi-Country Participatory Assessment of Structural Adjustment.* London: Zed Books.

Seo, S. N., R. Mendelsohn, A. Dinar, R. Hassan, and P. Kurukulasuriya. 2008. *A Ricardian Analysis of the Distribution of Climate Change Impacts on Agriculture across Agroecological Zones in Africa.* Policy Research Working Paper 4599. Washington, DC: World Bank.

Shiva, V. 1991. "The Green Revolution in the Punjab." *Ecologist* 21 (2): 57–60.

Wodon, Q., and H. Zaman. 2010. "Higher Food Prices in Sub-Saharan Africa: Poverty Impact and Policy Responses." *World Bank Research Observer* 25 (1): 157–176.

Woodhouse, P. 2009. "Technology, Environment and the Productivity Problem in African Agriculture: Comment on the World Development Report 2008." *Journal of Agrarian Change* 9 (2): 263–276.

World Bank. 1994. "Feasibility of Phosphate Rock Use as a Capital Investment in Sub-Saharan Africa: Issues and Opportunities." Unpublished report. Washington, DC.

———. 2007. World Development Report 2008: Agriculture for Development. Washington, DC.

Chapter 2

INTERTEMPORAL TRENDS IN AGRICULTURAL PRODUCTIVITY

Samuel Benin and Alejandro Nin-Pratt

Introduction

Changes or trends in agricultural productivity over time can shed light on the relative sources of agricultural growth as well as on resource and factor constraints to increasing agricultural production sustainably. Because improvements in agricultural productivity are important for reducing poverty and achieving other development objectives, it is essential to use the appropriate indicator and measure of agricultural productivity—partial factor productivity (PFP) or total factor productivity (TFP). However, because productivity embodies many different components, changes in productivity can catalyze a wide range of direct and indirect effects on the pathways to achieving different development objectives.

For example, output per worker or labor productivity, as indicators of PFP, may be better measures of productivity to identify linkages to nonagricultural growth, because it encapsulates the additional ways farm households earn income (Mellor 1999). Byerlee, Diao, and Jackson (2009) show that countries with the highest agricultural growth per worker experienced the greatest rate of rural poverty reduction. Other measures of PFP have been found to be significant determinants of poverty. Datt and Ravallion (1998), for example, find that higher land productivity (measured by agricultural output per unit area) had greater effect in reducing absolute poverty than in reducing the poverty gap or squared poverty gap, suggesting that the gains from higher land productivity were via rising average living standards, rather than improved distribution. Because changes in PFP could be caused by change from a variety of reasons, including change in the use of other inputs or change in output mix, the policy implications of changes in PFP measures are often unclear. Furthermore, changes in output and productivity also do not necessarily have similar impacts, and sometimes move in different directions, with differential consequences for poverty (Schneider and Gugerty 2011), and productivity gains, depending on the distribution of assets, may have limited impact on poverty reduction (Thirtle, Lin, and Piesse 2003).

Unlike these PFP measures, TFP measures provide a better sense of the changes in agricultural productivity that are attributable to technological change, which, for many policymakers and the Comprehensive Africa Agriculture Development Programme (CAADP), is a critical means of improving African agriculture. Known as the Solow residual, TFP measures the part of growth that is not accounted for by changes in conventional factors of production, such as land, labor, or capital. As a residual, however, the source of TFP growth is varied. Fan, Hazell, and Thorat (2000) found that investments in roads, agricultural research and development, and education were significant determinants of TFP, which in turn had a substantial effect on reducing poverty via reduced prices and increased wages, but at the cost of increased landlessness. Rahman and Salim (2013) show that the different sources can have different effects on the various components of TFP, which, similar to the case with the PFP measures, also suggests that the policy implications of changes in TFP can be complex.

Deciding what indicator and measure of agricultural productivity to use is complicated by knowledge gaps across several dimensions of the different components embodied in productivity, including

- Composition of agriculture—sector (all agriculture), subsector (crops, livestock, fisheries, forestry), commodity group (such as cereal, export crops, meat), and commodity (such as maize, rice, beef, tilapia);
- Type of factor (land, labor, capital), input (seed, fertilizer, feed), or husbandry (plant spacing, weeding, intensive livestock management);
- Measure of output and input—physical quantity or monetary value, which is important when aggregating across several subcomponents, because summing over weights or volumes may not be meaningful;
- Time (annual, long-term average, most recent years) and space (countries, regions, agroecologies, stage of development, endowment, etc.); and
- Level of aggregation (plot, farm, household, subnational, national, regional, continental).

The objective of this chapter is to assess changes over time and across different parts of Africa, in both partial and total measures of agricultural productivity, to understand the relative sources of productivity growth. We begin the next section with a presentation of the partial and total measures of productivity, in addition to the data used in estimating the indicators and in conducting the analysis. This is followed by trends analysis of the indicators and key drivers, and then conclusions and implications for using different measures and indicators.

Productivity Measures and Methodology

Partial factor productivity is a ratio of output to a subset of the inputs, usually one input, described as single-factor productivity. Two commonly used measures of PFP are land productivity (defined as the ratio of output to total harvested area) and labor productivity (the ratio of output to total number of hours worked). Obviously, these two PFP measures differ from one another by the variables they measure, as well as by the variables they exclude. PFP measures make it possible to focus on a given variable (that is, land or labor in the two examples above), to assess how that variable is changing relative to the output.

Total factor productivity, conceptually also a measure of output to inputs, is commonly measured as an index of the ratio of total agricultural outputs to total agricultural inputs. As such, TFP analysis can be seen as an extension of PFP analysis, since the variables used in measuring PFP are included in the variables used in measuring TFP. Use of TFP is favored in the analysis of productivity, because long-run agricultural growth depends on TFP growth, which can be decomposed into finer measures, including the three that are commonly estimated or presented in the literature: *technical change*, arising from movement of the technological frontier; *technical-efficiency change,* arising from movement of observations toward or away from the technological frontier; and *scale-efficiency change,* arising from movement of observations about the technological frontier to capture economies of scale.[1]

In principle, measuring PFP is straightforward, and the data requirements are not complicated. Measuring TFP, however, can be challenging, especially for developing countries that lack data on prices to use in aggregating outputs and inputs. Several methods for and approaches to measuring TFP are available, differing mainly in how outputs and inputs are aggregated. The methods can be classified into two broad groups: (1) nonparametric methods, including

1 Some studies have tried to decompose finer measures of efficiency change, distinguishing, for example, allocative efficiency change for inputs, allocative efficiency change for outputs, residual scale-efficiency change, etc. (for example, Rahman and Salim 2013). Details of the TFP decomposition in general, as well as the changes analyzed in the study (that is, technical-efficiency change and technical change), are presented in the appendix to this chapter.

index-based growth accounting (for example, the Törnqvist-Theil index) and data envelopment analysis (DEA); and (2) parametric methods, including econometric estimation of the technology, often by stochastic frontier analysis. (See, for example, Coelli, Prasada Rao, and Battese [1998] and Coelli and Prasada Rao [2001] for review of the different methods and measurement issues.) This study uses the Malmquist index approach, where the index is calculated by DEA, using one output and six inputs, and assuming sequential technology that rules out technological regression or negative growth rates in technical change. This method, referred to as the DEA-Malmquist index, is fully documented in Nin-Pratt and Yu (2010), so we present the main aspects of it in Appendix 2A of this chapter.

The literature has identified some issues with use of DEA methods to calculate distance functions. The major drawbacks include nonstochastic functions—that is, lack of including a random-error term to account for statistical noise; determination of implicit or shadow prices used in aggregating inputs; and dimensionality, or the number of inputs and outputs used relative to the number of observations in the cross-section. These issues and how they are dealt with in this study are also discussed in Appendix 2A.

To generate greater confidence in the findings associated with this method, however, we compare the results with those obtained using three other approaches that differently address the issues with DEA. These include two other versions of the DEA-Malmquist index: one is calculated by using two outputs to deal with the dimensionality issue, and the other is calculated by including lower and upper bounds on the shadow prices. The third method is the more conventional growth-accounting TFP index, where inputs are aggregated using fixed-input shares for all countries and periods. A brief comparison of the results is presented in Appendix 2B. Overall, the different methods yield similar TFP growth patterns, with the DEA-Malmquist–2-output-index giving higher growth rates, followed by the DEA-Malmquist–1-output-index, the DEA-Malmquist-bounds-index, and the growth-accounting-TFP-index.

Data and Sources of Data

The data used in the measurements of the different PFP and TFP indicators are drawn mostly from the Food and Agriculture Organization of the United Nations FAOSTAT database on agricultural production (FAO 2014), which covers the period 1961–2012. The data, which are detailed in Table 2.1, include one output (total agricultural production) and six inputs (land, labor, fertilizer, animal feed, crop capital, and livestock capital). The two PFP

TABLE 2.1 Description of variables and data used in estimating partial and total factor productivity

Variable	Description
Output	Value of gross crop and livestock production expressed in constant 2004–2006 international dollars. In the case of Nigeria, output for 2001–2012 was adjusted using agricultural value-added data from the World Development Indicators (World Bank 2014) to better reflect recent growth measured at the country level.
Land	Hectares of land, including land under temporary crops (doubled-cropped areas are counted only once); temporary meadows for mowing or pasture, such as land used permanently (five years or more) for herbaceous forage crops, either cultivated or growing wild (wild prairie or grazing land); land under market and kitchen gardens; land temporarily under fallow (less than five years); land cultivated with permanent crops, such as flowering shrubs (coffee), fruit trees, nut trees, and vines, but excluding land under trees grown for wood or timber.
Labor	Total economically active population engaged in or seeking work in agriculture, hunting, fishing, or forestry, whether as employers, their own account workers, salaried employees, or unpaid workers assisting in the operation of a family farm or business. This is an uncorrected measure of labor that does not account for actual hours worked or labor quality (education, age, experience, etc.). Data for Nigeria were adjusted following Fuglie (2011).
Fertilizer	Metric tons of nitrogen, phosphorus, and potassium nutrients consumed.
Animal feed	Metric tons (maize equivalent) of edible commodities (cereals, bran, oilseeds, oilcakes, fruits, vegetables, roots and tubers, pulses, molasses, animal fat, fish, meat meal, whey, milk, and other animal products) fed to livestock.
Crop capital	Sum of gross fixed capital stock in constant 2005 US$: • *Land development:* major improvements in the quantity, quality, or productivity of land to prevent its deterioration, including (1) on-field land improvement undertaken by farmers (includes work done on the field, such as making boundaries and irrigation channels); and (2) other activities, such as irrigation works, soil conservation works, and flood control structure, undertaken by government and other local bodies. • *Plantation crops:* trees yielding repeated products (including vines and shrubs) cultivated for fruits and nuts, sap and resin, bark and leaf products, etc. • *Machinery and equipment:* tractors (with accessories), harvesters and thrashers, and hand tools.
Livestock capital	Sum of gross fixed capital stock in constant 2005 US$: • *Animal stock:* stock of cattle and buffalo, camels, horses, mules, asses, pigs, goats, sheep, and poultry. • *Structures for livestock:* sheds constructed for housing cows, buffalo, horses, camels, and poultry. • *Milking machines:* machinery and related equipment used for milking animals.

Source: Authors' representation based on FAO (2014).

Notes: Crop and livestock capital cover 1975–2007. The values were developed by multiplying the quantity of physical assets in use by unit prices compiled from individual countries. Each asset held at a point in time is valued at the price at the same time, regardless of the age or actual condition of the asset.

measures of land productivity and labor productivity were obtained by dividing output by land and labor, respectively. TFP was calculated using the Malmquist index presented in Appendix 2A.

The results for each PFP and TFP indicator are presented at an aggregate level for the entire continent (which includes 45 of the 55 African countries because of data issues),[2] Africa south of the Sahara (SSA), and the five geographic regions of the African Union (central, eastern, northern, southern, and western). Table 2.2 presents the distribution of countries. The other type of aggregation or grouping used in the analysis and on which results are presented derives from the concept that different countries, depending on their resource endowments and stage of development, are on different trajectories toward achieving their development objectives (Diao et al. 2007). We use a four-category economic development typology based on three factors: agricultural potential, alternative (or nonagricultural) sources of growth, and income level (Benin et al. 2010; Appendix Table 2C.1).

We also present in Appendix Table 2C.2 the results of aggregation based on Regional Economic Communities (RECs). In general, we use the three different regional aggregations (geographic region, economic development classification, and REC) to generate greater confidence in the results of the analysis. However, the aggregations based on the latter two seem more rational, because they are based on economic criteria that are appropriate to the subsequent economic analysis of PFP and TFP. The aggregation based on geographic region seems more of a convenience or convention following the African Union. Therefore, apart from the political rationale and some economic ties in the case of western Africa (which also constitutes the Economic Community of West African States [ECOWAS] REC), the analyses of PFP and TFP that draw on a geographic perspective are relatively less significant. In all cases, the aggregated values of an indicator are estimated using the weighted sum approach, where the weight for each country is the share of that country's value of output relative to the total value of output for all countries in the region or group. To assess the effect of size and growth path as conditioning performance, we also present trends separately for the large and small agricultural economies and for the fast-growing and slow-growing agricultural economies (Appendix Table 2C.3).

2 Excludes Cape Verde, Comoros, Djibouti, Equatorial Guinea, Eritrea, Lesotho, Mayotte, São Tomé and Príncipe, Seychelles, and South Sudan for lack of data on all the relevant variables used in calculating TFP.

TABLE 2.2 Countries by geographic region and country's share in region's total agriculture value-added (%)

Central Africa (5.3)	East Africa (23.6)	North Africa (26.7)	Southern Africa (8.0)	West Africa (36.4)
Burundi (5.0)	Comoros (–)	Algeria (22.5)	Angola (21.0)	Benin (2.6)
Cameroon (35.7)	Djibouti (0.1)	Egypt (50.7)	Botswana (1.7)	Burkina Faso (3.6)
Central African Rep. (7.8)	Eritrea (–)	Libya (–)	Lesotho (0.8)	Cape Verde (0.1)
Chad (8.5)	Ethiopia (29.2)	Mauritania (1.5)	Malawi (9.4)	Côte d'Ivoire (5.3)
Congo, Dem. Rep. (37.4)	Kenya (13.7)	Morocco (18.3)	Mozambique (14.9)	Gambia, The (0.4)
Congo, Rep. (2.8)	Madagascar (5.1)	Tunisia (7.0)	Namibia (3.8)	Ghana (7.1)
Equatorial Guinea (2.6)	Mauritius (0.8)		South Africa (37.5)	Guinea (1.4)
Gabon (–)	Rwanda (3.6)		Swaziland (1.3)	Guinea-Bissau (0.4)
São Tomé & Príncipe (0.2)	Seychelles (0.0)		Zambia (9.6)	Liberia (0.6)
	Somalia (–)		Zimbabwe (–)	Mali (3.5)
	South Sudan (2.8)			Niger (2.4)
	Sudan (21.2)			Nigeria (67.4)
	Tanzania (15.3)			Senegal (2.2)
	Uganda (8.2)			Sierra Leone (1.3)
				Togo (1.6)

Source: Authors' calculations based on World Bank (2012).

Notes: The figures in parentheses are the regions' percentage shares in Africa's total agriculture value-added, and the countries' percentage shares in their respective regions' totals (2003–2010 annual average). Dashes indicate data are not available. Data for South Sudan and Sudan are based on 2008–2010 values.

Results

Annual trends in land and labor productivity

Annual trends in land and labor productivity are detailed in Tables 2.3a and 2.3b, Figure 2.1, and Appendix Figures 2C.1–2C.3 for the aggregations and for selected countries. The graphics are quite revealing and offer a quick overview of the comparative growth in land versus labor productivity or changes in land–labor intensity over time. There are three aspects to the plots in comparing the trends: their position in the quadrant space, their slope, and their length.

- The plot's *position* shows the magnitude, which is increasing in both land and labor productivity, going from the origin in a northeasterly direction.

- For a particular plot, the *slope* reflects the relative growth rates of labor and land productivity. With land productivity plotted on the *y*-axis and labor productivity on the *x*-axis, a slope steeper than the 45-degree line reflects a higher land productivity growth rate relative to the labor productivity growth rate and, therefore, a decreasing land-to-labor ratio. Conversely, a plot flatter than the 45-degree line means the labor productivity growth rate is higher than the land productivity growth rate and, therefore, has an increasing land-to-labor ratio. (This can be extended to compare different plots. For any two plots, the steeper one has a higher labor-to-land ratio, irrespective of the position of the plots in the quadrant.)
- The *length* of a plot reflects the magnitude of the combined growth rates, with a longer plot depicting a larger combined growth rate and a shorter plot depicting a smaller combined growth rate, again irrespective of the position of the plot in the quadrant.

AFRICA AND GEOGRAPHIC SUBREGIONS

Table 2.3a and Table 2.3b, Figure 2.1, and Appendix Figures 2C.1–2C.3 show that the trends in land and labor productivity are highly variable in different dimensions across different parts of the continent. For Africa as a whole, land productivity increased on average by 3.3 percent per year in 1961–2012, compared with a 2.0 percent increase per year for labor productivity, starting from 1961–1970 average levels of $256 per hectare (ha) and $892 per worker.[3] Compared with the entire continent, SSA realized lower growth rates in both land productivity (2.8 percent) and labor productivity (1.7 percent) over the same periods. These rates reflect the higher growth rates achieved in northern Africa, which experienced an annual average rate of growth of 3.2 percent in land productivity and 3.0 percent in labor productivity.

Northern and southern Africa have the highest annual average labor productivities, at $1,953 per worker in northern Africa and $3,333 per worker in southern Africa, compared with only $561 in central Africa, $612 in eastern Africa, and $999 in western Africa. Comparing the northern and southern Africa regions shows some significant differences, however. First, land productivity is much higher in northern Africa: $1,942/ha on average in 1961–2012, compared with only $92/ha in southern Africa over the same period. The relatively low land productivity in the southern region reflects the much higher

---

3 All currency is in constant 2004–2006 "international dollars," unless specifically noted as US dollars.

land-to-labor ratios associated with large plantations, with more mechanized agricultural operations. Second, whereas labor productivity has risen much faster than land productivity in southern Africa (with annual averages of 2.6 and 2.2 percent, respectively, in 1961–2012), land and labor productivities in northern Africa have risen at a roughly equal rate (3.2 and 3.0 percent, respectively). The trends observed in northern and southern Africa are driven by Egypt and South Africa, respectively: Egypt accounts for 51 percent of the total agriculture value-added in northern Africa, while South Africa accounts for about 44 percent in southern Africa (Table 2.2).

Figure 2.1 shows that the trends in central and eastern Africa are fairly similar, with land and labor productivity much lower than the levels for Africa as a whole. In 1961–2012, the annual average growth in land productivity in the two subregions was about 1.6 percent, compared with the range of 0.5–0.8 for labor productivity. The trend in western Africa is closer to the trend for Africa as a whole, although western Africa experienced a higher growth rate in land productivity (3.6 percent) and lower growth rate in labor productivity (1.8 percent) in 1961–2012.

Looking at the trends by subperiods (1961–1970, 1971–1980, 1981–1990, 1991–2000, and 2001–2012), Table 2.3b shows that labor productivity in Africa as a whole and in SSA increased more rapidly in 2001–2012 than in any of the preceding decades since 1961. While the patterns for the five geographic subregions differ, the general trend remains.

Many of the above results are consistent with the findings of other studies, but some are different. For example, whereas the findings of higher land than labor productivity growth rates in Africa as a whole are consistent with the results of previous studies on Africa (for example, Thirtle and Piesse 2008; Fuglie and Nin-Pratt 2013), the growth rates obtained in this chapter are larger, and are especially so for labor productivity. For example, Thirtle and Piesse (2008) estimate respective land and labor productivity growth rates at 2.0 and 0.4 percent for 1961–2003,[4] whereas Fuglie and Nin-Pratt (2013) estimate respective land and labor productivity growth rates at 2.3 and 0.6 percent for 1971–2009. Although we use more updated data in this study, country composition in this and other studies accounts for the bulk of the differences. In the Thirtle and Piesse (2008) study, for example, South Africa, which experienced a much higher growth rate in labor productivity than land productivity (Table 2.3b), was excluded. The results obtained here for the

4 The estimated average slope coefficient for the plots in Thirtle and Piesse (2008), however, is almost exactly unity, suggesting equal growth rates in land and labor productivity.

TABLE 2.3a Land and labor productivity (annual average level, 1961–2012)

	Subperiods			
	1961–1970		1971–1980	
Aggregations	Land	Labor	Land	Labor
Africa	256	892	323	1,102
Africa, south of the Sahara	104	863	124	1,063
Geographic location				
Central	136	488	152	518
Eastern	117	560	147	605
Northern	923	1,010	1,186	1,261
Southern	52	1,697	75	2,533
Western	115	768	125	736
Economic group				
LI–1	82	434	99	451
LI–2	98	466	131	506
LI–3	124	605	145	577
MI	362	1,167	460	1,516
Regional Economic Community				
CEN-SAD	379	807	484	886
COMESA	528	637	647	727
EAC	208	710	260	932
ECCAS	138	483	167	501
ECOWAS	115	768	125	736
IGAD	89	550	124	615
SADC	85	1,250	101	1,776
UMA	74	1,075	96	1,450
Other economic groups				
Large	374	1,061	476	1,357
Small	553	1,105	542	1,262
Fast-growing	125	700	145	632
Slow-growing	196	816	206	1,063
Selected countries				
Large				
Egypt	1,889	933	2,394	1,050
Ethiopia	57	335	62	299
Kenya	66	460	94	494
Morocco	94	924	107	1,019

Subperiods (continued)							
1981–1990		1991–2000		2001–2012		1961–2012	
Land	Labor	Land	Labor	Land	Labor	Land	Labor
456	1,271	685	1,557	935	2,078	547	1,407
142	1,189	208	1,315	328	1,771	187	1,261
177	526	202	536	266	706	189	561
149	553	175	575	236	742	168	612
1,685	1,580	2,390	2,414	3,261	3,242	1,942	1,953
90	3,347	100	3,663	137	5,077	92	3,333
155	770	275	1,086	460	1,531	235	999
123	451	135	417	143	434	117	437
135	451	181	436	249	482	162	469
174	574	163	586	220	687	167	609
669	1,811	965	2,155	1,287	2,856	769	1,938
692	1,006	981	1,465	1,296	1,975	787	1,257
938	804	1,450	1,234	1,966	1,687	1,139	1,044
267	921	309	1,195	401	1,237	293	1,008
207	495	221	502	303	641	211	529
155	770	275	1,086	460	1,531	235	999
115	555	152	597	204	813	139	633
118	2,189	131	2,372	160	3,349	121	2,232
123	1,857	164	2,562	219	3,294	138	2,095
697	1,601	1,035	1,905	1,360	2,585	811	1,736
511	1,422	545	1,534	592	1,779	550	1,434
186	655	305	1,023	522	1,510	267	927
230	1,081	263	1,457	289	1,770	239	1,258
3,398	1,275	4,425	2,263	5,901	3,191	3,690	1,798
75	279	121	224	203	274	107	282
130	507	161	457	239	521	142	489
137	1,258	172	1,576	252	2,447	156	1,483

(continued)

TABLE 2.3A (continued)

Aggregations	Subperiods 1961–1970		1971–1980	
	Land	Labor	Land	Labor
Nigeria	133	858	140	749
South Africa	58	2,586	79	3,939
Sudan	22	677	31	796
Tanzania	72	364	83	379
Small				
Botswana	6	844	7	957
Gabon	21	524	27	689
Gambia, The	176	574	164	450
Guinea-Bissau	66	394	65	353
Mauritius	1,933	2,244	1,975	2,331
Swaziland	78	982	128	1,508
Fast-growing				
Angola	17	503	17	432
Cameroon	154	525	196	668
Malawi	147	284	203	349
Mozambique	24	290	27	277
Nigeria	133	858	140	749
Rwanda	384	376	485	418
Sierra Leone	112	365	130	408
Zambia	18	340	27	410
Slow-growing				
Burundi	418	465	386	469
Congo, Dem. Rep.	100	455	119	447
Liberia	82	520	116	597
Mauritius	1,933	2,244	1,975	2,331
Namibia	9	2,001	11	2,330
Tunisia	114	1,536	173	2,397
Zimbabwe	73	576	106	708

Source: Authors' calculation and representation based on FAO (2014).

Notes: Land productivity is in constant 2004–2006 international dollars (I$) per hectare, and labor productivity is in constant 2004–2006 I$ per worker. LI–1 = low income, more favorable agriculture, and mineral rich; LI–2 = low income, more favorable agriculture, and nonmineral rich; LI–3 = low income and less favorable agriculture; MI = middle income. CEN-SAD = Community of Sahel-Saharan States; COMESA = Common Market for Eastern and Southern Africa; EAC = East African Community; ECCAS = Economic Community of Central African States; ECOWAS = Economic Community of West African States; IGAD = Intergovernmental Authority on Development; SADC = Southern African Development Community; UMA = Union du Maghreb Arabe. Large agricultural economies have at least 3.0 percent of Africa's total agricultural output; small agricultural economies have less than 0.1 percent of Africa's total agricultural output; fast-growing agricultural economies surpass the CAADP agricultural growth rate target of 6.0 percent per year; and slow-growing agricultural economies have an agricultural growth rate of less than 1.0 percent per year.

Subperiods (continued)							
1981–1990		1991–2000		2001–2012		1961–2012	
Land	Labor	Land	Labor	Land	Labor	Land	Labor
176	777	334	1,223	586	1,807	286	1,111
90	5,236	97	6,039	123	9,334	91	5,577
35	796	45	1,005	78	1,556	44	988
105	384	123	339	186	426	117	380
8	970	8	917	10	827	8	900
36	898	44	1,091	52	1,401	37	939
159	324	158	214	228	230	179	354
92	413	116	477	162	613	103	456
2,045	2,635	2,309	3,496	2,721	4,805	2,217	3,168
197	1,974	208	1,760	229	1,999	170	1,658
14	273	19	275	45	480	23	396
216	704	283	781	452	1,185	267	788
238	337	296	359	465	537	277	379
23	209	29	214	47	283	30	256
176	777	334	1,223	586	1807	286	1111
587	422	608	391	937	435	613	409
145	407	145	393	231	626	156	447
32	345	37	323	54	413	34	368
432	418	450	366	457	308	430	401
151	459	166	397	148	286	137	404
134	586	108	453	154	489	120	527
2,045	2,635	2,309	3,496	2,721	4,805	2,217	3,168
9	1,814	10	1,672	11	1,618	10	1,877
216	2,794	295	3,852	348	4,349	234	3,038
115	598	118	522	100	505	102	579

TABLE 2.3B **Land and labor productivity (%, annual average growth rate, 1961–2012)**

	Subperiods			
	1961–1970		1971–1980	
Aggregations	Land	Labor	Land	Labor
Africa	2.20	1.87	3.01	2.26
Africa, south of the Sahara	3.05	1.85	−0.08	2.49
Geographic location				
Central	1.82	0.66	0.32	0.22
Eastern	3.22	1.45	−0.82	−1.07
Northern	2.32	1.98	4.04	1.31
Southern	3.26	2.35	3.49	5.99
Western	3.29	1.70	−0.67	−1.50
Economic group				
LI–1	1.98	0.36	1.41	−0.18
LI–2	3.82	1.47	−0.75	−1.34
LI–3	2.48	1.13	0.95	−0.75
MI	2.01	2.05	3.73	3.17
Regional Economic Community				
CEN-SAD	1.91	1.85	3.60	0.01
COMESA	1.87	1.65	3.21	0.42
EAC	3.56	1.24	−0.93	−0.31
ECCAS	2.60	0.91	1.65	−0.38
ECOWAS	3.29	1.70	−0.67	−1.50
IGAD	4.83	1.82	−2.32	−1.26
SADC	1.39	2.10	2.02	5.03
UMA	1.89	2.16	1.56	1.66
Other economic groups				
Large	1.93	2.11	3.62	3.26
Small	−0.29	1.47	−0.25	0.78
Fast-growing	3.81	1.97	0.56	−2.69
Slow-growing	0.50	0.76	0.61	0.36
Selected countries				
Large				
Egypt	1.58	1.80	3.82	0.95
Ethiopia	1.97	0.06	1.07	−0.66
Kenya	2.96	0.38	3.78	0.77
Morocco	3.21	3.66	0.26	−0.62

Subperiods (continued)							
1981–1990		1991–2000		2001–2012		1961–2012	
Land	Labor	Land	Labor	Land	Labor	Land	Labor
3.86	1.25	3.53	2.14	2.16	3.18	3.31	2.04
3.16	0.32	3.27	1.65	5.20	3.32	2.82	1.67
2.53	0.17	−0.15	0.29	4.81	4.80	1.63	0.81
2.01	−0.18	1.30	1.07	2.96	1.59	1.59	0.54
2.99	3.93	3.45	3.02	1.63	3.19	3.20	2.96
1.24	0.78	2.64	1.27	3.36	3.58	2.22	2.56
4.73	2.28	4.60	2.66	5.75	4.02	3.59	1.81
2.51	0.38	−1.22	−1.74	2.35	1.98	1.37	−0.06
2.68	−0.03	2.99	0.17	2.75	1.15	2.21	−0.04
1.52	0.44	−2.20	0.67	4.81	2.30	1.26	0.32
3.73	1.13	3.21	1.90	1.89	3.15	3.25	2.14
3.48	2.98	3.04	2.84	1.88	3.19	3.13	2.30
4.22	2.31	3.76	4.44	1.84	2.38	3.41	2.47
1.74	1.74	0.94	0.27	2.86	1.48	1.49	1.29
1.83	0.00	−0.32	0.04	4.89	4.41	1.85	0.61
4.73	2.28	4.60	2.66	5.75	4.02	3.59	1.81
2.73	−0.22	2.34	1.80	2.58	1.74	1.89	0.79
1.85	0.32	0.62	1.92	2.49	3.51	1.52	2.27
4.63	3.89	1.18	1.69	4.03	3.71	2.67	2.77
3.89	0.90	3.51	2.24	1.69	3.17	3.32	2.13
1.66	2.02	0.10	−0.35	−0.70	1.50	0.16	1.13
3.63	3.20	4.20	2.68	5.90	4.70	3.63	2.06
2.44	1.74	−0.21	1.07	1.45	2.92	1.01	1.81
3.21	3.84	2.22	4.76	2.13	2.72	2.87	3.22
0.98	−2.74	6.66	−1.08	4.63	3.10	3.21	−0.60
3.88	0.92	1.25	2.04	4.67	2.85	3.11	0.20
5.76	5.71	1.10	0.73	4.83	5.65	2.48	2.43

(continued)

TABLE 2.3B (continued)

Aggregations	Subperiods 1961–1970 Land	1961–1970 Labor	1971–1980 Land	1971–1980 Labor
Nigeria	3.74	1.85	–1.20	–3.08
South Africa	3.63	2.49	2.86	5.97
Sudan	2.62	1.50	2.88	0.96
Tanzania	1.67	0.80	2.81	1.10
Small				
Botswana	3.32	2.72	–0.61	–1.31
Gabon	1.68	2.08	3.49	3.84
Gambia, The	2.21	0.25	–4.69	–6.66
Guinea-Bissau	–3.07	–2.88	2.78	0.45
Mauritius	–0.06	0.52	0.20	–0.46
Swaziland	4.21	4.39	5.11	3.61
Fast-growing				
Angola	2.97	1.97	–4.33	–6.24
Cameroon	3.17	2.60	0.42	1.11
Malawi	1.76	1.67	3.58	1.38
Mozambique	2.78	1.16	–1.76	–3.84
Nigeria	3.74	1.85	–1.20	–3.08
Rwanda	4.40	2.77	1.71	0.43
Sierra Leone	2.87	2.71	1.20	0.58
Zambia	2.78	1.13	3.10	0.46
Slow-growing				
Burundi	0.40	0.81	–0.77	–0.58
Congo, Dem. Rep.	1.72	–0.21	1.42	–0.58
Liberia	4.37	2.09	1.73	–0.38
Mauritius	–0.06	0.52	0.20	–0.46
Namibia	3.27	2.30	–1.32	–1.62
Tunisia	0.18	1.54	2.92	1.66
Zimbabwe	2.63	1.18	1.71	–0.23

Source: Authors' calculation and representation based on FAO (2014).

Notes: Annual average growth rates are calculated using the "LOGEST" function in Microsoft Excel. LI–1 = low income, more favorable agriculture, and mineral rich; LI–2 = low income, more favorable agriculture, and nonmineral rich; LI–3 = low income and less favorable agriculture; MI = middle income. CEN-SAD = Community of Sahel-Saharan States; COMESA = Common Market for Eastern and Southern Africa; EAC = East African Community; ECCAS = Economic Community of Central African States; ECOWAS = Economic Community of West African States; IGAD = Intergovernmental Authority on Development; SADC = Southern African Development Community; UMA = Union du Maghreb Arabe. Large agricultural economies have at least 3.0 percent of Africa's total agricultural output; small agricultural economies have less than 0.1 percent of Africa's total agricultural output; fast-growing agricultural economies surpass the CAADP agricultural growth rate target of 6.0 percent per year; and slow-growing agricultural economies have an agricultural growth rate of less than 1.0 percent per year.

Subperiods (continued)							
1981–1990		1991–2000		2001–2012		1961–2012	
Land	Labor	Land	Labor	Land	Labor	Land	Labor
5.40	3.58	4.91	2.67	6.15	4.66	3.87	2.03
0.93	1.05	1.25	2.36	2.57	4.81	1.73	3.00
−1.01	−0.61	4.89	3.63	3.62	2.53	2.87	1.91
2.37	−0.25	1.25	−1.23	4.09	2.75	2.30	0.25
1.14	1.15	−1.49	−4.76	3.71	2.13	1.03	−0.12
2.21	1.89	2.00	2.00	1.94	3.18	2.27	2.41
0.31	−3.48	4.12	0.81	0.17	−1.14	0.51	−2.42
3.39	2.15	2.44	2.12	3.86	2.50	2.37	1.20
1.84	4.33	0.78	1.73	1.02	2.64	0.84	1.93
2.38	0.39	−0.76	−1.42	1.64	2.33	2.57	1.54
0.89	−1.79	4.55	1.80	8.55	5.52	2.11	−0.35
1.75	0.13	3.33	2.08	5.67	5.96	2.54	1.81
0.61	−1.25	5.11	4.90	4.32	3.43	2.68	1.34
−0.84	−0.80	6.60	3.37	5.38	3.51	1.54	−0.18
5.40	3.58	4.91	2.67	6.15	4.66	3.87	2.03
0.55	−1.65	1.06	−1.76	4.92	3.01	2.05	0.25
1.65	0.02	−1.92	−1.38	9.59	8.40	1.57	1.08
4.00	0.35	0.83	−0.22	5.87	4.29	2.48	0.22
2.91	−0.52	−2.36	−2.31	1.56	−1.94	0.33	−1.07
2.85	0.62	−2.08	−4.31	0.43	−1.39	1.03	−1.09
0.22	−0.18	5.97	3.67	0.34	−2.11	1.24	−0.37
1.84	4.33	0.78	1.73	1.02	2.64	0.84	1.93
−0.21	−2.39	−0.34	−1.54	0.42	−0.16	0.44	−0.73
3.04	3.40	1.29	0.38	2.52	2.34	2.68	2.47
2.21	−0.86	2.02	2.14	−2.43	−1.35	0.67	−0.54

FIGURE 2.1 Line plots of land and labor productivity by geographic region (1961–2012)

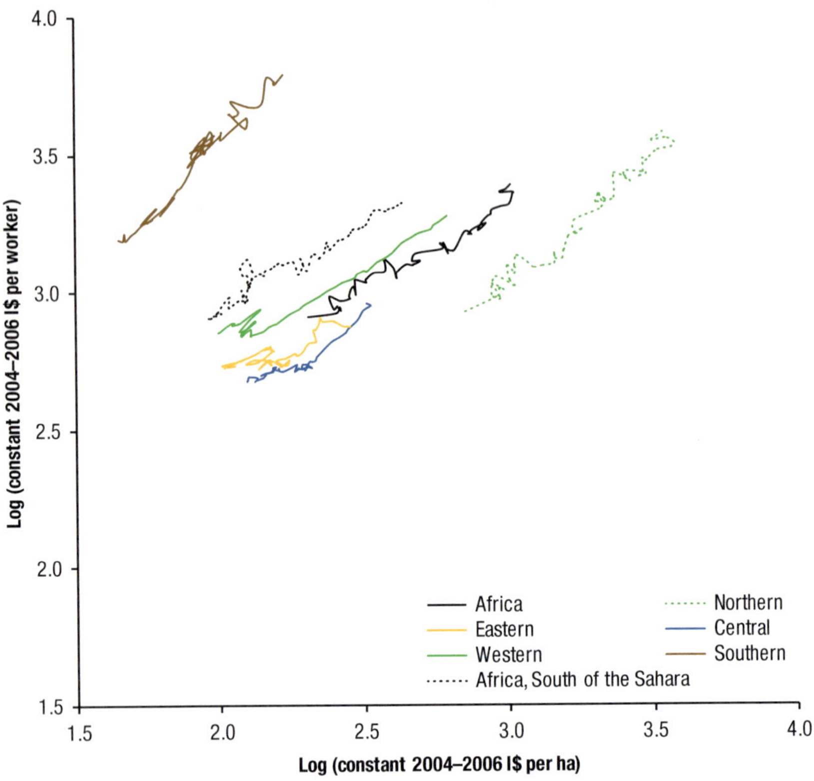

Source: Authors' calculation and representation based on FAO (2014).
Note: I$ = international dollar.

different subperiods are consistent with the finding of a decline in the poverty rate in Africa from the long-standing average of 50 percent to 47 percent in 2008, and also in the decline in the number of the extreme poor since 2005—the first time ever (World Bank 2012).

OTHER GROUPINGS

The trends in land and labor productivity analyzed by the other aggregations (that is, by economic classification, regional economic communities, or size and growth of the agricultural sector) are presented in Tables 2.3a and 2.3b and Appendix Figures 2C.1–2C.3. Looking at the trends by economic classification (Figure 2C.1), the *middle-income* (MI) category clearly outperformed the others in both measures of productivity. In the MI countries, land

and labor productivity increased at 3.2 percent and 2.1 percent, respectively, in 1961–2012. The performance of the MI group as whole is heavily influenced by the performance of Egypt and Nigeria, which accounted for about 24 and 27 percent, respectively, of the group's total agriculture value-added (Table 2.2). However, while the land productivity growth rate was higher in Nigeria (3.9 percent) than in Egypt (2.9 percent), the labor productivity growth rate was higher in Egypt (3.2 percent) than in Nigeria (2.0) in 1961–2012 (Table 2.3b).

The other three categories of countries are *low income, more favorable agriculture, and mineral rich* (LI–1); *low income, more favorable agriculture, and nonmineral rich* (LI–2); and *low income and less favorable agriculture* (LI–3). For these groups, we see negative or stagnant growth in labor productivity in the LI–1 and LI–2 groups and little increase in the LI–3 group for 1961–2012, compared with moderate increase in land productivity (annual average growth rate of 1.3–2.2 percent for the same period). Average land and labor productivity in the LI–1 group was the lowest, with a respective annual average of only $117/ha and $437/worker in 1961–2012. Note that the LI–1 group has favorable agriculture production potential and is also rich in minerals—dominated by the Democratic Republic of the Congo, which accounts for about 41 percent of the group's total agriculture value-added. The poor performance thus seems consistent with the "resource curse" thesis. The trends by subperiods (1961–1970, 1971–1980, 1981–1990, 1991–2000, and 2001–2012) reveal that, for all four economic categories, the growth rates in both land and labor productivity were generally lower (and negative in many cases) on average in the 1970s and 1990s than in the other three subperiods. Overall, labor productivity increased more rapidly in 2001–2012 than in any of the preceding decades since 1961.

Appendix Figure 2C.2 shows the trends by REC. Two of the RECs outperformed the others in land productivity: (1) the Common Market for Eastern and Southern Africa (COMESA) REC, dominated by Egypt in total agriculture value-added, shown in Appendix Table 2C.1, with an average level of $1,139/ha; and (2) the Community of Sahel-Saharan States (CEN-SAD) REC, dominated by Nigeria and Egypt, with an average level of $787/ha for the entire period (Table 2.3a). The Southern African Development Community (SADC), dominated by Tanzania and South Africa, and the Union du Maghreb Arabe (UMA), dominated by Algeria and Morocco, outperformed the other RECs in labor productivity, with an average of $2,232/worker for SADC and $2,095/worker for UMA. The lowest-performing RECs in levels of both land and labor productivity are the Economic Community of

Central African States (ECCAS) and the Intergovernmental Authority on Development (IGAD), with respective average land and labor productivity values in the range of $139–$211/ha and $529–$633/worker. With respect to the UMA REC, where labor productivity increased faster than land productivity in 1961–2012, land productivity increased relatively faster than labor productivity in the other RECs. The East African Community (EAC) and SADC RECs experienced the most variability in land and labor productivity, as reflected in the tortuous shape of their plots in Appendix Figure 2C.2.

SELECTED COUNTRIES

Turning now to the selected countries representing the large and small agricultural economies as well as the fast-growing and slow-growing agricultural economies in Africa, Appendix Figure 2C.3a shows the plots for the four different groups, and Appendix Figures 2C.3b and 2C.3c show the results for the individual countries. Two distinct characteristics stand out. First, the plots are longer for the large or fast-growing agricultural economies and shorter for the small or slow-growing agricultural economies. This indicates more rapid combined growth in land and labor productivities in the large or fast-growing agricultural economies, which is confirmed in the individual country plots in Figure 2C.3b and the results in Table 2.3b. Second, the plots are flatter for the large or fast-growing agricultural economies and seemingly steeper but tortuous for the small or slow-growing agricultural economies. This indicates a relatively higher land-to-labor productivity growth ratio in the large or fast-growing agricultural economies. The small or slow-growing agricultural economies are dominated by Mauritius, which has extremely high levels of land and labor productivity, and whose labor productivity increased more rapidly than land productivity in 1961–2012 (Tables 2.3a and b). Unfortunately, most of the countries in these two groups experienced average negative growth rates in labor productivity.

Looking at the performance of individual selected countries, we see that Egypt leads the group of countries in both levels and growth rates of land and labor productivity (Appendix Figure 2C.3b). Whereas Mauritius, Morocco, Namibia, Nigeria, South Africa, Swaziland, and Tunisia have similar or higher labor productivity values, averaging more than $2,000/worker in 1961–2012, Egypt clearly outperformed all of the other selected countries in land productivity, with an average of $3,690/ha in the same periods, compared with the next-highest levels of $2,217/ha in Mauritius and $613/ha in Rwanda (Table 2.3a). Egypt, South Africa, and Mauritius also stand out among the group of countries in terms of having higher average growth rate in labor productivity than in land productivity in 1961–2012.

Other countries with high average growth rates in land productivity in 1961–2012 include Kenya (3.1 percent), Ethiopia (3.2 percent), and Nigeria (3.9 percent), whereas others with high average growth rates in labor productivity include Gabon (2.4 percent), Morocco (2.4 percent), and South Africa (3.0 percent). Countries with the lowest average levels of land productivity (less than $50/ha) in 1961–2012 include Angola, Botswana, Gabon, Mozambique, Namibia, Sudan, and Zambia, whereas those with the lowest average levels of labor productivity (less than $400/worker) include Angola, Ethiopia, The Gambia, Malawi, Mozambique, and Zambia.

Countries with the lowest annual average growth rates in land productivity (less than 1.0 percent) in 1961–2012 include Burundi, The Gambia, Mauritius, Namibia, and Zimbabwe, whereas those with negative annual average growth rates in labor productivity include Angola, Botswana, Burundi, Democratic Republic of the Congo, Ethiopia, The Gambia, Liberia, Mozambique, Namibia, and Zimbabwe. Several of the countries classified recently as fast-growing agricultural economies (particularly Angola, Mozambique, and Rwanda—reflected in their high land and labor productivity growth rates in 2000–2012) show worse performance in the overall 1961–2012 trend, because of their initial low levels of land and labor productivity emerging from civil war.

It is clear from the results that high performance in one indicator of PFP does not mean equally high performance in other PFP indicators. South Africa, for example, is the top performer in labor productivity (with an average of $5,577/worker in 1961–2012), but has very low land productivity (with an average of only $91/ha in the same period). Figure 2.2 shows countries' relative rankings in the two indicators, using the average annual levels in 2000–2012 for illustration. Only Mozambique and Zambia have the same ranking in both measures, as the second- and third-lowest performers. Analyzing the trend by subperiods (1961–1970, 1971–1980, 1981–1990, 1991–2000, and 2001–2012), the 2000s saw strong positive growth in both land and labor productivity in many countries, headed by Sierra Leone and followed by Angola, Cameroon, Nigeria, and Morocco. These four countries experienced roughly equal average annual growth rates in land and labor productivity (Figure 2.3a and b).

SUMMARY OF FINDINGS ON TRENDS IN LAND AND LABOR PRODUCTIVITY

To summarize, we find that the trends in land and labor productivity are highly variable in different dimensions across different parts of Africa. High performance in one indicator does not necessarily mean equally high performance in the other indicator. Looking at the annual trends over the entire

FIGURE 2.2 Land and labor productivity for selected countries (average 2000–2012)

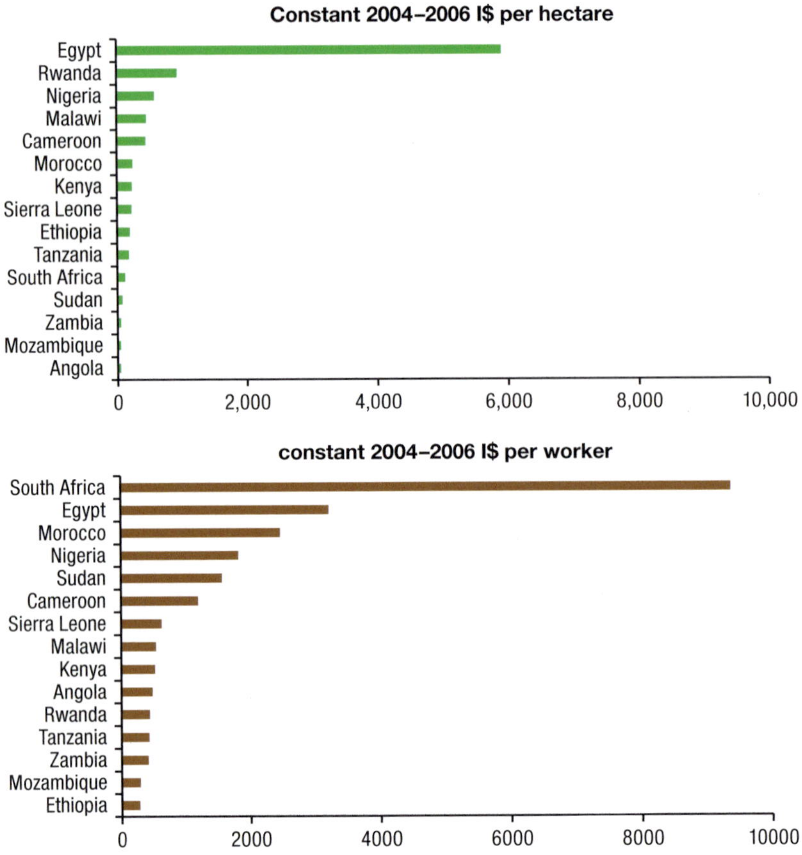

Source: Authors' calculation and representation based on FAO (2014).

Notes: Countries are large agricultural economies with at least 3 percent of Africa's total agricultural output, or fast-growing agricultural economies surpassing the CAADP agricultural growth rate target of 6 percent per year. I$ = international dollar.

1961–2012 period, we find that land productivity has risen much faster than labor productivity in Africa as a whole and in the majority of the subregions and countries analyzed. Looking at the trends by subperiods (1961–1970, 1971–1980, 1981–1990, 1991–2000, and 2001–2012), we find a mostly slower or declining rate of increase in both land and labor productivity in the 1970s and 1990s. The 2000s saw strong positive growth, especially in labor productivity, which is consistent with the finding of a decline in the poverty rate in Africa from the long-standing average of 50 percent to 47 percent in 2008 (World Bank 2012).

FIGURE 2.3 **Growth rate in land and labor productivity for selected countries (annual average 2000–2012)**

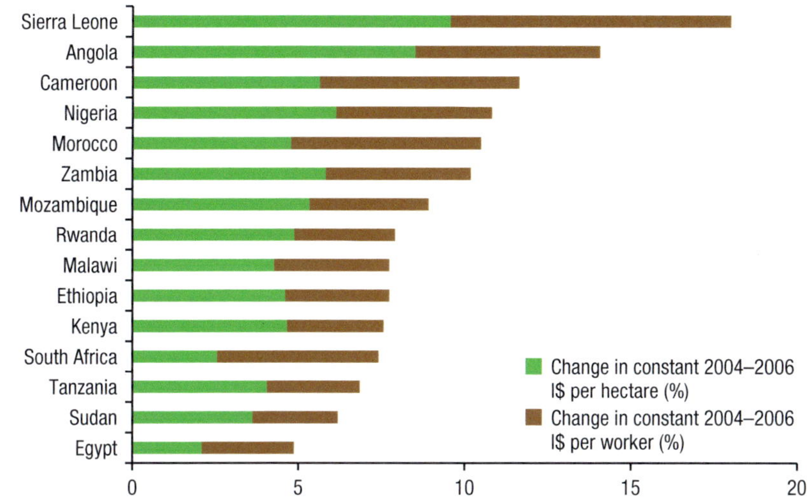

Source: Authors' calculation and representation based on FAO (2014).

Notes: Countries are large agricultural economies with at least 3 percent of Africa's total agricultural output, or fast-growing agricultural economies surpassing the CAADP agricultural growth rate target of 6 percent per year. I$ = international dollar.

However, the analysis for Africa as a whole hides significant differences across different subregions and countries. For example, the northern and southern regions have the highest annual labor productivities. In Egypt, Gabon, Mauritius, and South Africa (and many of the subregions and groups to which these countries belong), labor productivity has risen much faster than land productivity. Egypt stands out as a high performer in both level and growth of land and labor productivity. These differences reflect differences in factors that are not analyzed in the PFPs, including high capital endowment and use of other inputs (for example, fertilizer and irrigation) in those countries and subregions, compared with the largely rainfed average observed in many parts of Africa. Such shortcomings with the PFP measures are addressed in the TFP measures in the next section.

Trends in total factor productivity

By accounting for all factors and inputs used in production, TFP better captures the overall performance of agricultural production than PFP. In this study, TFP growth is decomposed into *technical-efficiency change*, or movement of observations toward the technological frontier, and *technical change*, or movement of the technological frontier. The results are shown in Table 2.4, Figure 2.4, and Appendix Figures 2C.4–2C.6 for the different aggregations and selected countries. Table 2.4 shows the average annual growth rates of TFP and its decomposed parts for the period 1961–2012 and for the same five subperiods (1961–1970, 1971–1980, 1981–1990, 1991–2000, and 2001–2012). Although useful from a quantitative perspective, because the annual averages shown in the table can hide significant variations across time, Figure 2.4 and Appendix Figures 2C.4–2C.6 based on the plots of the underlying TFP index (indexed at 1961=1) give a bird's-eye view of such variations.

TFP GROWTH AT THE AGGREGATE LEVELS

For Africa as a whole, TFP increased at an annual average growth rate of 0.71 percent in 1961–2012 (Table 2.4). For SSA, TFP increased at an annual average growth rate of 0.5 percent during the same period. These findings are consistent with previous estimates, including Headey, Alauddin, and Prasada Rao (2010), who estimated TFP growth for SSA to be less than 0.5 percent in 1970–2001, and Fuglie and Nin-Pratt (2013), who estimated it to be 0.5 percent in 1971–2009. Compared with other developing regions of the world, SSA has an annual average TFP growth rate similar to that of Latin America and the Caribbean, higher than that of South and Southeast Asia, and lower than that of China and Northeast Asia (Fuglie and Nin-Pratt 2013). Headey, Alauddin, and Prasada Rao (2010) show that the estimate for SSA is lower than those for other developing regions when the DEA method is used, but comparable when the stochastic frontier analysis is used.

The above analysis, however, hides the significant variation in TFP growth over time and across different parts of the continent. For Africa as a whole, for example, the results show that TFP remained stagnant until the mid-1980s, when it started to rise steadily at an annual average rate of 0.8 percent in 1981–1990, 1.4 percent in 1991–2000, and 2.0 percent in 2001–2012. The trend for SSA is similar, where TFP also stagnated until the mid-1980s, and then started to rise steadily at an annual average rate of 0.5 percent in 1981–1990, 1.1 percent in 1991–2000, and 2.0 percent in 2001–2012. The rapid growth in TFP in 2001–2012 is consistent with the earlier finding of rapid growth, especially in labor productivity within the same period.

Looking at the trends in the index for the different subregions—geographic region (Figure 2.4), income group (Appendix Figure 2C.4), and REC (Appendix Figure 2C.5)—we find some significant differences across different parts of Africa. We can distinguish three broad categories in terms of the pattern of TFP growth: (1) TFP, as observed for Africa as a whole, stagnated initially until the mid-1980s and then increased—observed in the majority of the subregions and groups; (2) TFP declined initially and then increased and has caught up with or surpassed the 1961 initial level in northern and western Africa, in the LI–3 and MI economic groups, and in the ECOWAS and UMA RECs; and (3) TFP consistently increased, rising slowly initially, in southern Africa and in the COMESA, EAC, and SADC RECs.

TFP GROWTH DECOMPOSITION AT THE AGGREGATE LEVELS

Decomposition of TFP growth into technical-efficiency change (or simply, efficiency change) and technical change, presented in Figure 2.5, shows that the stagnation of or decline in TFP observed in most parts of Africa prior to the mid-1980s was the result of negative efficiency change (Figure 2.5a). This is typically characterized by using more inputs to obtain the same amount of output—as occurs, for example, when inputs are distributed freely to farmers to replace lost harvest (Irz and Thirtle 2004). The negative efficiency change was largest in central and western Africa and in the ECOWAS and UMA RECs, averaging more than –1.4 percent per year. In general, the estimated negative efficiency change associated with the periods prior to the mid-1980s is consistent with the low overall economic growth in the continent in the 1970s and 1980s, and when agricultural output in SSA, for example, grew by only 1.0 percent per year on average (see Chapter 1 of this book).

From the mid-1980s onward, both efficiency change and technical change contributed positively to TFP growth. Figure 2.5b shows that technical change accounted for about 50 percent of the growth in TFP in many parts of the continent, with the contribution being more than 70 percent in northern Africa, the LI–3 economic group, and the EAC and COMESA RECs. The TFP growth decomposition results for the LI–1 economic group and the EAC REC for 1985–2012 stand out: in the LI–1 economic group, technical change accounted for only 7 percent of the growth in TFP; and in the EAC REC, efficiency change was negative.

TRENDS IN TFP AND TFP GROWTH DECOMPOSITION AT THE COUNTRY LEVEL

Appendix Figure 2C.6 shows considerable variation in the trends in levels of TFP, efficiency, and technology across the selected countries, representing the large or small agricultural economies and fast-growing or slow-growing agricultural economies. Nigeria and Egypt are the top two largest agricultural economies. For Nigeria, TFP was stagnant initially, and then declined rapidly until the mid-1980s when it began to rise, and is only recently catching up with the 1961 initial level. Egypt, however, has realized consistent increase in TFP, with more rapid growth since the late 1980s.

TABLE 2.4 Total factor productivity growth, efficiency change, and technical change (%, annual average, 1961–2012)

	Subperiods					
	1961–1970			1971–1980		
Aggregations	TFP	Eff	Tech	TFP	Eff	Tech
Africa	−0.07	−0.88	0.83	−0.29	−1.10	0.83
Africa, south of the Sahara	0.13	−0.71	0.86	−0.37	−1.24	0.91
Geographic location						
Central	−0.44	−1.25	0.82	−0.59	−2.24	1.85
Eastern	0.37	−0.68	1.10	0.86	−0.22	1.07
Northern	−0.87	−1.68	0.75	−0.02	−0.58	0.51
Southern	0.14	−0.37	0.52	1.43	0.99	0.42
Western	0.05	−0.78	0.84	−3.47	−4.10	0.67
Economic group						
LI-1	−0.17	−0.86	0.70	−0.12	−1.90	2.07
LI-2	0.16	−0.75	0.97	0.21	−0.67	0.88
LI-3	−0.45	−1.93	1.47	1.25	0.27	0.90
MI	−0.11	−0.81	0.70	−0.74	−1.35	0.63
Regional Economic Community						
CEN-SAD	−0.14	−0.97	0.84	−1.00	−1.63	0.62
COMESA	0.28	−0.80	1.11	0.78	−0.33	1.19
EAC	0.89	−0.07	0.96	1.26	−0.44	1.84
ECCAS	−0.11	−0.98	0.87	−0.39	−2.02	1.71
ECOWAS	0.05	−0.78	0.84	−3.47	−4.10	0.67
IGAD	0.07	−1.13	1.26	1.28	0.08	1.21
SADC	0.25	−0.27	0.54	0.73	−0.08	0.95
UMA	−2.40	−3.10	0.63	−0.85	−1.52	0.56
Other economic groups						
Large	0.09	−0.70	0.83	−0.46	−0.91	0.50
Small	1.01	−0.24	1.26	0.37	−1.21	1.71
Fast-growing	0.47	−0.28	0.74	−3.71	−4.34	0.69
Slow-growing	−0.01	−1.06	1.08	−0.30	−1.99	1.88
Selected countries						
Large						
Egypt	0.48	−0.38	0.86	0.57	0.11	0.46
Ethiopia	−1.07	−3.08	2.08	−0.67	−0.73	0.06
Kenya	0.62	−0.05	0.67	1.68	1.42	0.26
Morocco	−1.75	−2.32	0.58	−0.99	−1.42	0.43

	Subperiods (continued)									All		
	1981–1990			1991–2000			2001–2012			1961–2012		
	TFP	Eff	Tech	TFP	Eff	Tech	TFP	Eff	Tech	TFP	Eff	Tech
	0.77	0.24	0.44	1.38	0.56	0.77	1.98	1.16	0.82	0.71	−0.02	0.70
	0.47	0.05	0.32	1.11	0.76	0.37	1.98	1.18	0.73	0.50	−0.09	0.56
	0.73	0.54	0.18	1.16	1.11	0.00	1.71	0.61	1.03	0.57	−0.01	0.62
	1.11	0.63	0.39	0.97	1.03	0.17	1.91	0.28	1.45	0.85	0.20	0.67
	1.67	0.82	0.84	2.02	−0.09	2.02	2.18	1.19	1.22	1.36	0.21	1.13
	0.73	0.05	0.60	0.95	−0.20	1.17	2.27	2.12	0.27	0.94	0.16	0.75
	0.26	0.07	0.21	2.02	1.51	0.38	2.28	1.81	0.45	0.12	−0.33	0.42
	0.58	0.35	0.18	1.01	0.94	−0.07	0.66	0.75	0.26	0.36	−0.15	0.54
	1.10	0.61	0.34	−0.13	−0.02	0.08	2.03	0.81	1.08	0.50	−0.01	0.51
	0.52	0.11	0.40	1.16	0.64	0.42	1.73	0.47	1.15	0.55	−0.27	0.76
	0.71	0.10	0.54	1.93	0.72	1.13	2.04	1.36	0.73	0.81	0.00	0.77
	0.89	0.37	0.46	1.93	0.93	0.90	2.05	1.07	0.96	0.75	−0.02	0.71
	1.28	0.48	0.70	1.81	0.90	0.98	1.81	0.18	1.55	1.18	0.17	1.01
	1.77	1.02	0.69	−0.48	−1.23	0.84	1.91	0.27	1.36	0.95	−0.15	1.13
	0.63	0.29	0.32	1.43	1.13	0.26	2.41	1.52	1.12	0.79	0.07	0.76
	0.26	0.07	0.21	2.02	1.51	0.38	2.28	1.81	0.45	0.12	−0.33	0.42
	1.37	0.77	0.45	1.07	1.40	0.03	2.14	0.34	1.60	0.99	0.32	0.70
	0.59	0.10	0.41	0.91	0.18	0.68	1.52	1.29	0.24	0.67	0.03	0.62
	1.68	1.72	0.13	0.95	−0.52	1.34	2.81	2.97	0.04	0.94	0.31	0.64
	0.73	0.03	0.58	1.82	0.74	0.98	1.99	1.13	0.85	0.82	0.03	0.73
	0.90	0.41	0.31	−0.96	−1.13	0.23	0.87	0.41	0.44	0.28	−0.35	0.63
	−0.36	−0.66	0.30	2.10	2.00	0.09	2.99	2.57	0.44	0.06	−0.34	0.40
	0.93	0.76	0.20	0.90	0.31	0.59	−0.62	−1.06	0.36	0.35	−0.42	0.78
	1.72	0.14	1.57	2.23	−0.04	2.27	2.08	−0.05	2.13	1.57	0.08	1.49
	−0.65	−0.65	0.00	1.71	1.57	0.13	2.76	1.98	0.77	0.57	0.36	0.21
	1.41	0.69	0.72	−0.06	−0.25	0.19	2.81	−0.02	2.83	1.34	0.45	0.80
	3.27	3.27	0.00	−0.15	−0.96	0.82	4.10	4.06	0.04	1.15	0.76	0.39

(continued)

TABLE 2.4 (continued)

	Subperiods					
	1961–1970			1971–1980		
Aggregations	TFP	Eff	Tech	TFP	Eff	Tech
Nigeria	0.50	−0.29	0.79	−6.44	−7.25	0.88
South Africa	0.17	−0.23	0.41	2.61	2.05	0.55
Sudan	−0.64	−1.70	1.07	0.95	0.32	0.63
Tanzania	0.82	0.65	0.17	−1.14	−1.19	0.06
Small						
Botswana	1.47	0.78	0.68	−4.20	−4.93	0.76
Gabon	1.11	0.00	1.11	0.23	−0.19	0.42
Gambia, The	2.11	−0.40	2.52	−6.13	−6.13	0.00
Guinea-Bissau	−3.03	−3.76	0.76	0.51	0.51	0.00
Mauritius	1.34	−0.34	1.69	3.09	−2.08	5.29
Swaziland	1.14	0.30	0.84	2.37	2.05	0.31
Fast-growing						
Angola	−0.42	−0.46	0.04	−3.50	−3.61	0.11
Cameroon	0.66	−0.18	0.84	−1.69	−1.81	0.13
Malawi	−0.14	−0.79	0.66	0.97	0.97	0.00
Mozambique	−0.23	−0.24	0.01	−3.13	−3.13	0.00
Nigeria	0.50	−0.29	0.79	−6.44	−7.25	0.88
Rwanda	3.59	1.31	2.25	1.70	0.71	0.98
Sierra Leone	0.08	−0.61	0.69	−0.42	−0.44	0.02
Zambia	−0.93	−1.25	0.32	1.88	1.80	0.08
Slow-growing						
Burundi	−0.27	−1.01	0.75	−0.73	−2.10	1.40
Congo, Dem. Rep.	−0.09	−0.93	0.85	−0.20	−3.49	3.40
Liberia	1.44	−0.06	1.50	−0.18	−0.50	0.33
Mauritius	1.34	−0.34	1.69	3.09	−2.08	5.29
Namibia	2.10	0.49	1.61	−1.31	−2.63	1.36
Tunisia	−1.24	−2.12	0.90	−1.00	−1.45	0.46
Zimbabwe	0.14	−1.69	1.85	0.27	0.27	0.00

Source: Authors' calculation based on the Malmquist index model results.

Notes: TFP = total factor productivity growth; Eff = efficiency change; Tech = technical change. LI-1 = low income, more favorable agriculture, and mineral rich; LI-2 = low income, more favorable agriculture, and nonmineral rich; LI-3 = low income and less favorable agriculture; MI = middle income. CEN-SAD = Community of Sahel-Saharan States; COMESA = Common Market for Eastern and Southern Africa; EAC = East African Community; ECCAS = Economic Community of Central African States; ECOWAS = Economic Community of West African States; IGAD = Intergovernmental Authority on Development; SADC = Southern African Development Community; UMA = Union du Maghreb Arabe. Large agricultural economies have at least 3.0 percent of Africa's total agricultural output; small agricultural economies have less than 0.1 percent of Africa's total agricultural output; fast-growing agricultural economies surpass the CAADP agricultural growth rate target of 6.0 percent per year; and slow-growing agricultural economies have an agricultural growth rate of less than 1.0 percent per year.

Subperiods (continued)									All		
1981–1990			1991–2000			2001–2012			1961–2012		
TFP	Eff	Tech	TFP	Eff	Tech	TFP	Eff	Tech	TFP	Eff	Tech
0.26	0.00	0.25	3.01	2.97	0.04	2.73	2.69	0.04	–0.16	–0.50	0.33
0.92	–0.01	0.94	1.24	–0.85	2.10	2.50	1.84	0.65	1.40	0.22	1.18
–0.11	–0.11	0.00	5.28	5.27	0.01	2.35	–0.52	2.89	1.43	0.77	0.65
0.96	0.81	0.15	0.62	0.59	0.02	0.62	–0.11	0.73	0.18	0.04	0.14
–0.56	–0.56	0.00	–2.61	–2.61	0.00	2.19	2.18	0.01	–0.65	–0.91	0.27
0.18	0.09	0.09	1.36	0.25	1.11	1.00	0.05	0.94	0.69	–0.08	0.77
–0.64	–0.64	0.00	1.09	1.09	0.00	–1.42	–2.42	1.02	–1.24	–1.49	0.26
2.09	2.07	0.02	1.02	0.91	0.11	2.51	0.32	2.19	0.66	0.38	0.27
0.85	0.85	0.00	–2.33	–2.33	0.00	0.33	0.30	0.03	0.25	–0.82	1.09
1.84	0.31	1.53	–1.69	–2.40	0.73	0.59	0.35	0.23	1.02	0.10	0.91
–1.46	–1.46	0.00	3.09	3.09	0.00	5.23	5.02	0.20	0.40	0.33	0.07
0.57	0.57	0.00	0.06	0.02	0.05	4.67	1.85	2.76	0.72	0.36	0.36
–0.26	–0.26	0.00	2.83	2.32	0.50	1.11	0.04	1.07	0.77	0.30	0.47
–2.46	–2.46	0.00	2.80	2.80	0.00	3.69	3.46	0.22	–0.61	–0.63	0.02
0.26	0.00	0.25	3.01	2.97	0.04	2.73	2.69	0.04	–0.16	–0.50	0.33
0.89	–0.49	1.38	3.72	1.06	2.63	2.18	0.42	1.76	2.07	0.13	1.94
–0.37	–0.39	0.01	–0.40	–0.40	0.00	6.15	2.29	3.77	0.37	–0.11	0.48
1.65	1.65	0.00	–0.05	–0.12	0.07	4.38	4.21	0.16	0.76	0.69	0.07
1.48	0.74	0.73	0.58	–0.25	0.83	–3.97	–4.28	0.32	–0.14	–1.04	0.91
0.71	0.62	0.09	1.92	1.66	0.26	–1.90	–2.21	0.32	0.44	–0.42	0.86
2.16	0.16	1.99	1.78	1.69	0.09	–1.69	–1.71	0.02	0.12	–0.70	0.83
0.85	0.85	0.00	–2.33	–2.33	0.00	0.33	0.30	0.03	0.25	–0.82	1.09
–0.63	–0.63	0.00	–0.83	–0.83	0.00	0.34	0.16	0.19	–0.11	–0.63	0.53
0.85	0.69	0.16	–0.16	–2.10	1.99	1.68	1.50	0.18	0.59	–0.31	0.90
1.75	1.75	0.00	1.20	1.20	0.00	–1.18	–1.23	0.05	–0.07	–0.18	0.12

FIGURE 2.4 Levels of total factor productivity, efficiency, and technology by geographic region (1961–2012: indexed at 1961=1)

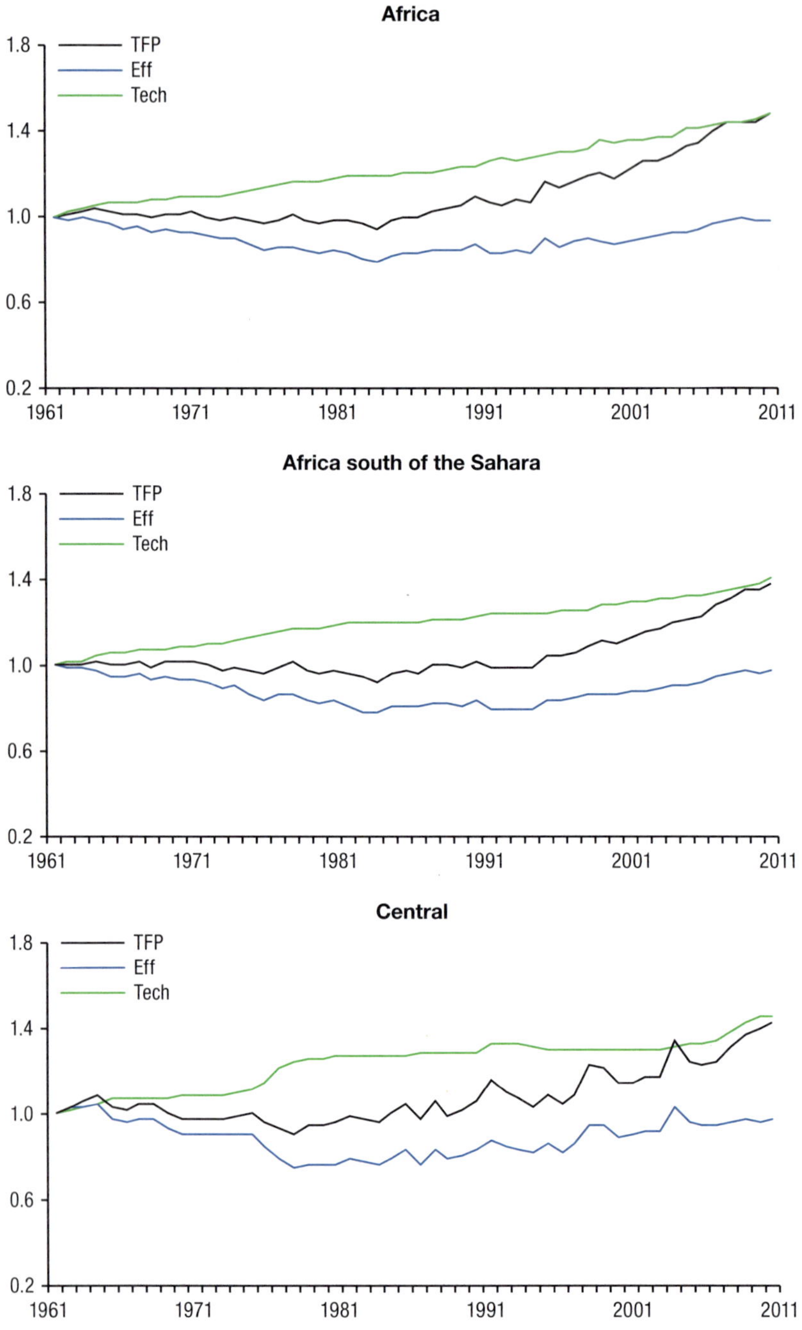

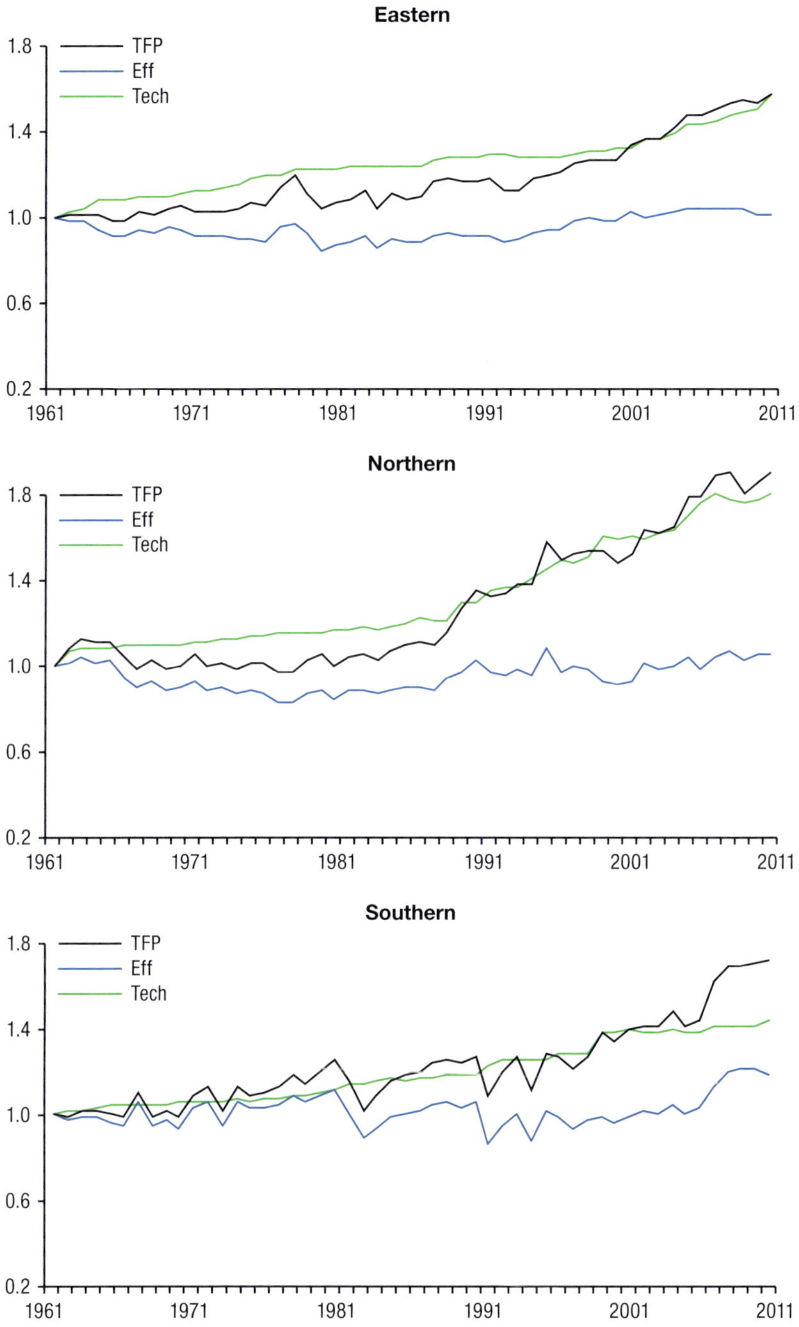
Eastern
1.8
1.4
1.0
0.6
0.2
TFP
Eff
Tech
1961
1971
1981
1991
2001
2011
Northern
1.8
1.4
1.0
0.6
0.2
TFP
Eff
Tech
1961
1971
1981
1991
2001
2011
Southern
1.8
1.4
1.0
0.6
0.2
TFP
Eff
Tech
1961
1971
1981
1991
2001
2011

(continued)

FIGURE 2.4 (continued)

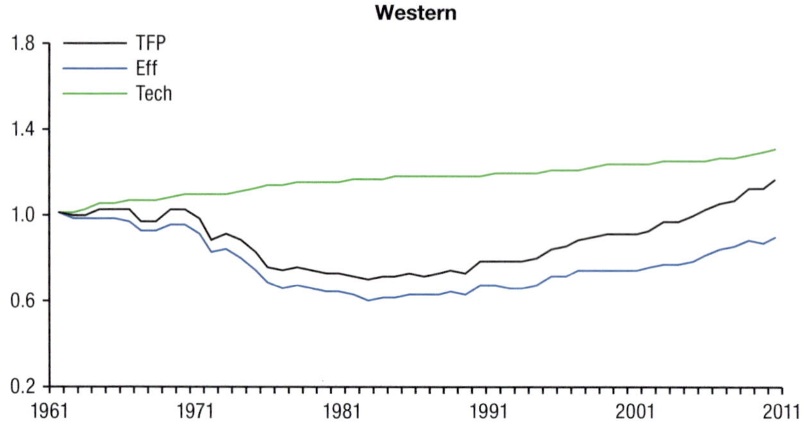

Source: Authors' calculation and illustration based on TFP model results.

Notes: TFP = total factor productivity; Eff = efficiency; Tech = technology.

FIGURE 2.5A Total factor productivity growth decomposition by group (%, annual average 1961–1985)

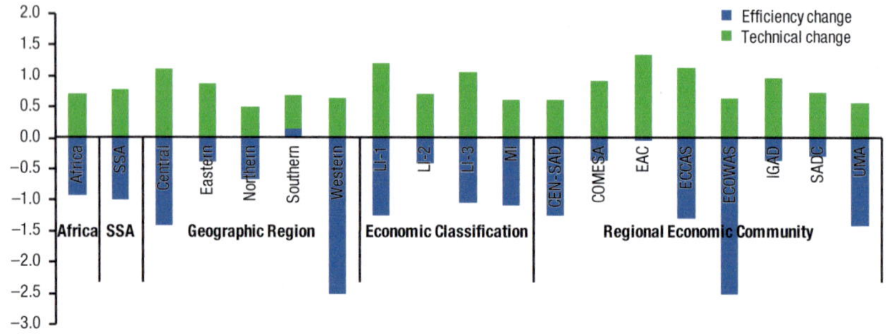

Source: Authors' calculation and illustration based on TFP model results.

Notes: LI–1 = low income, more favorable agriculture, and mineral rich; LI–2 = low income, more favorable agriculture, and nonmineral rich; LI–3 = low income and less favorable agriculture; MI = middle income. CEN-SAD = Community of Sahel-Saharan States; COMESA = Common Market for Eastern and Southern Africa; EAC = East African Community; ECCAS = Economic Community of Central African States; ECOWAS = Economic Community of West African States; IGAD = Intergovernmental Authority on Development; SADC = Southern African Development Community; SSA = Africa south of the Sahara; UMA = Union du Maghreb Arabe.

As most of the observations are on or close to the technological frontier, TFP growth is dominated by technical change. Malawi and Angola are the fastest-growing agricultural economies in the recent past decade in terms of overall agricultural growth. Whereas there has been little technical change in Angola, its remarkable agricultural growth starting in the mid-1990s reflects its emergence

FIGURE 2.5B Total factor productivity growth decomposition by group (%, annual average, 1985–2012)

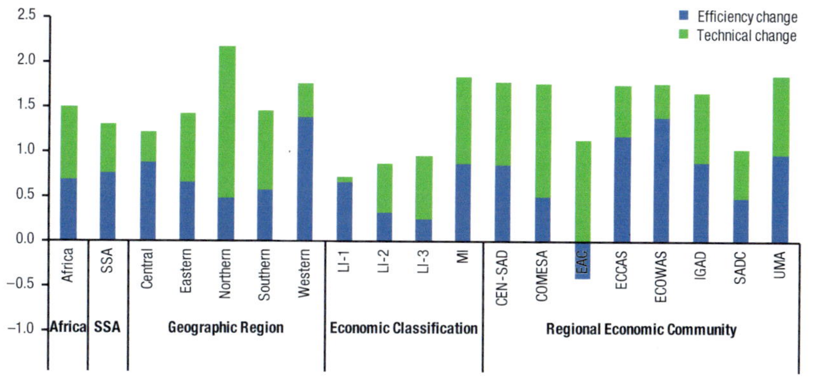

Source: Authors' calculation and illustration based on TFP model results.

Notes: LI–1 = low income, more favorable agriculture, and mineral rich; LI–2 = low income, more favorable agriculture, and non-mineral rich; LI–3 = low income and less favorable agriculture; MI = middle income. CEN-SAD = Community of Sahel-Saharan States; COMESA = Common Market for Eastern and Southern Africa; EAC = East African Community; ECCAS = Economic Community of Central African States; ECOWAS = Economic Community of West African States; IGAD = Intergovernmental Authority on Development; SADC = Southern African Development Community; SSA = Africa south of the Sahara; UMA = Union du Maghreb Arabe.

from war and catching up rapidly with and surpassing the initial 1961 level. In Malawi, TFP remained at the initial 1961 level until the late 1990s, when it increased as a result of both positive technical change and efficiency change.

The trends in The Gambia and Botswana, representing Africa's two smallest agricultural economies, are very similar, with TFP declining initially and then becoming stagnant over time in the remaining periods. Mauritius and Namibia, the two slowest-growing agricultural economies, also have similar trends in TFP growth, increasing initially, declining toward the 1961 level, and then remaining stagnant or fluctuating around the 1961 level.

Several of our findings are consistent with previous estimates. For example, Headey, Alauddin, and Prasada Rao (2010) find acceleration (2.5–5.5 percent) in TFP growth in 1985–2001 for Angola, Egypt, and Malawi, compared with sluggish growth in 1970–1985 for Egypt and Malawi (0.3–0.5 percent) or negative growth for Angola (−0.5 percent). Our finding for Nigeria, however, differs from that of Headey, Alauddin, and Prasada Rao (2010), who find positive TFP growth (0.5 percent) for Nigeria in 1970–1985 compared with our negative growth rate in the 1970s (−6.4 percent) and sluggish growth rate in the 1980s (0.3 percent) (Table 2.4 and Appendix Figure 2C.6). Our finding of rapid TFP growth for Nigeria from 1985 onward is consistent with that of Headey, Alauddin, and Prasada Rao (2010). Irz and Thirtle (2004) find negative TFP

FIGURE 2.6 **Total factor productivity growth decomposition at country level (%, annual average 1985–2012)**

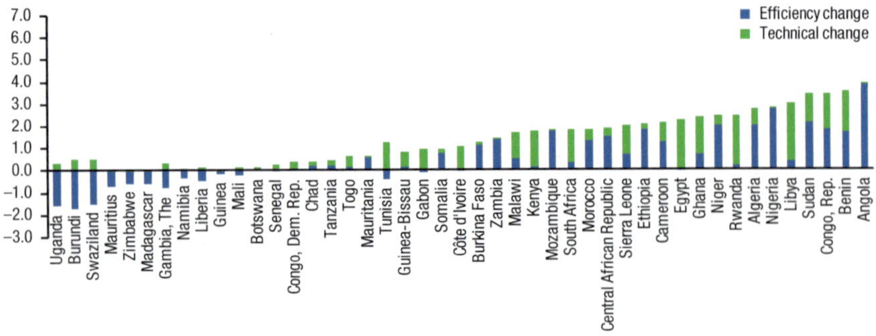

Source: Authors' calculation and illustration based on TFP model results.

growth (−2.3 percent) in Botswana's traditional agriculture sector in 1979–1996, but positive TFP growth (1.2 percent) in the commercial agriculture sector. They also find significant technological regression (−2.9 percent) in Botswana's traditional agriculture sector—a finding that is fundamentally different from the findings in this study because of differences in methodologies used.[5]

Whereas the above analysis shows the patterns in TFP growth over the entire 1961–2012 period considered here, the patterns in more recent years better reflect the current trajectory of the countries in agricultural transformation. We analyze two subperiods: 1985–2012, representing the general period following the recovery or turnaround in the decline in TFP growth (Figure 2.6), and 2000–2012 (Figure 2.7). For the first subperiod, the year 2000 is when African countries signed the Millennium Declaration that

5 The DEA-Malmquist index used in this study assumes *sequential technology* instead of *contemporaneous technology* in the sense that there is dependence between the production sets across time. This approach is based on the assumption that "production units can always do what they did before" in the production process, ruling out the possibility of technological regression or negative technical change (see Appendix 2A for details). This is captured in the efficiency change component, which in Botswana we find to be negative in the 1970s (−4.9 percent), 1980s (−0.6 percent), and 1990s (−2.6 percent) (Table 2.4 and Appendix Figure 2C.6). Therefore, our overall result of declining TFP in Botswana in 1971–2000 is consistent with the "technological regression" of Irz and Thirtle (2004). In general, differences in the decomposition from using the sequential versus contemporaneous technology are more pronounced in the 1970s and 1980s for SSA countries with a low-capital agriculture sector (particularly Angola, Botswana, Ethiopia, Mozambique, and Namibia). This is reflected by the largely zero technical change and negative efficiency change for those countries in those periods (Table 2.4). The technology frontier collapses in those periods, which would result in technological regression or negative technical change when the contemporaneous technology assumption is used in calculating the Malmquist index.

FIGURE 2.7 Total factor productivity growth decomposition at country level (%, annual average 2000–2012)

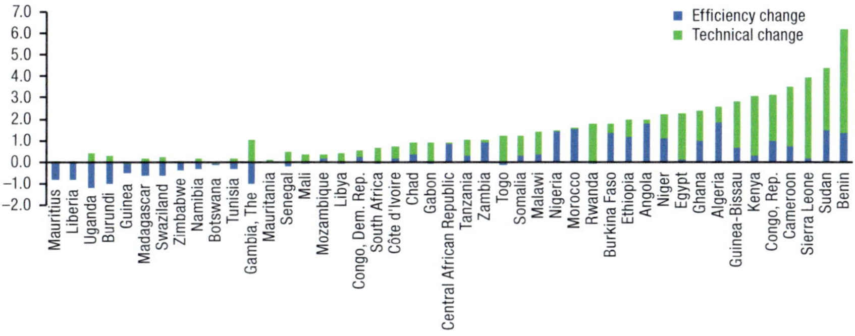

Source: Authors' calculation and illustration based on TFP model results.

defined the Millennium Development Goals; it also represents the start of the most recent decade of rapid growth in labor productivity seen earlier.

During 1985–2012, Figure 2.6 shows that about one-third of the 45 countries achieved an annual average TFP growth rate of at least 2.0 percent, with Angola in front (at 3.9 percent), followed by Benin, Republic of the Congo, Sudan, and Libya (at 3.0 percent or higher). Technical change accounted for less than one-half of the TFP growth in the majority of all of the countries. For the period 2000–2012, Figure 2.7 also shows that about one-third of the countries achieved an average annual TFP growth rate of at least 2.0 percent. However, the ranking of countries shifts, with Benin, Sudan, Sierra Leone, Cameroon, Republic of the Congo, and Kenya taking over the lead, with at least a 3.0 percent annual average growth rate in TFP. Many more countries also show positive and large rates of technical change.

SUMMARY OF FINDINGS ON TFP TRENDS AND TFP GROWTH DECOMPOSITION

For Africa as a whole, TFP increased at an annual average growth rate of 0.71 percent in 1961–2012, and in SSA by 0.5 percent. These figures are consistent with previous estimates, but hide the significant variation in TFP growth over time and across different parts of the continent. For Africa as a whole, TFP remained stagnant between 1961 and the mid-1980s, when it started to rise steadily at an annual average rate of 0.8 percent in 1981–1990, 1.4 percent in 1991–2000, and 2.0 percent in 2001–2012. The rapid growth in TFP in 2001–2012 is consistent with the earlier finding of rapid growth, especially in labor productivity, within the same period.

TABLE 2.5 Input and capital per worker and technical change, annual average (1995–2012)

Technical change (average %)	Land (ha)	Crop capital (2005 US$)	Livestock capital (2005 US$)	Fertilizer (kg)	Feed (kg)
High (2.0%)	21.4	6.9	2.3	94.6	1.6
Medium (0.6%)	14.5	3.4	3.4	83.0	1.0
Low (0.1%)	33.2	0.8	2.8	12.0	0.3

Source: Authors' calculation and illustration based on TFP model results.
Note: ha = hectare; kg = kilogram.

TFP growth decomposition shows that the widespread stagnation or decline in TFP observed prior to the mid-1980s was due to loss in efficiency or negative efficiency change, also consistent with previous findings. The rate of negative efficiency change was largest in central and western Africa and in the ECOWAS and UMA RECs, averaging more than –1.4 percent per year. From the mid-1980s onward, however, efficiency change and technical change contributed positively and equally to TFP growth. During the periods of recovery and turnaround, technical change contributed more than 70.0 percent of TFP growth in northern Africa, the LI–3 economic group, and the EAC and COMESA RECs, but only 7.0 percent in the LI–1 economic group.

At the country level, Nigeria and Egypt, which are the two largest agricultural economies in Africa in terms of their share of Africa's total agriculture value-added, show distinct TFP growth paths, particularly prior to the mid-1980s. For example, compared with the U-shaped pattern observed in Nigeria, Egypt realized a consistent increase in TFP, with more rapid growth since the late 1980s; however, as most of the observations are on or close to the technological frontier, TFP growth in Egypt is dominated by technical change. Considering the period 1985–2012, about one-third of the 45 countries analyzed achieved annual average TFP growth rates of at least 2 percent, with countries in the lead, including Angola, Benin, Congo, Sudan, and Libya (at 3 percent in 1985–2012), and Benin, Sudan, Sierra Leone, Cameroon, Congo, and Kenya taking over the lead in the more recent periods of 2000–2012.

A key question then is, what is driving the strong role of technical change in TFP growth in northern Africa, the LI–3 economic group, and the EAC and COMESA RECs compared with, for example, countries in the LI–1 economic group? In calculating the Malmquist index, countries with similar input and capital intensities below and at the frontier are compared, which results in different speeds of frontier expansion for the different groups of comparable countries. We find that the frontier for low-input and low-capital countries, mostly in SSA, has been moving slowly or not at all (under the sequential-technology assumption) or collapsing (under the contemporary-technology assumption). On the other hand, the frontiers for the high-input and high-capital countries have been moving steadily and faster. Table 2.5 shows that countries with larger endowments of crop capital and using more fertilizer and feed per worker (likely also those with more commercial-oriented agriculture) are those experiencing rapid technical change.

Correlation between PFP and TFP measures

From the above analysis of PFP and TFP measures, we have seen that whereas TFP measures can provide a better sense of the changes in agricultural productivity, measuring TFP can be challenging (especially for developing countries), compared with measuring PFP, which is straightforward and has uncomplicated data requirements. A key question, therefore, concerns not which measure to use instead of the other, but how they complement each other.

Figure 2.8 shows that the patterns of growth in the PFP and TFP measures are quite different. Growth in land productivity increased from an initial average rate of 2.2 percent achieved in the 1960s to 3.0 percent in the 1970s, reached a high of 3.9 percent in the 1980s, and then declined in the 1990s and 2000s to the average rate achieved in the 1960s. Growth in labor productivity also increased from its initial average rate of 1.9 percent achieved in the 1960s to 2.3 percent in the 1970s, then dropped rapidly to 1.2 in the 1980s, and then increased to 2.1 percent in the 1990s and 3.2 percent in

FIGURE 2.8 Land, labor, and total factor productivity growth in Africa (%, annual average 1961–2012)

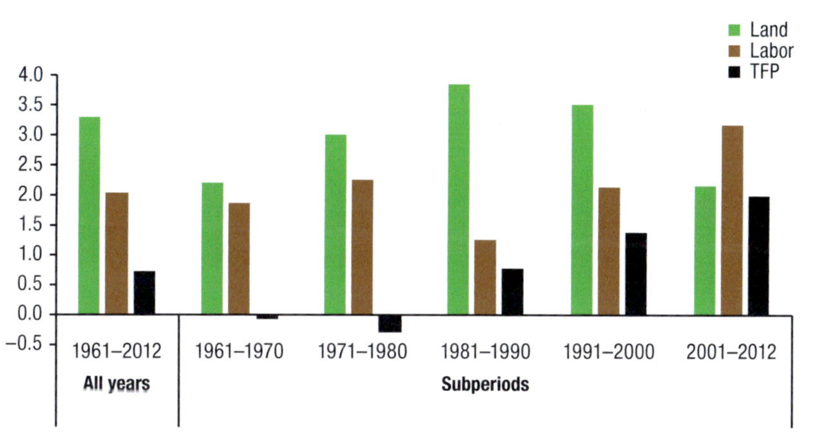

Source: Authors' calculation and illustration based on FAO (2014) and TFP model results.
Note: TFP = total factor productivity.

TABLE 2.6 Correlation coefficients between land, labor, and total factor productivity (TFP) growth by technical change and input intensity (1961–2012)

			1961–1985			
Coefficients	Land	Labor	Land	TFP	Labor	TFP
All Africa	0.98	***	0.84	***	0.82	***
Technical change (%)						
Low (0.00)	0.98	***	0.84	***	0.88	***
Medium (0.07)	0.97	***	0.86	***	0.84	***
High (2.94)	0.98	***	0.72	***	0.68	***
Input per worker						
Land (ha)						
Low (2.04)	0.98	***	0.86	***	0.85	***
Medium (6.87)	0.99	***	0.84	***	0.83	***
High (52.66)	0.98	***	0.83	***	0.81	***
Crop capital (2005 US$)						
Low (350)	0.96	***	0.86	***	0.84	***
Medium (900)	0.98	***	0.77	***	0.74	***
High (6,160)	0.98	***	0.87	***	0.86	***
Livestock capital (2005 US$)						
Low (420)	0.97	***	0.80	***	0.79	***
Medium (1,260)	0.98	***	0.83	***	0.81	***
High (5,190)	0.98	***	0.85	***	0.84	***
Fertilizer (kg)						
Low (0.85)	0.97	***	0.68	***	0.65	***
Medium (6.23)	0.98	***	0.85	***	0.84	***
High (142.71)	0.98	***	0.91	***	0.91	***

Source: Authors' calculation based on FAO (2014) and TFP model results (2011).

Notes: *** = significant at 1% level; ha = hectare; kg = kilogram. Figures in parentheses are average values for the respective category, where low, medium, and high are terciles of the indicator.

the 2000s. Annual average TFP growth rates were negative in the first two decades, and then increased to 0.8 percent in the 1980s, 1.4 percent in the 1990s, and 2.0 percent in the 2000s As such, for Africa as a whole from 1961 to 1980, we observe a U-shaped trend for growth in land and labor productivity, but a declining trend for TFP growth. From the 1980s onward, there was an increasing trend for growth in labor productivity and TFP, but a declining trend for growth in land productivity.

	1986–2012					
Coefficients	Land	Labor	Land	TFP	Labor	TFP
All Africa	0.96	***	0.74	***	0.74	***
Technical change (%)						
Low (0.00)	0.96	***	0.81	***	0.80	***
Medium (0.07)	0.94	***	0.82	***	0.84	***
High (2.94)	0.96	***	0.58	***	0.57	***
Input per worker						
Land (ha)						
Low (2.04)	0.94	***	0.64	***	0.66	***
Medium (6.87)	0.98	***	0.87	***	0.87	***
High (52.66)	0.97	***	0.78	***	0.77	***
Crop capital (2005 US$)						
Low (350)	0.95	***	0.82	***	0.84	***
Medium (900)	0.96	***	0.54	***	0.54	***
High (6,160)	0.97	***	0.85	***	0.84	***
Livestock capital (2005 US$)						
Low (420)	0.97	***	0.72	***	0.72	***
Medium (1,260)	0.92	***	0.60	***	0.61	***
High (5,190)	0.98	***	0.89	***	0.88	***
Fertilizer (kg)						
Low (0.85)	0.97	***	0.57	***	0.57	***
Medium (6.23)	0.94	***	0.85	***	0.87	***
	0.97	***	0.85	***	0.85	***

Looking at the correlation between the growth rates of the three productivity measures, the results presented in Table 2.6 show differences in the correlation coefficients, which differ by periods—for example, from 1961 to 1985 (during the periods of TFP decline) and from 1986 to 2012 (during the periods of TFP recovery and increase). In general or for Africa as a whole, the coefficients for the correlation between land and labor productivity growth (which are close to 1) are larger than those for the correlations between land productivity and TFP growth and between labor productivity and TFP growth. These patterns hold for the different periods, as well as for different observations grouped according to technical change and input and capital intensities.

Furthermore, the coefficients for the correlation between land and labor productivity growth are the same across the different groups of technical change and input and capital intensities. There are differences, however, in the coefficients for the correlations between land productivity and TFP growth and between labor productivity and TFP growth across the different groups of technical change and input and capital intensities. For example, going from low to high within any group, the coefficients are declining for technical change, increasing for land and fertilizer use per worker, and U-shaped for crop and livestock capital per worker.

Together, the above results suggest that analysis of agricultural productivity in Africa involving analysis of both PFP and TFP measures will provide strong complementarity. Because growth in land and labor productivity is strongly correlated (which is confirmed by the plots in Figure 2.1 and Appendix Figures 2C.1–2C.3 for different parts of the continent), using either of the PFP measures for a rapid assessment of changes in agricultural productivity is likely to be acceptable. But, because there are differences in the productivity effects of different factors and inputs, analysis of which are excluded in PFP measures, the policy implications of PFP analysis are not likely to be strong. To get a better sense of the long-term changes in agricultural productivity that are attributable to technological change, such as required for CAADP, analysis of TFP and TFP decomposition will be necessary and critical.

Conclusions and Implications

This chapter assessed changes in indicators of both PFP and TFP (using the DEA-Malmquist index) measures over time (1961–2012) and across different parts of Africa at the aggregate, subregional, and country levels. The results shed light on the relative sources of agricultural growth, on the resource and factor constraints for increasing agricultural production sustainably, and on the relative usability of the different indicators in strategic monitoring of agriculture sector performance. Between 1961 and 2012, we find that for Africa as a whole, land productivity increased the fastest, at a 3.3 percent annual average, followed by labor productivity at an annual average of 2.0 percent, and then TFP at an annual average of 0.7 percent, with technical change accounting for nearly all of the TFP growth. These findings are consistent with the literature, but hide significant differences across different subregions and countries, as well as over different subperiods of time.

Looking at differences across different parts of Africa, we found that the southern region, for example, had relatively high labor productivities but relatively low land productivities compared with other geographic regions. This finding is consistent with the high land–labor and capital–labor intensities associated with large plantations and more mechanized, commercial agricultural operations that take place there. Land productivities in northern Africa are as high as in southern Africa, but whereas labor productivity has risen much faster than land productivity in southern Africa, land and labor productivities in northern Africa have risen at roughly equal rates. The trends observed in western Africa are closer to those observed for Africa as a whole, compared with the generally lower average levels and growth rates observed in central and eastern Africa. For TFP, three broad patterns of growth were found: (1) TFP, as observed for Africa as a whole, stagnated initially until the mid-1980s and then increased, as observed in the majority of the subregions and groups; (2) TFP declined initially and then increased and has caught up with or surpassed the 1961 initial level in northern and western Africa, the LI–3 and MI economic groups, and the ECOWAS and UMA RECs; and (3) TFP consistently increased, rising slowly initially, in southern Africa and the COMESA, EAC, and SADC RECs.

Looking at differences over different subperiods of time, we found that from 1961 until the mid-1980s there was a U-shaped trend for growth in land and labor productivity, but a declining trend for TFP growth. From the mid-1980s onward, there was an increasing trend for growth in labor productivity and TFP, but a declining trend for growth in land productivity. TFP growth decomposition shows that the widespread stagnation or decline in TFP observed prior to the mid-1980s was the result of negative efficiency change, especially in central and western Africa and in the ECOWAS and UMA RECs. From the mid-1980s onward, however, efficiency change and technical change contributed positively and equally to TFP growth for Africa as a whole, although technical change contributed more than 70 percent of TFP growth in northern Africa, the LI–3 economic group, and the EAC and COMESA RECs, and only 7 percent in the LI–1 economic group.

The findings from both the PFP and the TFP analyses suggest that different policies and investments will be needed in different parts of the continent to increase and sustain high agricultural productivity and growth. However, only the TFP analysis sheds light on the relative sources of agricultural productivity growth to help inform specific strategies to accelerate the expansion of Africa's technical frontier and improve efficiency in production systems. For example, we found that the technological frontier for countries with relatively low-input and low-capital intensities have been moving slowly or not at all, compared with the faster-moving frontier for those with relatively high-input and high-capital intensities. In particular, countries with larger endowment of crop capital and using more fertilizer and feed per worker (likely also those with more commercial-oriented agriculture) are experiencing rapid technical change. Therefore, policies and investments that help farmers to intensify and capitalize their agricultural production processes will be critical for increasing and sustaining high technological advancement in the sector. This support will be particularly important in places with a slowdown in land availability, to help improve rural incomes and further reduce poverty.

Depending on data availability, one important area for additional work that could help sharpen the policy implications of the TFP analysis is using data at the firm or farm level, rather than at the country level as done here and in the literature, which loses the heterogeneity of production systems and decisionmaking units or farms within the country. In general, considering the data and analytical challenges associated with measuring TFP compared with the relatively easy requirements for measuring PFP, analysis of changes in agricultural productivity that involves analysis of both PFP and TFP measures will provide strong complementarity. This will be especially important when comparing production units or systems with different input and factor use intensities, as their respective patterns of growth in PFP and TFP measures are likely to be different.

Appendix 2A: Measuring Total Factor Productivity: The Malmquist Index

The Malmquist index, pioneered by Caves, Christensen, and Diewert (1982) and based on distance functions, became extensively used in the measure and analysis of productivity, after Färe et al. (1994) showed that the index can be estimated using data envelopment analysis (DEA), a nonparametric approach. The nonparametric Malmquist index has been especially popular because it is easy to compute and does not require information about input or output prices or assumptions regarding economic behavior, such as cost minimization and revenue maximization. This ease of use is attractive in the context of African agriculture, where usually market prices for the inputs are either nonexistent or insufficiently reported to provide any meaningful information for land, labor, and livestock. In addition, the nonparametric approach can be applied in a multiple-input, multiple-output setting. Also important is its ability to decompose productivity growth into two mutually exclusive and exhaustive components: changes in technical efficiency over time (catching up) and shifts in technology over time (technical change).

We adopt the following notations and definitions: t=1, ...,T is time period in years; j=1, ..., J is an index of production points or units or countries; m=1, ... , M is an index of outputs; n=1, ..., N is an index of inputs; x_j is a column vector of inputs used by production unit j $(x_{j1}, x_{j2}, ..., x_{jN})$; y_j is a column vector of outputs of production unit j $(y_{j1}, y_{j2}, ..., y_{jM})$; k is the number of country groups k=1, ..., K, where each group corresponds to an agroecological zone; z_j is a row vector of nonnegative weights; and Θ is a scalar "contraction" or "shrinking" factor.

To calculate the output-based Malmquist index, we define the production possibility set (PPS), which contains all the correspondences of input and output vectors that are feasible and within which the production units operate. In our analysis, we will refer to these production units as countries. Denote the PPS for a particular period t (t=1, ..., T) as S^t, such that:

$$S^t = \left\{(x^t, y^t) \in \Re_+^{n+m} \mid x^t \text{ can produce } y^t\right\} \tag{2A.1}$$

The PPS contains all feasible correspondence of inputs $x^t \in \Re_+^N$ capable of producing output levels $y^t \in \Re_+^M$. The set S^t is also referred to as the production technology and can also be represented from the input or output perspective:

$$L^t(y) = \left\{x^t \middle| (x^t, y^t) \in S^t\right\} \tag{2A.2}$$

$$P^t(x) = \left\{y^t \middle| (x^t, y^t) \in S^t\right\} \tag{2A.3}$$

These are alternative ways of describing the possibilities for the transformation of inputs x into outputs y. Figure 2A.1 illustrates the technology in the form of an input possibility set (for periods t and t+1), as defined in equation 2A.2. This is the set of input vectors that can produce output vector y, for the technology in period t and with $x^t \in \Re_+^N$ inputs and $y^t \in \Re_+^M$ outputs.

In the figure, the frontier is defined by two production points (B and C) representing efficient combinations of inputs x_1 and x_2 used by production points B and C in period t (B^t, C^t) and in period t+1 (B^{t+1}, C^{t+1}) to produce, respectively, y^t and y^{t+1}. The frontier of the input possibilities for a given output vector in a particular period is defined as the input vector that cannot be reduced by a uniform factor without leaving the set. Formally, the frontier in input space is represented by the isoquant, such that:

$$I(y) = \left\{x \middle| L(y), \Theta x \notin L(y), \Theta < 1\right\} \tag{2A.4}$$

FIGURE 2A.1 Input possibility set, periods *t* and *t*+1

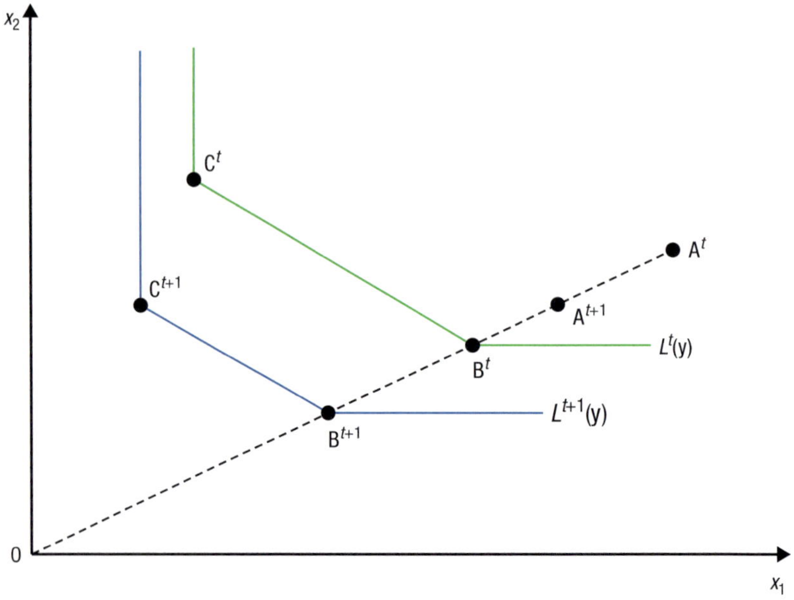

Source: Authors' illustration based on literature review.

In Figure 2A.1, the input set $L^t(y)$ is the space to the right of and above the isoquant defined by B^t and C^t. The efficient subset for this technology is the segment of the isoquant between points B^t and C^t: efficiency is attained at the technological frontier, when a decrease in any input requires an increase in at least another input.

In Figure 2A.1, the technical efficiency of country A is the distance from the production point A to the frontier and can be expressed as the ratio $TE^t(x^t,y^t) = OB^t/OA^t<1$. This is a measure of how far the production point A is from the frontier in period *t*. We can also define efficiency between the production point in *t*+1 and the frontier in *t* $TE^t(x^{t+1}, y^{t+1})$ as the ratio OB^t/OA^{t+1}. In the same way, the efficiency of the production point in *t* with respect to the frontier in *t*+1 is calculated as $TE^{t+1}(x^t, y^t) = OB^{t+1}/OA^t$. Finally, the distance from the production point in *t*+1 to the frontier in *t*+1 is $TE^{t+1}(x^{t+1}, y^{t+1}) = OB^{t+1}/OA^{t+1}$. The efficiency measure equals 1 when the production point in period *t* is on the frontier for period *t*, as is the case for point B in Figure 2A.1. (When evaluating the distance function for a production point *t* relative to some other period's frontier, the distance measure can exceed 1.)

The input-oriented measures of efficiency defined using Figure 2A.1 can be expressed in terms of input-distance functions, which are used to define the Malmquist productivity index. The efficiency of a production unit with respect to the frontier in t is defined using distance functions as:

$$T_i^t(x^t, y^t) = \frac{1}{D_i^t(x^t, y^t)} = 1/sup\{\Theta: [x_j^t/\Theta, y_j^t] \in L^t(y)\} \quad (2A.5)$$

where θ is the coefficient representing the maximum feasible contraction of the input vector x^t at period t given y^t for production unit j, and i indicates that this is an input-oriented distance measure. The analysis that follows will use only input-oriented measures for ease of notation and thus drop the i index in the equations. Distances for points in t with respect to the frontier in t+1 or points in t+1 with respect to the frontier in t are defined similarly. Depending on the technology used as reference, we can define a period t-based or a period t+1-based input-oriented Malmquist index. The period t-based Malmquist index is defined as:

$$M^t = \frac{D^t(x^{t+1}, y^{t+1})}{D^t(x^t, y^t)} \quad (2A.6)$$

Using the technology at t+1 as the reference, the period t+1-based Malmquist index is defined as:

$$M^t = \frac{D^{t+1}(x^{t+1}, y^{t+1})}{D^{t+1}(x^t, y^t)} \quad (2A.7)$$

We can apply these definitions to measure productivity growth in the frontier country B in Figure 2A.1, recalling that, for a frontier country, $D^t(x^t, y^t) = D^{t+1}(x^{t+1}, y^{t+1}) = 1$. For this particular case, the Malmquist indexes defined in equations 2A.6 and 2A.7 are $M_i^t = (OB^t/OB^{t+1}) > 1 = M_i^{t+1} = 1/(OB^{t+1}/OB^t) > 1$. An index greater than 1, as in this example, means that productivity is growing. For a frontier country like B, productivity growth is equivalent to a shift in the frontier. A shift in the frontier upward and to the right, as in Figure 2A.1, can indicate technical progress. In the particular example presented here, the period t-based and period t+1-based Malmquist indexes both result in the same estimate of productivity growth.

The two Malmquist indexes in equations 2A.6 and 2A.7 give the same result only if, as pointed out by Färe et al. (1997), either of the conditions in (i) holds in conjunction with any of those in (ii):

1. $y^{t+1} = \lambda y^t, \lambda > 0$, or technology exhibits implicit Hicks output neutral technical change;

2. $x^{t+1} = x^t$ or $x^{t+1} = \lambda x^t, \lambda > 0$ and technology exhibits constant returns to scale, or technology exhibits constant returns to scale and implicit Hicks-input neutral technical change.

As the choice between either of the two indexes is arbitrary, Färe et al. (1994) defined their Malmquist index as the geometric mean of M_i^t and M_i^{t+1}:

$$M = [M^t \times M^{t+1}]^{1/2} = \left[\frac{D^t(x^{t+1},y^{t+1})}{D^t(x^t,y^t)} \times \frac{D^{t+1}(x^{t+1},y^{t+1})}{D^{t+1}(x^t,y^t)}\right]^{1/2} \tag{2A.8}$$

Growth decomposition

Färe et al. (1994) showed that the Malmquist index could be decomposed into a technical-efficiency change (or simply, efficiency change) component and a technical change component, and that these results applied to the different period-based Malmquist indexes. It follows that:

$$M = \frac{D^{t+1}(x^{t+1},y^{t+1})}{D^t(x^t,y^t)} \left[\frac{D^t(x^{t+1},y^{t+1})}{D^t(x^t,y^t)} \times \frac{D^{t+1}(x^{t+1},y^{t+1})}{D^{t+1}(x^t,y^t)}\right]^{1/2} \tag{2A.9}$$

The ratio outside the square brackets measures the efficiency change between periods t and t+1, or the change in how far observed input is from the minimum potential input that can be used to produce y between periods t and t+1. The technical change component captures the shift of technology (the frontier) between the two periods.

Note that if we decompose the two Malmquist indexes in equations 2A.6 and 2A.7 into their efficiency change and technical change components, the efficiency change index will be the same for both, but they will differ in the way they measure the shift in the frontier (technical change). The index M^t measures the shift in the frontier along a ray through the origin and the production point in t+1. The index M^{t+1} measures the shift in the frontier through the production point in t. The technical change component of the Malmquist index in equation 2A.9 is just the geometric mean of the technical change components in M^t and M^{t+1}.

A value of the efficiency change component of the Malmquist index greater than 1 means that the production unit is closer to the frontier in period t+1 than it was in period t: the production unit is catching up to the frontier. A value of less than 1, which is negative efficiency change, indicates efficiency regress. The same range of values is valid for the technical change component, meaning technical progress when the value is greater than 1 and technical regress when the index is less than 1. Notice that in our example of Figure 2A.1, the efficiency change component of the productivity indexes for country B equals 1, because this country is on the frontier in periods t and in

t+1, implying that there is no change in efficiency. In that case, the Malmquist indexes for country B are equivalent to the technical change component, measuring the shift in the frontier.

The Malmquist index owes some of its popularity to these productivity change decompositions. However, the Malmquist index represents a correct measure of productivity only if the reference technology exhibits constant returns to scale (CRS), as assumed here for country comparisons. This relates to two main problems with the initial definition of the Malmquist index by Caves, Christensen, and Diewert (1982). First, the Malmquist index did not comply with the definition of an "adequate" measure of productivity change because it did not fulfill the property of proportionality. This property states that if outputs are increased in the same proportion from one period to the next while inputs remain the same (that is, output-oriented measure of productivity), then the productivity index should increase in the same proportion. In the case of the Malmquist index, this property requires that the distance functions be linearly homogeneous of degree +1 in outputs and –1 in inputs, which means that the benchmark technology is characterized by CRS.

If the technology exhibits variable returns to scale (VRS), then the Malmquist index does not comply with the proportionality properly; however, most important, it is an inaccurate measure of productivity change, because it ignores the contribution of scale change to productivity change (Grifell-Tatjé and Lovell 1995). In the case of equation 2A.9, if VRS is present, then the efficiency change and technical change components are not actual measures of "pure" efficiency change and technical change, but they both include a scale-efficiency change. There is a discussion in the literature about how to correctly introduce and measure the effect of scale economies in the decomposition of the Malmquist index. A good summary of this discussion and the conceptual interpretation of different decompositions of the Malmquist index can be found in Zofio (2007). Decomposition of technical change into biased and magnitude components is proposed in Färe et al. (1998), for which several variations have been proposed (for example, Balk [2001]).

Because a common CRS technology is assumed for all African countries in calculating the Malmquist index, we are not able to identify efficiency change or technical change resulting from differences in structural characteristics of production in different regions, like natural resource quality and agroecologies. These effects can be obtained by measuring the Malmquist index and decomposition separately for different groups of countries that are grouped according to geographic or agroecological location or other relevant criteria. These effects can then be compared with the results obtained from pooling all countries in a metafrontier Malmquist index (MM). By representing the distance to the frontier of the kth group of a country in this group as $D_k^t(x^t, y^t)$ and the distance of this same country to the metafrontier as $D_F^t(x^t, y^t)$, we can define a technology gap ratio (TGR) at period t as the ratio of the two technical efficiencies. Following Rambaldi, Prasada Rao, and Dolan (2007), the TGR for group k is:

$$TGR_k^t\left(x^t, y^t\right) = \frac{D_F^t(x^t, y^t)}{D_k^t(x^t, y^t)} \qquad (2A.10)$$

As the metafrontier envelops the group frontiers, we have that $D_F^t(x^t, y^t) \leq D_k^t(x^t, y^t)$, which means that $TGR_k^t \leq 1$. Using the distance functions calculated with respect to the group and the metafrontier, Rambaldi, Prasada Rao, and Dolan (2007) show that the efficiency change and technical change components of the Malmquist index can be decomposed as follows:

$$E_F^{t,t+1} = E_k^{t,t+1} \times \frac{TGR_k^{t+1}(x^{t+1}, y^{t+1})}{TGR_k^t(x^t, y^t)} \qquad (2A.11)$$

$$TC_F^{t,t+1} = TC_k^{t,t+1} \times \left[\frac{TGR_k^t}{TGR_k^{t+1}} \times \frac{TGR_k^t(x^{t+1}, y^{t+1})}{TGR_k^{t+1}(x^t, y^t)}\right]^{1/2} \qquad (2A.12)$$

$E_F^{t,t+1}$ and $TC_F^{t,t+1}$ are measures of efficiency change and technical change between t and t+1, respectively, measured with respect to the metafrontier (as represented in equation 2A.9). Equation 2A.11 shows that efficiency change for a particular country relative to the metafrontier is equal to efficiency change within the kth group, times the change in the technology gap between group k and the metafrontier. Similarly, equation 2A.12 shows that technical change for a particular country relative to the metafrontier is equal to the technical change relative to the group frontier, times the geometric mean of the inverse of the technology gap growth index evaluated at (x^t, y^t) with respect to period t+1 technology, and at (x^{t+1}, y^{t+1}) with respect to period t technology. According to Rambaldi, Prasada Rao, and Dolan (2007), this term can be interpreted as the inverse of the relative improvement in the technology gap of a specific country between t and t+1.

In sum, productivity growth measures that result from assuming a common technology will be the same as those obtained by using multiple technologies. The difference is that, with the single-technology assumption, we cannot separate the efficiency change and technical change effects related to changes between the different technologies. With the assumption of different technologies, technical change for countries located in southern Africa, for example, could be decomposed into growth of the southern Africa technology frontier and a reduction in the gap between the southern Africa frontier and the metafrontier. The same decomposition could be applied for countries located in the different geographic locations (Table 2.2), economic groups (Appendix Table 2C.1), or agroecological zones or farming systems (Chapters 3 and 4 of this book). With a single technology, we cannot observe these differential effects, which are beyond the scope of this study.

Estimation by data envelopment analysis

To measure the Malmquist index and decompose efficiency change and technical change based on the concepts of PPS presented earlier, we use DEA. In DEA, the PPS is deduced from observed input–output correspondences by making assumptions as to the nature of the PPS. These assumptions are included as constraints in the different linear programs used to estimate four different distance functions. As presented in Charnes, Cooper, and Rhodes (1978), frequent assumptions made are:

(i) Convexity of the PPS: If $(x,y) \in S$ and $(x',y') \in S$, then $(\lambda(x,y) + (1-\lambda)(x',y')) \in S$

(ii) Monotonicity or strong disposability of inputs and outputs:
  a. If $(x,y) \in S$ and $x' \geq x$, then $(x',y) \in S$
  b. If $(x,y) \in S$ and $y' \leq y$, then $(x,y') \in S$

(iii) CRS: If $(x,y) \in S$ and $(x',y') \in S$, then $(\lambda x, \lambda y) \in S$ for any $\lambda \geq 0$.

Notice that under CRS and efficient production, scaling of inputs by a certain factor leads to the outputs being scaled by the same factor. Because of this, when assuming CRS, we obtain the same results using input- or output-oriented distance functions. In what follows, we define input-oriented problems under CRS.

We need to solve four different linear programming (LP) problems to determine the distance functions needed to calculate the Malmquist index for a particular production point (country) C between t and t+1. The distance of production point C in t to the frontier in t is:

$$D_c^t(x^t, y^t) = \max \Theta^c \; s.t. \; y_c^{t,m} \leq \sum_{j=1}^{J} z_j^t y_j^{t,m} \text{ and}$$
$$(1/\Theta^c) x_c^{t,n} \geq \sum_{j=1}^{J} z_j^{t,n} x_j^{t,n}, \text{ and } z_j^t \geq 0 \qquad (2A.13)$$

Where c is one of the j production units: j = 1,..,c,...,J; Similarly, the distance of production point c in t+1 to the frontier in t is:

$$D_c^t(x^{t+1}, y^{t+1}) = \max \Theta^c \; s.t. \; y_c^{t+1,m} \leq \sum_{j=1}^{J} z_j^t y_j^{t,m} \text{ and}$$
$$(1/\Theta^c)\, x_c^{t+1} \geq \sum_{j=1}^{K} z_j^t x_j^t, \text{ and } z_j^t \geq 0 \qquad (2A.14)$$

Computation for $D_c^{t+1}(x^{t+1}, y^{t+1})$ is like in $D_c^t(x^t, y^t)$, but with t+1 substituted for t. Finally, $D_c^{t+1}(x^t, y^t)$ is calculated as in $D_c^t(x^{t+1}, y^{t+1})$ (equation 2A.13), but the t and t+1 subscripts are transposed (Färe et al. 1994). All these problems assume CRS. To impose VRS, we need to include one more constraint: $\sum_{j=1}^{J} z_j^t = 1$.

Problems with DEA

OUTLIERS

Several problems have been pointed out in the literature that result from the use of DEA methods to calculate distance functions. One of these problems is that the DEA frontier defined in the linear problems above is not stochastic—that is, it does not contain a random-error term to account for statistical noise. This means that the efficiency of a production unit measured using DEA methods is typically defined by a small proportion of the observations—those at the frontier of the PPS. In practice, some of the frontier units are atypical, either because of a much stronger performance than other units in the sample or as the result of an atypical mix of inputs and outputs (Thanassoulis, Portela, and Despić 2008). For this reason, it is important that the data for these particular units be reliable.

To detect outliers in our sample of countries, we use the method suggested by Tran, Shively, and Preckel (2010), based on two scalar measures. The first is the relative frequency with which an observation appears in the construction of the frontier when testing the efficiency of other observations. The second measure is the cumulative weight of an observation in the construction of the frontier. For example, using constraints in the dual-optimization problem (equation 2A.10), we define z-count (C_j) as the number of times an observation appears during the construction of the DEA hull (the DEA problem is solved $_J$ times, the number of production units that define the PPS):

$$C_j = \sum_{j\, if\, z_j^n > 0} 1 \qquad (2A.15)$$

We define z-sum (S_j) as the cumulative weight of an observation in all constructed efficient sets (when solving the LP problem for a particular country C). It is computed as:

$$S_j = \sum_j z_j^t \qquad (2A.16)$$

The DEA model yields nonzero values for z-count and z-sum for all efficient observations (the ones that appear with values z_j>0 in the solution to the LP problems), while all inefficient firms have zero values of both z-count and *z*-sum. We followed the procedure suggested by Tran, Shively, and Preckel (2010) to detect outliers. First, and based on the values of C_j and S_j, we identify potential outliers: observations in the dataset that exert an especially strong influence on the construction of the efficient frontier. After identifying observations with a high frequency or level for their weights, we drop these observations, and with the remaining observations we repeat the DEA to obtain new values for C_j and S_j, exclusive of the dropped observations. We drop observations in an iterative fashion, and the process stops once we reach a desired degree of convergence in the observed weights. Given that we work with a limited sample of countries, we do not drop observations identified as likely outliers. These observations are not included in the sample when we calculate the distance for other observations, so reported results are not influenced by these observations. However, we still calculate distance functions separately for these potential outliers, and report these results with results for other countries.

INAPPROPRIATE SHADOW PRICES

One of the reasons for the popularity of the DEA method approach to international comparisons of productivity is that it does not require market prices as weights (normally not available) to obtain an index of total inputs or outputs to measure total factor productivity (TFP). However, even though a priori price information is not needed, the DEA approach still uses implicit price information derived from the shape of the production surface, which allows the estimation of efficiency measures and nonparametric Malmquist indexes. This implicit determination of shadow prices entails potential problems, because these methods are susceptible to the effect of data noise, and shadow prices can prove to be inconsistent with prior knowledge or accepted views on relative prices or cost shares. This is the case when linear programming problems used in DEA methods to calculate distance functions assign a zero or close-to-zero price to some factors because of the particular shape of the production possibility set. As a consequence, inputs considered important a priori could be all but ignored in the analysis, or could end up being dominated by inputs of secondary importance (Pedraja-Chaparro, Salinas-Jimenez, and Smith 1997).

We check our results for the incidence of zero shadow input prices in the standard estimation of the nonparametric Malmquist index, and use a modified procedure to calculate the index that constrains the values of shadow prices in the DEA approach, introducing a priori information on the expected values of shadow input shares.

Constraints to implicit shadow prices are introduced by using the dual-optimization problem in equation 2A.10. This dual problem can be thought of as minimizing shadow cost subject to the constraint that shadow revenue is normalized to 1, and subject to the constraints that when these multipliers are assigned to all producers in the sample, no producer earns positive shadow profit (Thanassoulis, Portela, and Despić 2008). This dual problem is defined for a particular production unit c as:

$$D_c^t(x^t, y^t) = \min\left[\sum_{n=1}^{N} w^n x_c^{t,n}\right], \text{ s.t. } \sum_{m=1}^{M} \rho^m y_c^{t,m} = 1, \left(\sum_{n=1}^{N} w^n x_j^{t,n} - \sum_{m=1}^{M} \rho^m y^{t,m}\right) \geq 0 \quad (2A.17)$$

With ρ^m, $w^n \geq 0$ being shadow prices of outputs and inputs, respectively, and the set of production units j, outputs m, and inputs n, as defined above.

The optimization problem shown in equation 2A.10 and its dual counterpart in 2A.17 allow for total flexibility in choosing shadow prices. To define suitable limits to the value that input shares take, we introduce additional constraints to the original formulation in 2A.14 that set upper and lower bounds (a^n, b^n) to the input share. We define the standard distance function, where ρ and w are, respectively, the output and input shadow prices, and $w^{t,n} \times x_c^{t,n}$ (the input shadow prices multiplied by the input quantities) is equal to the implicit input shares, as shown in Coelli and Prasada Rao (2001). Then, constraints to shadow shares are expressed as:

$$b_c^{t,n} \leq w_c^{t,n} x_c^{t,n} \leq a_c^{t,n} \quad (2A.18)$$

Restricted and unrestricted models will provide the same results only if all the additional restrictions imposed are nonbinding. In general, the narrower the imposed bounds, the larger the expected differences between the outcomes of each model. To define the bounds for the input shares, we first solve the model to obtain average shadow shares for each input, and then define a range of two standard deviations around the mean within which we allow solutions to the LP problems. In this way, we still take advantage of the flexibility of the DEA approach to define shadow prices, while controlling for extreme and zero values in the solution.

THE CURSE OF DIMENSIONALITY

Suhariyanto and Thirtle (2001) called attention to two main difficulties that may result from dimensionality, or the number of inputs and outputs relative to the number of observations in the cross-section, when using DEA for international comparisons: (1) the greater the number of input and output variables, the higher the probability that a particular decisionmaking unit will appear as efficient; and (2) the technology frontiers may be unstable, with the frontier for different periods intersecting and introducing unlikely levels of technological regression.

Estimation approach

In this study we follow Suhariyanto and Thirtle (2001) and Nin, Arndt, and Preckel (2003), who suggest the use of a *sequential technology* instead of the *contemporaneous technology* frequently used in DEA analysis. A contemporaneous technology is the technology defined by the equation 2A.2: $L^t(y) = \{x^t | (x^t, y^t) \in S\}$. With this definition, successive production sets are essentially unrelated to one another—that is, they may or may not overlap in any possible way. The sequential production set, on the other hand, assumes that there is some form of dependence between the production sets across time. This dependence stems from the assumption that "production units can always do what they did before in the production process." For each time period t=1, ..., T, rewrite the technology as:

$$L^{(1,t)}(y) = \{x^{t\text{-}g} | (x^{t\text{-}g}, y^{t\text{-}g}) \in S\} \tag{2A.19}$$

With this technology, the input–output mix used in previous years (t-g) is always available and is part of the technology in period t, which means that successive sequential reference production sets are nested into one another. Using this definition of technology instead of the contemporaneous technology definition, we increase the number of observations defining the PPS, reducing the dimensionality problem while ruling out the possibility of technical regress: contractions of the frontier are not allowed.

We calculate a Malmquist index for one output and six inputs, as described in the main text, using the sequential technology, thus ruling out the possibility of technical regress. Constraints to implicit shadow prices are introduced by using the dual-optimization problem, as described above. Before calculating the different components of the Malmquist index, the method suggested by Tran, Shively, and Preckel (2010) is used to detect outliers.

Sensitivity analysis

To generate greater confidence with the findings associated with the one-output, six-inputs constrained or bounded DEA-Malmquist index method used here, we compare our results with those obtained using three other approaches that differently address the problems with the DEA discussed above: (1) DEA-Malmquist index calculated using two-outputs—crops and livestock; (2) DEA-Malmquist index calculated including lower and upper bounds on the shadow prices; and (3) growth-accounting TFP index, where inputs are aggregated using fixed-input shares for all countries and periods. A brief comparison of the results is presented in Appendix 2B. These results are based on the data from 1971 to 2012, which are complete for all the relevant variables required for the different models.

Appendix 2B: Comparative Analysis of Alternative Index-based Methods

This appendix compares the overall results produced by the four different methods described above: (1) our preferred method, referred to as the "DEA-Malmquist–1-output-index"; (2) the index calculated using two outputs, referred to as the "DEA-Malmquist–2-outputs-index"; (3) the index calculated including lower and upper bounds to shadow prices, referred to as the "DEA-Malmquist-bounds-index"; and (4) the index using the more conventional growth accounting method, referred to as "TFP-fixed-shares-index." These shares are average shadow shares for all countries obtained from the DEA, which are 0.10 for land, 0.20 for labor, 0.07 for fertilizer, 0.22 for feed, 0.22 for crop capital, and 0.18 for livestock capital. To rule out the possibility of zero shadow prices, upper and lower bounds are used by adding +1 and –1 standard deviation to the shares. Without the bounds, the incidence of zero input prices is shown in Figure 2B.1.

About 30 percent of countries on average per year show zero shadow prices for land and labor, and 27 percent show zero shadow prices for livestock capital. In contrast, the percentage of countries with zero crop capital and fertilizer shadow prices is much lower (18 and 11 percent, respectively), whereas only 3 percent of countries show zero shadow prices for feed. This suggests that zero shadow prices in our sample of countries are related to unusual combinations of inputs—for example, large values for labor relative to capital in crop production. With zero shadow price, input substitution is not defined and, continuing with the example, a reduction of labor will have no effect on productivity, given that its shadow price is zero, which means that labor in this case will not be considered for estimating efficiency.

Looking now at the respective results, Table 2B.1 shows the average TFP growth rates and their components for Africa as a whole, and Figure 2B.2 shows the growth paths over time. The results show that the more flexibility one allows in the calculation of the DEA-Malmquist index, the higher the estimated TFP growth (Table 2B.1) and the higher the TFP level in

FIGURE 2B.1 Percentage of zero shadow prices for different inputs, annual average (1971–2012)

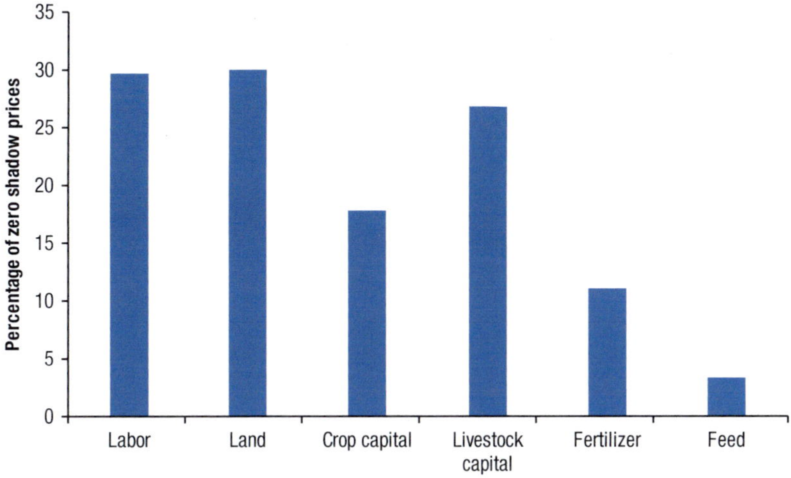

Source: Authors' calculation and illustration based on DEA-Malmquist index method.

2012 compared with that in the initial 1971 period (Figure 2B.2). The DEA-Malmquist–2-outputs-index produces the highest TFP annual average growth rate of 1.8 percent in 1995–2012, compared with 1.3 percent resulting from the TFP-fixed-shares-index. On the other hand, TFP growth paths, as well as the improved performance that started in 1995, are similar for all indexes, although we find larger differences occurring during the first half of the analyzed periods, when the region experienced low or negative TFP growth. Although we did not carry out statistical tests on differences in the growth rates obtained with the different methods, many of the differences are small. We find high-correlation coefficients among the results (Figure 2B.3 and Figure 2B.4).

Differences in the methods are more enhanced for some countries. These are also reflected in Figures 2B.3 and 2B.4. Each point in the figures represents a country, and the coordinates of each point are the TFP growth rates of the TFP indexes being compared. Points on or close to the 45-degree line are those for which the method used has no or very little effect on TFP estimates. The parallel lines that bound the 45-degree line are calculated as the value of the growth rate in the 45-degree line minus (lower bound) or plus (upper bound) two standard deviations measured as the distance between the country points and the 45-degree line. The figures show that countries cluster along the 45-degree line with the highest variability for the DEA-Malmquist–2-outputs-index, as observed in the comparison of average

TABLE 2B.1 Annual average TFP growth rates for Africa using different TFP index methods, 1971–2012

	Malmquist index							
	1-output		2-outputs		Bounds		TFP fixed shares	
Index method	1971–1994	1995–2012	1971–1994	1995–2012	1971–1994	1995–2012	1971–1994	1995–2012
TFP	0.3	1.7	0.5	1.8	0.1	1.6	−0.2	1.3
Efficiency	−0.4	0.7	−0.4	0.5	−0.4	0.7	n.a.	n.a.
Technical Change	0.7	1.0	1.0	1.3	0.6	0.8	n.a.	n.a.

Source: Authors' calculations based on DEA-Malmquist and TFP index methods.
Note: n.a. = not applicable; TFP = total factor productivity.

FIGURE 2B.2 Average TFP indexes for Africa using different index methods, 1971–2012

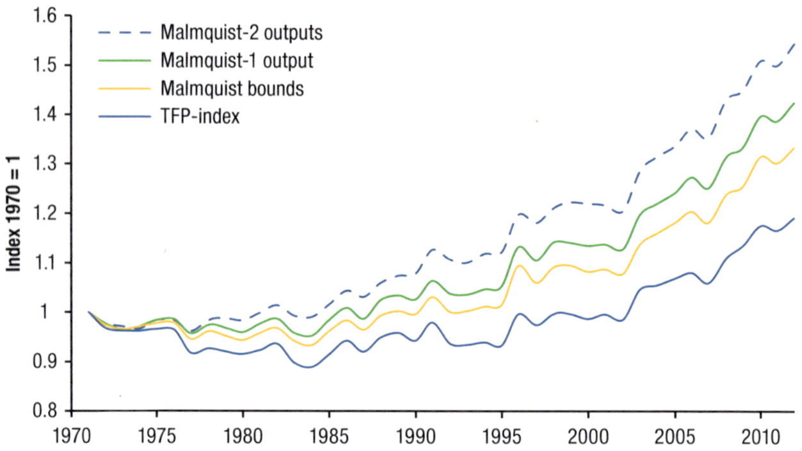

Source: Authors' calculation and illustration based on DEA-Malmquist and TFP index methods.
Note: TFP = total factor productivity.

indexes. Ranking country performance using the different methods will result in a similar order of countries. Note that with two exceptions, all countries are within the range of two standard deviations from the 45-degree line. Also note that the two exceptions are the countries with the highest TFP growth calculated using the DEA-Malmquist index.

For further analysis at the country level, Table 2B.2 shows the estimated annual average growth rate obtained with the different methods, as well as the absolute value of the difference between the DEA-Malmquist indexes and the TFP-fixed-share-index. Countries with the largest variation include

FIGURE 2B.3 Scatter plots of TFP growth rates from different index methods for Africa south of the Sahara (annual averages, 1995–2012)

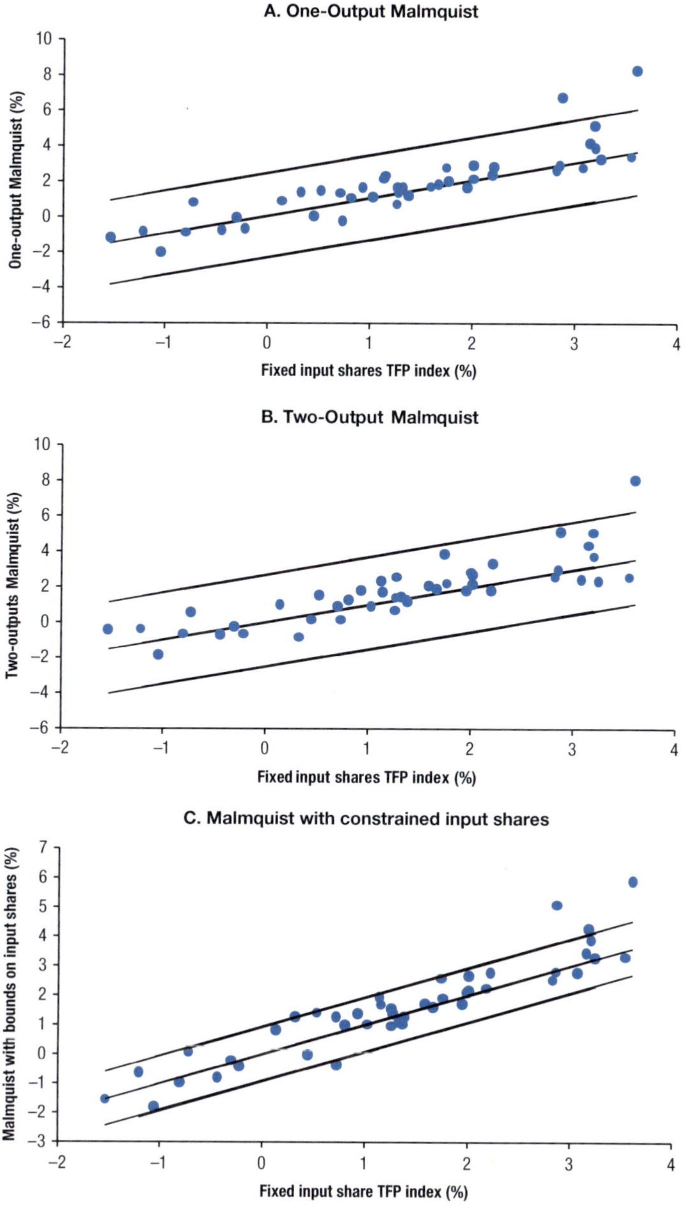

Source: Authors' calculation and illustration based on DEA-Malmquist and TFP index methods.

Note: If methods deliver exactly the same total factor productivity (TFP) growth rate, points should be on the 45-degree line in the figure: distance to the line reflects differences in TFP estimates by the different methods. Upper and lower bounds are calculated as plus and minus two standard deviations, respectively.

FIGURE 2B.4 Scatter plots of TFP growth rates from different DEA-Malmquist index methods (annual averages, 1995–2012)

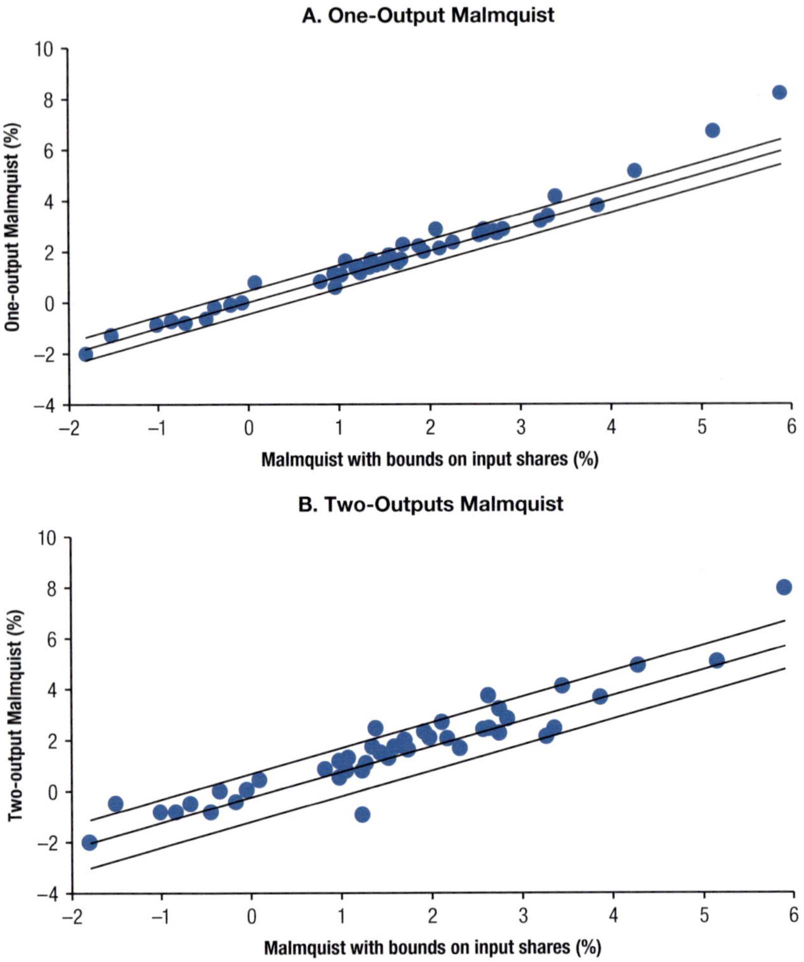

Source: Authors' calculation and illustration based on different DEA-Malmquist index methods.

Note: If methods deliver exactly the same total factor productivity (TFP) growth rate, points should be on the 45-degrree line in the figure: distance to the line reflects differences in TFP estimates by the different methods. Upper and lower bounds are calculated as plus and minus two standard deviations, respectively.

TABLE 2B.2 Annual average TFP growth rates for African countries using different TFP index methods, 1971–2012

	DEA-Malmquist index			TFP fixed-input	Difference between DEA-Malmquist indexes and TFP fixed-input-shares index (absolute value)			
Countries	1 output[a]	2 outputs[b]	Bounds[c]	shares[d]	1 output[a]	2 outputs[b]	Bounds[c]	Average
Benin	8.2	8.1	5.9	3.6	4.6	4.5	2.3	3.8
Rwanda	6.7	5.2	5.1	2.9	3.8	2.3	2.2	2.8
Congo, Rep.	5.1	5.1	4.3	3.2	1.9	1.9	1.1	1.6
Sudan	2.7	3.9	2.6	1.7	1.0	2.2	0.8	1.3
Congo, Dem. Rep.	0.7	0.6	0.1	−0.7	1.5	1.3	0.8	1.2
Liberia	1.3	−0.8	1.2	0.3	1.0	1.1	0.9	1.0
Ghana	2.1	2.4	1.9	1.1	0.9	1.3	0.8	1.0
Mauritania	1.4	1.6	1.4	0.5	0.9	1.1	0.9	1.0
Madagascar	−0.3	0.2	−0.4	0.7	1.0	0.5	1.1	0.9
Burundi	−2.1	−1.8	−1.8	−1.0	1.0	0.8	0.8	0.9
Sierra Leone	4.1	4.4	3.4	3.1	0.9	1.3	0.3	0.8
Libya	2.2	1.8	1.7	1.1	1.1	0.6	0.6	0.8
Guinea	0.8	1.1	0.8	0.1	0.7	0.9	0.7	0.8
Kenya	2.8	3.4	2.7	2.2	0.6	1.2	0.5	0.7
Ethiopia	2.8	2.7	2.6	2.0	0.8	0.7	0.6	0.7
Gabon	1.6	1.9	1.4	0.9	0.7	1.0	0.4	0.7
Mozambique	3.8	3.8	3.8	3.2	0.6	0.6	0.7	0.6
Egypt	2.8	2.9	2.1	2.0	0.8	0.9	0.1	0.6
Gambia, The	−0.9	−0.3	−0.7	−1.2	0.3	0.9	0.5	0.6
Central Afr. Rep.	1.4	2.6	1.4	1.3	0.1	1.4	0.1	0.5
Swaziland	−1.3	−0.3	−1.5	−1.5	0.3	1.2	0.0	0.5
Tanzania	0.7	0.7	1.0	1.3	0.6	0.5	0.3	0.5
Burkina Faso	1.3	1.0	1.2	0.7	0.6	0.3	0.5	0.4
Angola	3.3	2.6	3.3	3.5	0.2	0.9	0.2	0.4
Zambia	2.7	2.5	2.7	3.1	0.4	0.6	0.3	0.4
Mali	0.0	0.2	−0.1	0.5	0.5	0.2	0.5	0.4
Zimbabwe	−0.7	−0.6	−0.5	−0.2	0.5	0.4	0.2	0.4
Namibia	−0.8	−0.6	−0.9	−0.4	0.4	0.2	0.4	0.3
Botswana	1.0	1.4	1.0	0.8	0.2	0.5	0.2	0.3
Algeria	3.2	2.4	3.2	3.2	0.0	0.8	0.0	0.3
Malawi	1.9	2.3	1.9	1.8	0.2	0.5	0.2	0.3
Senegal	1.6	1.5	1.5	1.3	0.3	0.2	0.2	0.2

(continued)

TABLE 2B.2 (continued)

Countries	DEA-Malmquist index			TFP fixed-input shares[d]	Difference between DEA-Malmquist indexes and TFP fixed-input-shares index (absolute value)			
	1 output[a]	2 outputs[b]	Bounds[c]		1 output[a]	2 outputs[b]	Bounds[c]	Average
Tunisia	1.6	1.9	1.7	2.0	0.4	0.1	0.3	0.2
Togo	1.6	1.5	1.1	1.3	0.2	0.2	0.3	0.2
Somalia	1.6	2.2	1.7	1.6	0.0	0.6	0.1	0.2
Cameroon	2.6	2.6	2.6	2.8	0.2	0.2	0.3	0.2
Morocco	1.1	1.3	1.0	1.4	0.2	0.1	0.3	0.2
Uganda	−1.0	−0.6	−1.0	−0.8	0.2	0.2	0.2	0.2
Guinea-Bissau	1.1	1.3	1.3	1.4	0.3	0.1	0.1	0.2
Niger	2.3	1.9	2.3	2.2	0.1	0.3	0.1	0.2
Côte d'Ivoire	1.8	1.9	1.6	1.7	0.1	0.3	0.1	0.2
Mauritius	−0.1	−0.2	−0.2	−0.3	0.2	0.1	0.1	0.1
South Africa	2.1	2.2	2.1	2.0	0.1	0.2	0.1	0.1
Nigeria	2.8	3.0	2.8	2.9	0.0	0.2	0.0	0.1
Chad	1.0	1.0	1.0	1.0	0.0	0.0	0.0	0.0

Source: Authors' calculation based on different DEA-Malmquist and TFP index methods.

Notes: [a] Malmquist index with one output and six inputs; [b] two outputs (crop and livestock) and six inputs; [c] one output and six inputs, but upper and lower bounds (plus and minus one standard deviation from the mean, respectively) imposed to shadow prices; [d] fixed shares are average-input shadow shares from linear programming problems used to calculate distance functions. TFP = total factor productivity.

Benin and Rwanda. Republic of the Congo, Sudan, and Democratic Republic of the Congo also show relatively large differences. Comparing the results of Malmquist-DEA methods with those obtained from a Törnqvist-Theil index for 93 countries, Coelli and Prasada Rao (2001) concluded that the observed differences between estimates could result from poorly estimated shadow prices for some countries because of the dimensionality problem in DEA. Or, if shadow shares are well estimated, problems could arise from some countries differing significantly from the sample average, because of country-specific factors, such as land scarcity and labor abundance.

Appendix 2C: Country Groupings and Plots of Partial and Total Factor Productivity Levels

TABLE 2C.1 Countries by economic development classification and country's share in group's total agriculture value-added

		Low income	Middle income (MI) (69.5)
More favorable agricultural conditions	Mineral rich (LI–1) (4.4)	Central African Republic (9.5)	Algeria (8.6)
		Congo, Dem. Rep. (45.4)	Angola (2.4)
		Guinea (11.9)	Botswana (0.2)
		Liberia (4.7)	Cameroon (2.7)
		Sierra Leone (10.9)	Cape Verde (0.0)
		Zambia (17.6)	Congo, Rep. (0.2)
	Nonmineral rich (LI–2) (22.0)	Benin (4.3)	Côte d'Ivoire (2.8)
		Burkina Faso (6.0)	Djibouti (0.0)
		Ethiopia (31.4)	Egypt (19.4)
		Gambia, The (0.7)	Equatorial Guinea (0.2)
		Guinea-Bissau (0.7)	Gabon (–)
		Kenya (14.7)	Ghana (3.7)
		Madagascar (5.5)	Lesotho (0.1)
		Malawi (3.4)	Libya (–)
		Mozambique (5.4)	Mauritius (0.3)
		Tanzania (16.4)	Morocco (7.0)
		Togo (2.6)	Namibia (0.4)
		Uganda (8.8)	Nigeria (35.3)
		Zimbabwe (–)	São Tomé & Príncipe (0.0)
Less favorable agricultural conditions (LI–3) (4.1)		Burundi (6.5)	Senegal (1.1)
		Chad (11.1)	Seychelles (0.0)
		Comoros (–)	South Africa (4.3)
		Eritrea (–)	South Sudan (1.0)
		Mali (31.0)	Sudan (7.2)
		Mauritania (9.8)	Swaziland (0.2)
		Niger (21.0)	Tunisia (2.7)
		Rwanda (20.6)	
		Somalia (–)	

Source: Authors' calculations based on Diao et al. (2007) and World Bank (2012).

Notes: The figure in parentheses is the region's percentage share in Africa's total agriculture value-added, or the country's share in the region's total (2003–2010 annual average). Dashes mean data are not available. Data for South Sudan and Sudan are based on 2008–2010 values. LI–1 = low income, more favorable agriculture, and mineral rich; LI–2 = low income, more favorable agriculture, and non-mineral rich; LI–3 = low income and less favorable agriculture.

TABLE 2C.2 Countries by Regional Economic Community (REC) and country's share in REC's total agriculture value-added

CEN-SAD (66.8)	COMESA (37.4)	EAC (8.2)	ECCAS (7.9)
Benin (1.4)	Burundi (0.7)	Burundi (3.3)	Angola (21.4)
Burkina Faso (2.0)	Comoros (–)	Kenya (39.6)	Burundi (3.4)
Central African Rep. (0.6)	Congo, Dem. Rep. (5.3)	Rwanda (10.3)	Cameroon (24.2)
Chad (0.7)	Djibouti (0.0)	Tanzania (23.0)	Central African Rep. (5.3)
Comoros (–)	Egypt (36.1)	Uganda (23.8)	Chad (5.8)
Côte d'Ivoire (2.9)	Eritrea (–)		Congo, Dem. Rep. (25.4)
Djibouti (0.0)	Ethiopia (18.4)		Congo, Rep. (1.9)
Egypt (20.2)	Kenya (8.6)		Equatorial Guinea (1.7)
Gambia, The (0.2)	Libya (–)		Gabon (–)
Ghana (3.9)	Madagascar (3.3)		Rwanda (10.8)
Guinea (0.8)	Malawi (2.0)		São Tomé & Príncipe (0.1)
Guinea-Bissau (0.2)	Mauritius (0.5)		
Kenya (4.8)	Rwanda (2.3)		
Liberia (0.3)	Seychelles (0.0)		
Libya (–)	South Sudan (1.8)		
Mali (1.9)	Sudan (13.4)		
Mauritania (0.6)	Swaziland (0.3)		
Morocco (7.3)	Uganda (5.2)		
Niger (1.3)	Zambia (2.1)		
Nigeria (36.7)	Zimbabwe (–)		
São Tomé & Príncipe (0.0)			
Senegal (1.2)			
Sierra Leone (0.7)			
Somalia (–)			
South Sudan (–)			
Sudan (8.5)			
Togo (0.9)			
Tunisia (2.8)			

Source: Authors' calculations based on World Bank (2012).

Notes: CEN-SAD = Community of Sahel-Saharan States; COMESA = Common Market for Eastern and Southern Africa; EAC = East African Community; ECCAS = Economic Community of Central African States; ECOWAS = Economic Community of West African States; IGAD = Intergovernmental Authority for Development; SADC = Southern Africa Development Community; and UMA = Union du Maghreb Arabe. The figure in parentheses is the region's percentage share in Africa's total agriculture value-added, or the country's share in the region's total (2003–2010 annual average). The shares across the RECs do not add up to 100 percent, as the constituent countries are not mutually exclusive. Dashes mean data are not available. Data for South Sudan and Sudan are based on 2008–2010 values.

ECOWAS (36.4)	IGAD (17.8)	SADC (15.0)	UMA (13.2)
Benin (2.6)	Djibouti (0.1)	Angola (11.2)	Algeria (45.6)
Burkina Faso (3.6)	Eritrea (–)	Botswana (0.9)	Libya (–)
Cape Verde (0.1)	Ethiopia (38.8)	Congo, Dem. Rep. (13.3)	Mauritania (3.0)
Côte d'Ivoire (5.3)	Kenya (18.2)	Lesotho (0.4)	Morocco (37.1)
Gambia, The (0.4)	Somalia (–)	Madagascar (8.1)	Tunisia (14.3)
Ghana (7.1)	South Sudan (3.7)	Malawi (5.0)	
Guinea (1.4)	Sudan (28.2)	Mauritius (1.2)	
Guinea-Bissau (0.4)	Uganda (10.9)	Mozambique (8.0)	
Liberia (0.6)		Namibia (2.0)	
Mali (3.5)		Seychelles (0.1)	
Niger (2.4)		South Africa (20.0)	
Nigeria (67.4)		Swaziland (0.7)	
Senegal (2.2)		Tanzania (24.0)	
Sierra Leone (1.3)		Zambia (5.1)	
Togo (1.6)		Zimbabwe (–)	

TABLE 2.C3 Countries by size and growth of agriculture sector

Size of agriculture sector		Growth of agriculture sector	
Large	**Small**	**Fast-growing**	**Slow-growing**
Egypt	Botswana	Angola	Burundi
Ethiopia	Gabon	Cameroon	Congo, Dem. Rep.
Kenya	Gambia, The	Malawi	Liberia
Morocco	Guinea-Bissau	Mozambique	Mauritius
Nigeria	Mauritius	Nigeria	Namibia
South Africa	Swaziland	Rwanda	Tunisia
Sudan		Sierra Leone	Zimbabwe
Tanzania		Zambia	

Source: Authors' calculations based on FAO (2014).

Notes: Large-agricultural economies have at least 3.0 percent of Africa's total agricultural output; small agricultural economies have less than 0.1 percent of Africa's total agricultural output; fast-growing agricultural economies surpass the CAADP agricultural growth rate target of 6.0 percent per year; and slow-growing agricultural economies have an agricultural growth rate of less than 1.0 percent per year.

FIGURE 2C.1 Line plots of land and labor productivity by economic classification (1961–2012)

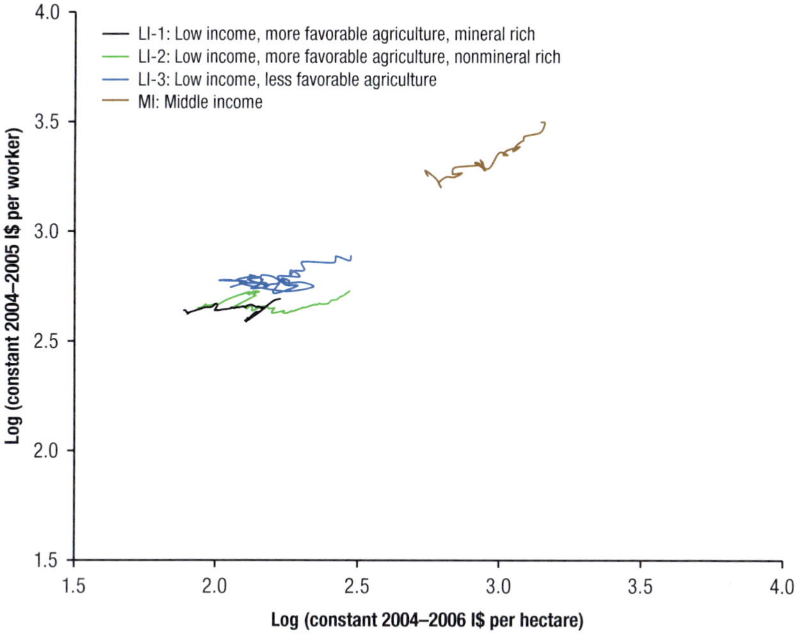

Source: Authors' calculation and representation based on FAO (2014).

Note: I$ = international dollar.

FIGURE 2C.2 Line plots of land and labor productivity by Regional Economic Community (1961–2012)

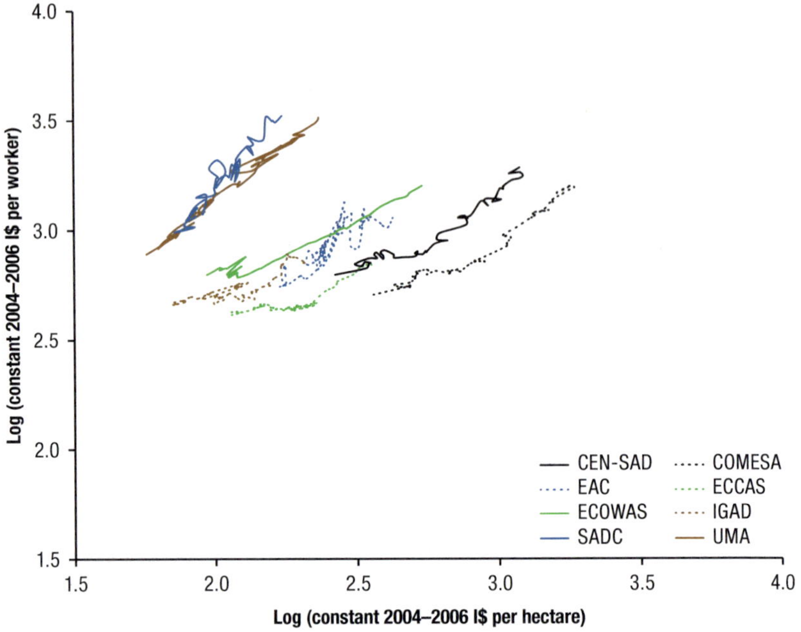

Source: Authors' calculation and representation based on FAO (2014).

Notes: CEN-SAD = Community of Sahel-Saharan States; COMESA = Common Market for Eastern and Southern Africa; EAC = East African Community; ECCAS = Economic Community of Central African States; ECOWAS = Economic Community of West African States; IGAD = Intergovernmental Authority for Development; SADC = Southern Africa Development Community; and UMA = Union du Maghreb Arabe; I$ = international dollar.

FIGURE 2C.3A Line plots of land and labor productivity by size or rate of growth of agriculture sector (1961–2012)

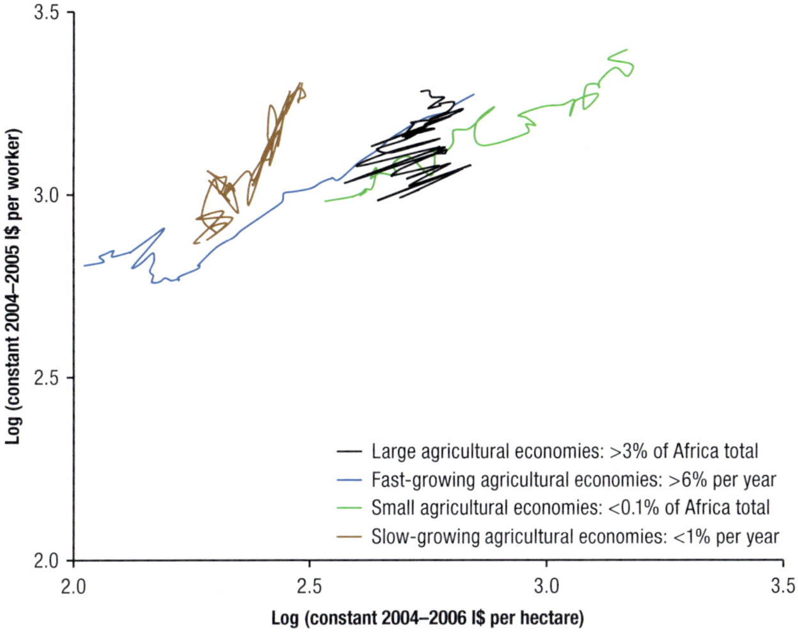

Source: Authors' calculation and representation based on FAO (2014).

Note: I$ = international dollar.

FIGURE 2.C3B Line plots of land and labor productivity for selected countries by size or rate of growth of agriculture sector (1961–2012)

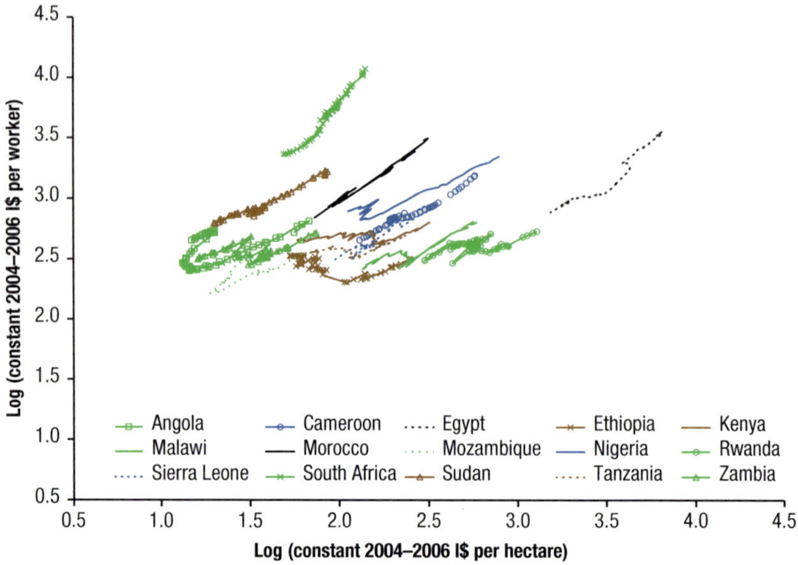

Source: Authors' calculation and representation based on FAO (2014).

Notes: Countries with at least 3 percent of Africa's total agricultural output are large economies. Countries that surpass the CAADP agricultural growth rate target of 6 percent per year are fast-growing agricultural economies. I$ = international dollar.

FIGURE 2C.3C Line plots of land and labor productivity for selected countries by size or rate of growth of agriculture sector (1961–2012)

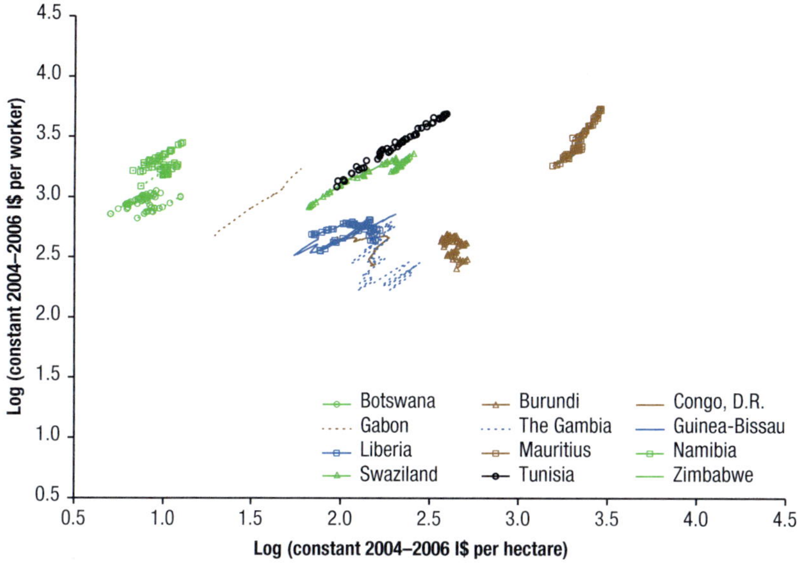

Source: Authors' calculation and representation based on FAO (2014).

Notes: Countries with less than 0.1 percent of Africa's total agricultural output are small agricultural economies. Countries with agricultural growth rates of less than 1.0 percent per year are slow-growing agricultural economies. I$ = international dollar.

FIGURE 2C.4 Levels of total factor productivity, efficiency, and technology by economic classification (1961–2012: indexed at 1961=1)

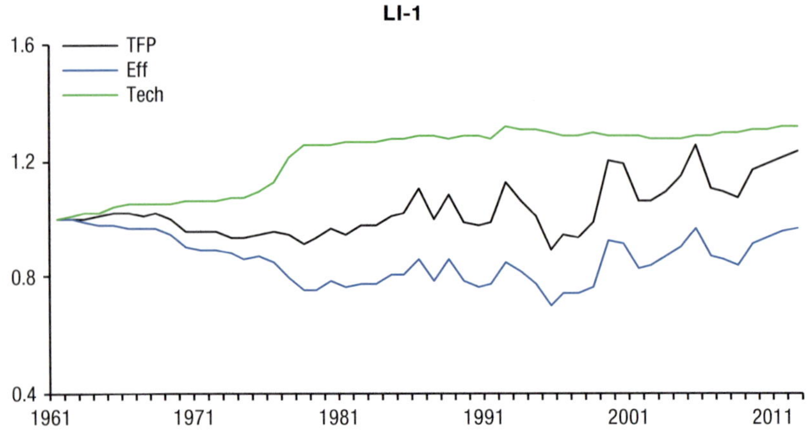

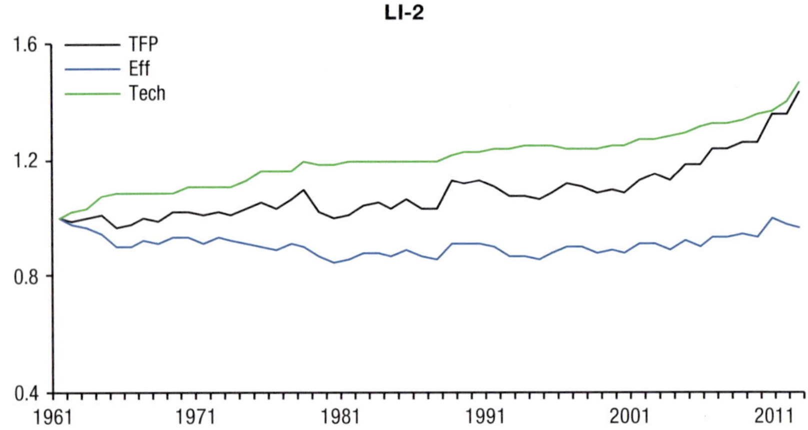

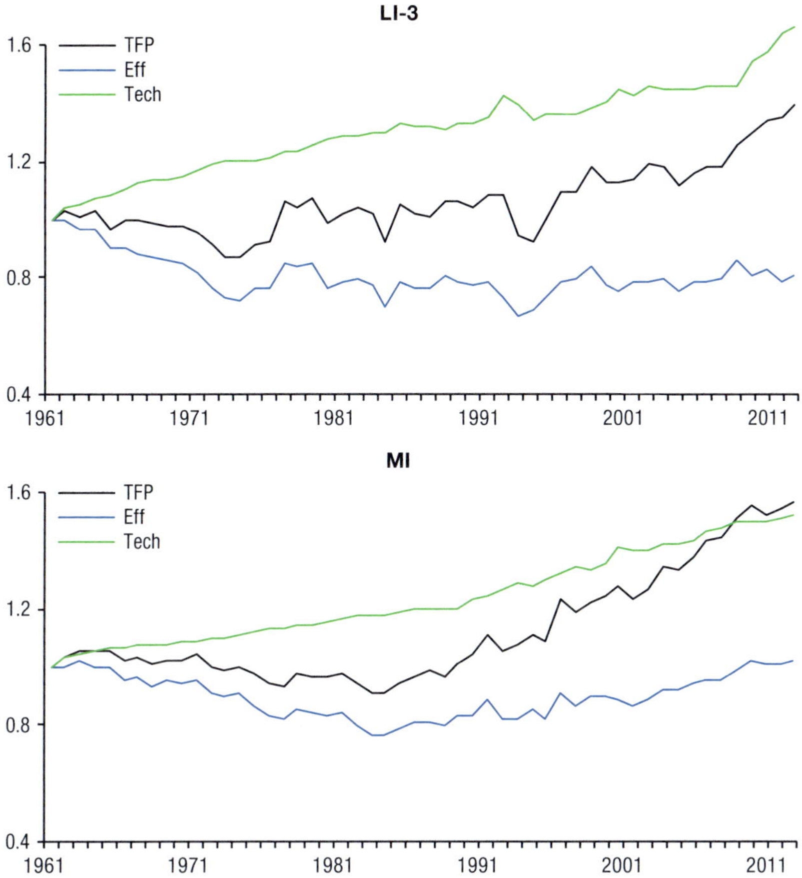

Source: Authors' calculation and illustration based on TFP model results.

Notes: TFP = total factor productivity; Eff = efficiency; = Tech = technology. LI–1 = low income, more favorable agriculture, and mineral rich; LI–2 = low income, more favorable agriculture, and nonmineral rich; LI–3 = low income and less favorable agriculture; MI = middle income.

FIGURE 2C.5 Levels of total factor productivity, efficiency, and technology by Regional Economic Community (1961–2012: indexed at 1961=1)

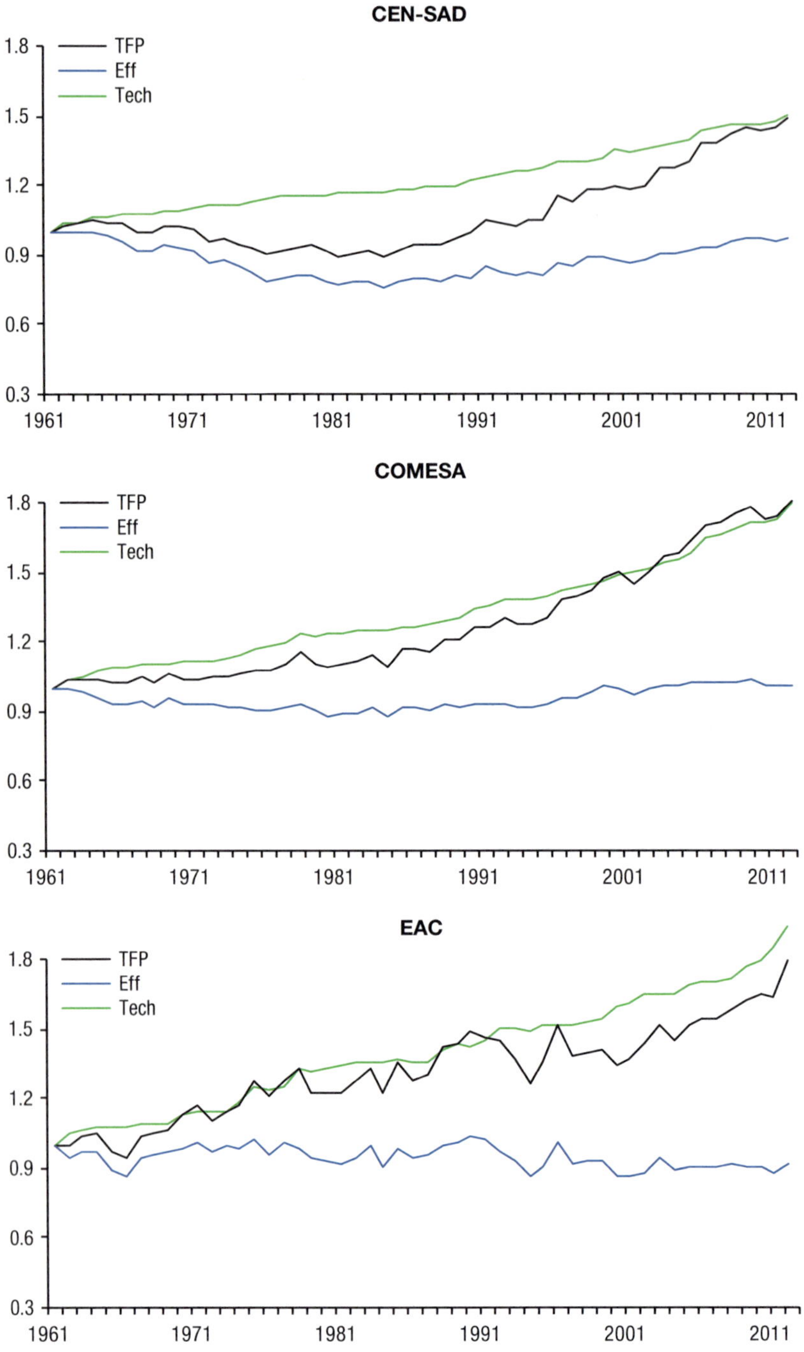

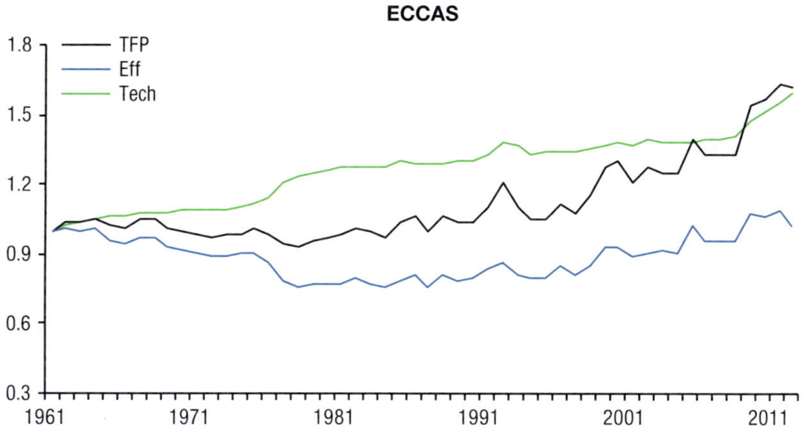
ECCAS
TFP
Eff
Tech
1.8
1.5
1.2
0.9
0.6
0.3
1961
1971
1981
1991
2001
2011

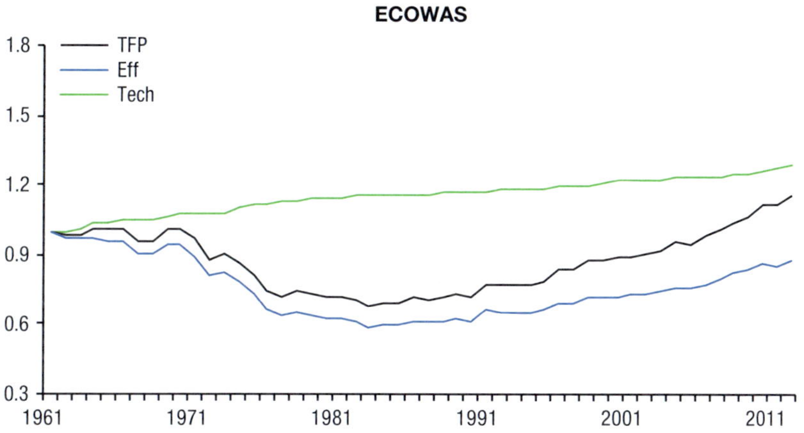
ECOWAS
TFP
Eff
Tech
1.8
1.5
1.2
0.9
0.6
0.3
1961
1971
1981
1991
2001
2011

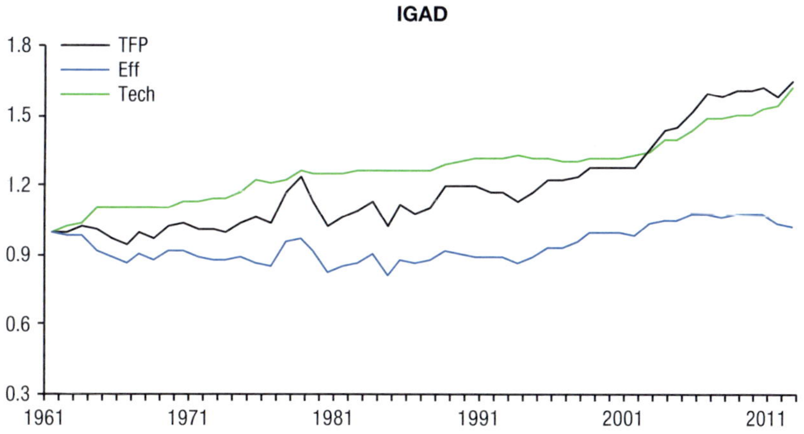
IGAD
TFP
Eff
Tech
1.8
1.5
1.2
0.9
0.6
0.3
1961
1971
1981
1991
2001
2011

(continued)

FIGURE 2C.5 (continued)

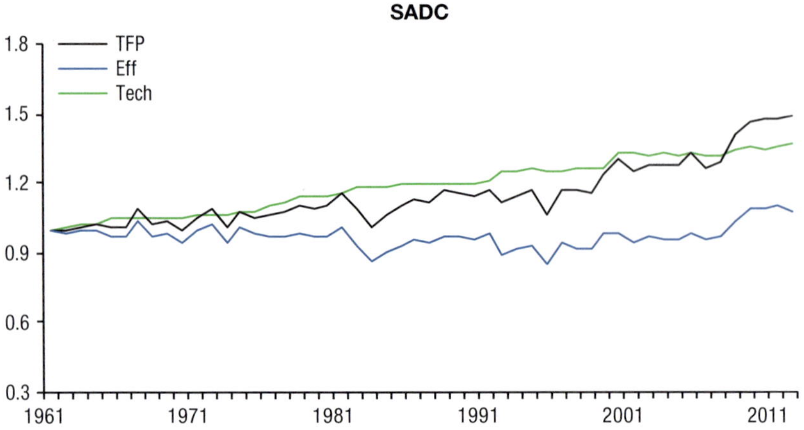

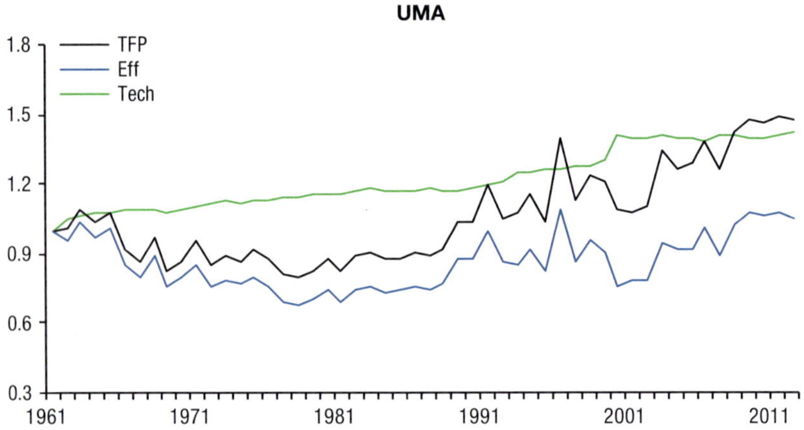

Source: Authors' calculation and illustration based on TFP model results.

Notes: TFP = total factor productivity; Eff = efficiency; Tech = technology; CEN-SAD = Community of Sahel-Saharan States; COMESA = Common Market for Eastern and Southern Africa; EAC = East African Community; ECCAS = Economic Community of Central African States; ECOWAS = Economic Community of West African States; IGAD = Intergovernmental Authority for Development; SADC = Southern Africa Development Community; and UMA = Union du Maghreb Arabe.

FIGURE 2C.6 Levels of total factor productivity, efficiency, and technology for selected countries (1961–2012: indexed at 1961=1)

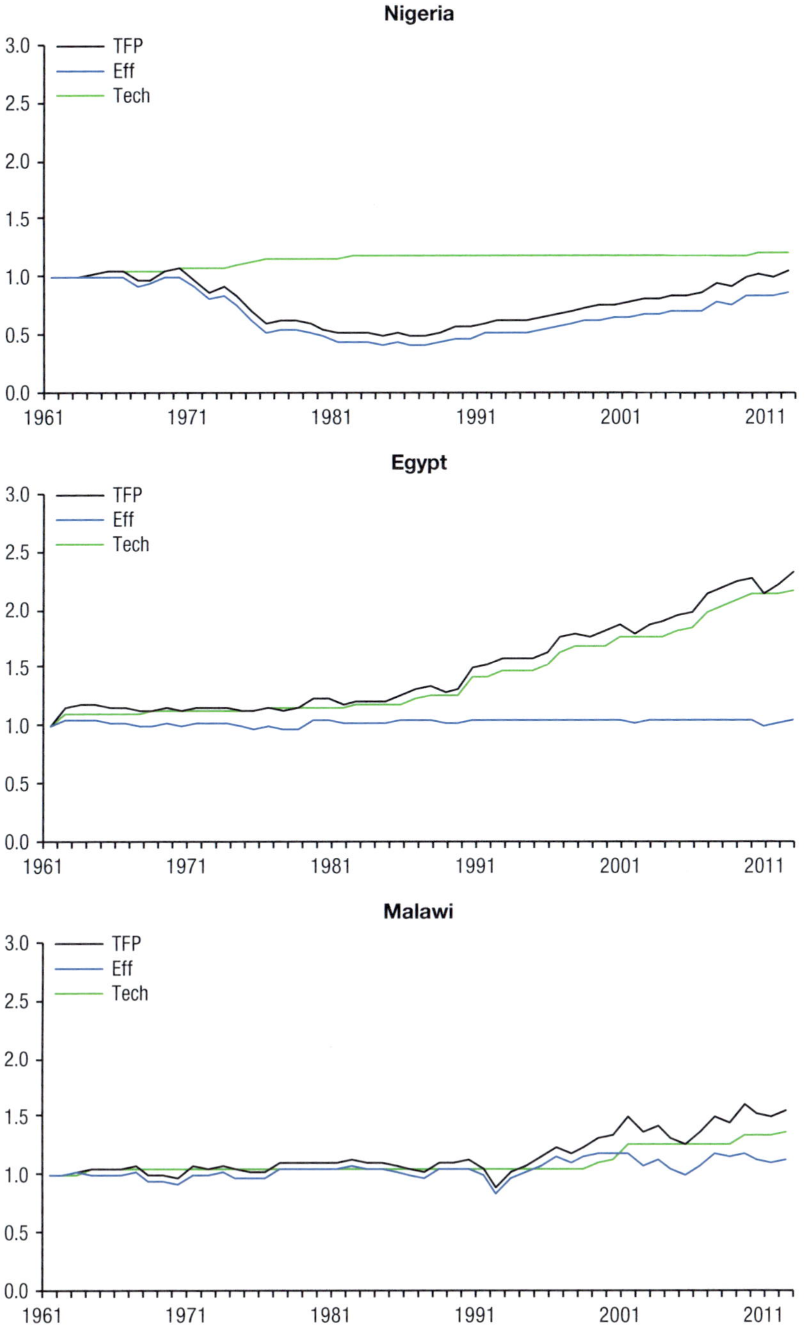

(continued)

FIGURE 2C.6 (continued)

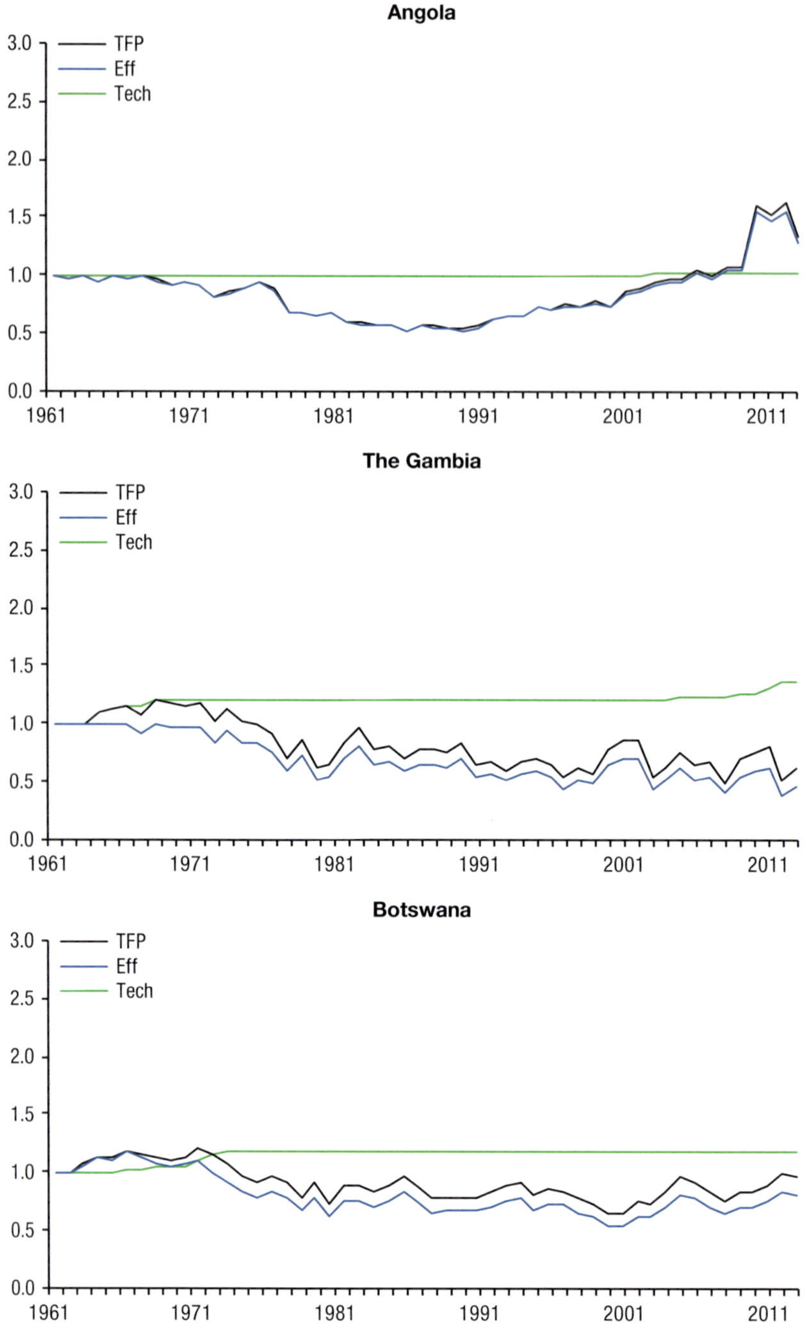

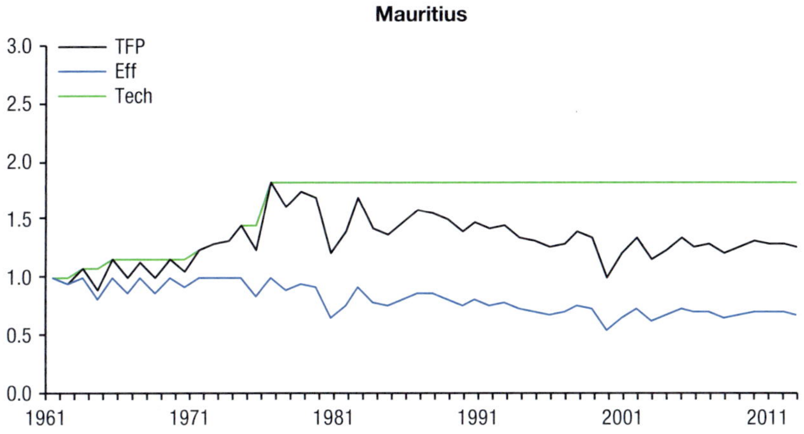

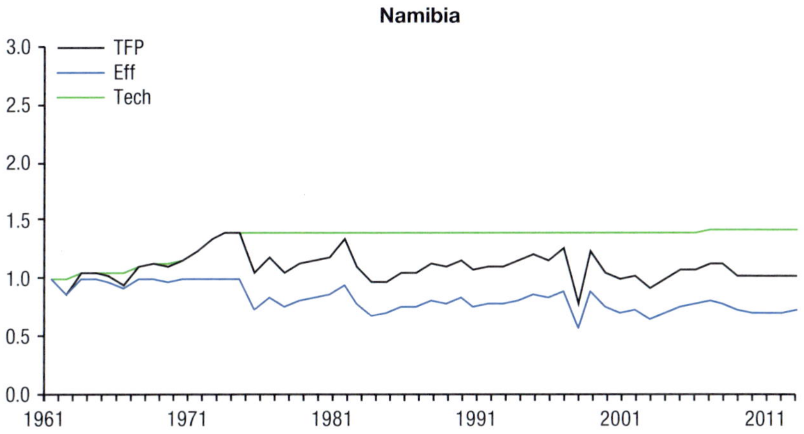

Source: Authors' calculation and illustration based on TFP model results.

Notes: TFP = total factor productivity; Eff = efficiency; Tech = technology. The selected countries are the top two largest agricultural economies in terms of percentage share in Africa's total agricultural output—Egypt and Nigeria; the top two fastest-growing agricultural economies—Angola and Malawi; the bottom two smallest agricultural economies—The Gambia and Botswana; and the bottom two slowest-growing agricultural economies—Mauritius and Namibia.

References

Balk, B. M. 2001. "Scale Efficiency and Productivity Change." *Journal of Productivity Analysis* 15 (3): 159–183.

Benin, S., A. Kennedy, M. Lambert, and L. McBride. 2010. *Monitoring African Agricultural Development Processes and Performance: A Comparative Analysis.* ReSAKSS Annual Trends and Outlook Report 2010. Washington, DC: International Food Policy Research Institute.

Byerlee, D., X. Diao, and C. Jackson. 2009. *Agriculture, Rural Development, and Pro-Poor Growth: Country Experiences in the Post-Reform Era*. Agriculture and Rural Development Discussion Paper 21. Washington, DC: World Bank.

Caves, D. W., L. R. Christensen, and W. E. Diewert. 1982. "The Economic Theory of Index Numbers and the Measurement of Input, Output, and Productivity." *Econometrica* 50: 1393–1414.

Charnes, A., W. W. Cooper, and E. Rhodes. 1978. "Measuring the Efficiency of Decision Making Units." *European Journal of Operational Research* 2: 429–444.

Coelli, T. J., and D. S. Prasada Rao. 2001. *Implicit Value Shares in Malmquist TFP Index Numbers.* Centre for Efficiency and Productivity Analysis Working Papers No. 4/2001. Armidale, New South Wales: University of New England, School of Economic Studies.

Coelli, T. J., D. S. Prasada Rao, and G. E. Battese. 1998. *An Introduction to Efficiency and Productivity Analysis.* Boston: Kluwer Academic.

Datt, G., and M. Ravallion. 1998. "Farm Productivity and Rural Poverty in India." *Journal of Development Studies* 34 (4): 62–85.

Diao, X., P. Hazell, D. Resnick, and J. Thurlow. 2007. *The Role of Agriculture in Development: Implications for Sub-Saharan Africa*. IFPRI Research Report 153. Washington, DC: International Food Policy Research Institute.

Fan, S., P. Hazell, and S. Thorat. 2000. "Government Spending, Agricultural Growth and Poverty in Rural India." *American Journal of Agricultural Economics* 82 (4): 1038–1051.

FAO (Food and Agriculture Organization of the United Nations). 2014. FAOSTAT database. "Production and Resource Domains." Accessed September. http://faostat.fao.org/default.aspx?lang=en.

Färe, R., S. Grosskopf, M. Norris, and P. Roos. 1998. "Malmquist Productivity Indexes: A Survey of Theory and Practice." In *Index Numbers: Essays in Honour of Sten Malmquist,* edited by R. Färe, S. Grosskopf, and R. Russell, 127–190. Boston, London, and Dordrecht: Kluwer Academic.

Färe, R., S. Grosskopf, M. Norris, and Z. Zhang. 1994. "Productivity Growth, Technical Progress, and Efficiency Change in Industrialized Countries." *American Economic Review* 84: 66–83.

Fuglie, K. 2011. "Agricultural Productivity in Sub-Saharan Africa." In *The Food and Financial Crisis in Africa*, edited by D. L. Lee and M. Ndulo, 122–153. Oxfordshire, UK: Commonwealth Agricultural Bureau International.

Fuglie, K., and A. Nin-Pratt. 2013. "Agricultural Productivity: A Changing Global Harvest." In *2012 Global Food Policy Report*. Washington, DC: International Food Policy Research Institute.

Grifell-Tatjé, E., and C. A. K. Lovell. 1995. "A Note on the Malmquist Productivity Index." *Journal of Economics Letters* 47: 169–175.

Headey, D., M. Alauddin, and D. S. Prasada Rao. 2010. "Explaining Agricultural Productivity Growth: An International Perspective." *Agricultural Economics* 41: 1–14.

Irz, X, and C. Thirtle. 2004. "Dual Technological Development in Botswana Agriculture: A Stochastic Input Distance Function Approach." *Journal of Agricultural Economics* 55 (3): 455–478.

Mellor, J. 1999. *Faster, More Equitable Growth: The Relation between Growth in Agriculture and Poverty Reduction*. Agricultural Policy Development Project Research Report No. 4. Washington, DC: United States Agency for International Development. http://www.abtassociates.com/reports/19998977199472.pdf.

Nin, A., C. Arndt, and P. V. Preckel. 2003. "Is Agricultural Productivity in Developing Countries Really Shrinking? New Evidence Using a Modified Nonparametric Approach." *Journal of Development Economics* 71 (2): 395–415.

Nin-Pratt, A., and B. Yu. 2010. "Getting Implicit Shadow Prices Right for the Estimation of the Malmquist Index: The Case of Agricultural Total Factor Productivity in Developing Countries." *Agricultural Economics* 41 (3–4): 349–360.

Pedraja-Chaparro, F., J. Salinas-Jimenez, and P. Smith. 1997. "On the Role of Weight Restrictions in Data Envelopment Analysis." *Journal of Productivity Analysis* 8: 215–230.

Rahman, S., and R. Salim. 2013. "Six Decades of Total Factor Productivity Change and Sources of Growth in Bangladesh Agriculture (1948–2008)." *Journal of Agricultural Economics* 64 (2): 275–294.

Rambaldi, A. N., D. Prasada Rao, and D. Dolan. 2007. "Measuring Productivity Growth Performance Using Metafrontiers with Applications to Regional Productivity Growth Analysis in a Global Context." Paper presented at the 2007 Australasian Meeting of the Econometric Society, Queensland, Australia, July 4–6.

Schneider, K., and M. K. Gugerty. 2011. "Agricultural Productivity and Poverty Reduction: Linkages and Pathways." *The Evans School Review* 1 (1): 56–74.

Suhariyanto, K., and C. Thirtle. 2001. "Asian Agricultural Productivity and Convergence." *Journal of Agricultural Economics* 52 (3): 96–110.

Thanassoulis, E., M. C. Portela, and O. Despić. 2008. "Data Envelopment Analysis: The Mathematical Programming Approach to Efficiency Analysis." In *The Measurement of Productive Efficiency and Productivity Growth*, edited by H. Fried, C. A. Knox Lovell, and S. S. Schmidt, 251–420. New York: Oxford University Press.

Thirtle, C., L. Lin, and J. Piesse. 2003. "The Impact of Research-Led Agricultural Productivity Growth on Poverty Reduction in Africa, Asia and Latin America." *World Development* 31 (12): 1959–1975.

Thirtle, C., and J. Piesse. 2008. "Factor Endowments, Biased Technological Change, Wages and Poverty Reduction: Can Genetically Modified Crops Bring a Green Revolution to SSA?" *International Journal of Biotechnology* 10 (2/3): 184–206.

Tran, N. A., G. Shively, and P. Preckel. 2010. "A New Method for Detecting Outliers in Data Envelopment Analysis." *Applied Economics Letters* 17 (4): 313–316.

World Bank. 2012. *The State of the Poor: Where Are the Poor and Where Are They Poorest?* Poverty Reduction and Equity Department Note. Washington, DC.

———. 2014. "World Development Indicators." Accessed September. http://data.worldbank.org/indicator.

Zofio, J. L. 2007. "Malmquist Productivity Index Decompositions: A Unifying Framework." *Applied Economics* 39 (18): 2371–2387.

Chapter 3

SPATIAL PATTERNS OF AGRICULTURAL PRODUCTIVITY

Stanley Wood, Zhe Guo, and Ulrike Wood-Sichra

Introduction

The previous chapter examined several measures of productivity, primarily in terms of their evolution over time, and reported those changes by countries and subregions. However, the conditions under which agriculture is practiced and specific production systems predominate are highly diverse spatially, even within a single country. This chapter, therefore, examines patterns of agricultural productivity not only at a greater spatial resolution but also in terms of the spatial distribution of specific production systems. We first summarize some of the reasons for growing interest in the spatial dimensions of agriculture, briefly review the general characteristics of the spatial datasets used, and then describe the specific production system schema underpinning the analyses presented in this and subsequent chapters. In the following sections we describe the spatial variability of key factors shaping the productivity of production systems, examine the overall value of (crop) production and associated spatial patterns of land and labor productivity, and briefly discuss the projected effects of climate change spatially. In the final section, we summarize our findings and their implications for prioritizing and targeting interventions, especially in the context of planning for knowledge and technology spillover across domains, countries, and subregions.

How a Spatial Perspective Helps

Development researchers and practitioners are increasingly recognizing the value of better understanding the spatial dimensions of productivity and, in particular, how that understanding can significantly improve the targeting and design of agricultural policies, investments, and farm service provision (Wood, Sebastian, and Chamberlin 2003; Dorosh et al. 2010; Benin et al. 2011). Several factors are shaping this growing awareness. First is the simple observation that distinct spatial patterns of endowments, cultures, and histories have conditioned the evolution of different agricultural development pathways in different geographic locations across Africa (Stock 2012). Second, whether planned or unplanned, spatial patterns of endowments, constraints, and interventions have led over time to the emergence of growth poles and lagging regions within and across countries that exacerbate regional inequalities and that can foster destabilizing social and political conditions (Stock 2012). Third, the footprint of agriculture is not fixed. Its spatial dynamics are witnessed, for example, in the expansion of the agricultural frontier into the forests of central Africa and into the more marginal savanna lands of West Africa, as well as in changing local climate patterns, the abandonment of degraded agricultural lands, and large-scale foreign direct investments in African farmland (FAO and World Bank 2009; Thornton et al. 2009; Denninger and Byerlee 2011). Fourth is the widely held perspective that, compared with the preconditions of the Green Revolution in Asia, successful agricultural development in Africa has been inhibited by greater spatial heterogeneity in agroecological, cultural, and socioeconomic conditions. Greater spatial diversity makes the development and diffusion of agricultural innovations more costly and complicates the scaling of local successes. It implies greater need to invest in more locally adapted innovations, rather than rely on more cost-effective strategies in which knowledge and technology can more readily spill over across multiple locations and production contexts.

Another implication of the complexity of knowledge transfer in Africa is the greater need for well-functioning mechanisms and institutional arrangements that can advocate for and accelerate knowledge spillover across ecologies and countries (Wood and Anderson 2009). Subregional research for development networks established and expected to catalyze and promote cross-border collaboration in agricultural research in different parts of Africa include the North African Sub-Regional Organization, the Association for Strengthening Agricultural Research in Eastern and Central Africa, Conseil Ouest Africain pour la Recherche et le Développement Agricoles/West and Central African Council for Agricultural Research and Development, and the Center for Coordination of Agricultural Research and Development in Southern Africa.

Last are the greater opportunities for and falling costs of exploring the spatial dimensions of agricultural development and economic growth. The data, tools, and human capacity surrounding the acquisition, management, and application of georeferenced data have improved substantially over the past 20 years. These advances have been spurred by rapid expansion in the range and accessibility of geographic information system tools; remote-sensing products (primarily satellite-based, but more recently from a growing range of low-cost, unmanned aerial vehicle platforms); and web and mobile device services that integrate Global Positioning System capabilities, satellite imagery, and mapping into a range of freely accessible consumer services (such as Google Earth and Bing maps). One important application of these new tools will be in monitoring and learning more about the spatial and temporal patterns of environmental change, including climate change, land and vegetation quality dynamics, water resource exploitation, and biodiversity, and their relationships to agricultural productivity growth and development. Access to spatially explicit information on, for example, changes in land use, the adoption of yield-enhancing technologies, soil erosion, diversity of organisms, and water extraction can help provide more locally tailored assessments of the drivers of and responses to environmental change. Although such analysis is beyond the scope of this chapter, we later on highlight some of the spatial data, tools, and models that are increasingly being applied in examining climate change scenarios and potential effects in Africa.

Opportunities for and challenges to using spatial data

Before describing the spatial analytical approach and results, it is worthwhile to reflect on the construction and general characteristics of the spatial data used and, in particular, on data quality implications and the ultimate level of confidence that may be placed in the type of regional spatial analysis presented.

The spatial datasets used throughout this chapter comprise a range of demographic, biophysical, and agricultural production-related variables that are conformed to a standard 5 arc-minute grid cell across the entire continent. Since the grid cells are described in spheroid geometry units, each grid cell represents a different physical area on the ground that approximates to a 9 × 9-kilometer (km) grid cell at the equator. Grid cell areas are constant at any given latitude, but decrease in proportion to distance from the equator. These latitudinal differences in grid cell area are accounted for by keeping grid cell area as a specific grid attribute for the purposes of appropriately area-weighting any subsequent grid-based analytical operation. The variables populating each of the 371,566 5 arc-minute grid cells[1] that comprise the African

1 The 5 arc-minute gridded database for Africa comprises 371,566 grid cells. The analysis presented in this chapter, however, covers only the 291,892 grid cells of Africa south of the Sahara, since one of the key spatial data inputs—the SPAM crop production database (You, Wood, and Wood-Sichra 2009)—is limited to this areal extent.

continent are derived from one of three approaches. First, and typically most reliable in capturing spatial variation, are grid cells derived from spatially continuous data collection by satellite-based, remote-sensing observation platforms. In this category fall such variables as elevation and land use. Second are grid cell variables constructed by spatial-interpolation techniques operating on a network of point observations, including weather- and climate-related variables that are spatially interpolated from observations made at recording stations within a meteorological network. And third are grid cell values estimated by spatial downscaling of statistical data from larger, typically administrative, areas (for example, estimating grid cell-specific variation in crop production using district-level statistics that are spatially decomposed using independent grid-scale covariates).

Table 3.1 summarizes the core spatial variables used in this chapter. Of note is that all datasets are themselves products of prior spatial modeling. While the different modeling efforts are associated with a range of peer review, none provides confidence intervals or formal reliability estimates for the spatial datasets themselves. Some rely solely on expert-based classification rules applied to the overlay of spatial variables, rather than on formal analytical techniques, such as the assignment of individual grid cells to specific farming-system categories using an expert-determined classification logic (Dixon, Gulliver, and Gibbon 2001). Another feature is the scarcity of relevant spatio-temporal datasets at the regional scale. This reflects the enormous logistical challenge of compiling and harmonizing time series of subnational statistical data in order to construct consistent, spatially disaggregated time series across all African states. As is typified by this volume, therefore, regional development studies typically rely on *either* time-series analysis of national-scale data, as presented in Chapter 2, for around 50 spatial units (countries) that span an enormous range of physical extent from some 235 million hectares (Mha; Democratic Republic of the Congo) to around 0.4 Mha (Cape Verde); *or* more highly spatially disaggregated explicit analysis using thousands of grid cells, but only for a single or limited number of points in time. The notable exception is for satellite-derived data sources, where thematic datasets for medium (250–500-meter [m])- to high (3–30-m)-resolution grids are capable of delivering complete spatial coverage of Africa every 10–30 days.[2] While variables derived from satellite data are, therefore, much better suited to supporting

2 In practice, cloud cover in the tropics, particularly during the growing season, severely limits the temporal availability of usable imagery for agricultural applications from most (optical) satellite-based sensors. Furthermore, highly specialized and scaled computing resources are needed to support image processing and interpretation to generate time-series products.

spatial time-series analysis, the thematic range of such remote-sensed data products is limited primarily to biophysical attributes, such as vegetation condition (for example, the vegetation index used in Chapter 4), major land cover change, fire detection, rainfall, and flooding (ESA 2012).

For the purposes of spatially explicit productivity analysis, however, we seek spatially disaggregated assessments of variables, such as cropland area, farm labor proxies, and area and yield of individual crops. These variables are more specialized spatial data products that demand a higher degree of interpretation, ancillary data, modeling, and validation and that typically take several years to construct (although the trend is for those lag times to decrease). To support this study, the most recent set of relevant and consistent regional spatial databases (consistent both in the sense of period of observation and in the choice and use of common ancillary data in their construction) were for 2000–2001. As described in the following section, the grid cell values for these datasets were all preprocessed (rescaled) to ensure they were consistent with reported 2005–2007 national means and totals for analysis and reporting purposes. Our strong assumption is that while cropland and rural population densities, and the harvested areas and yields of individual crops, all changed between 2000 and the 2005–2007 period for which we report productivity results, the spatial pattern of their relative values remained consistent. We substantiate this assumption by noting that between 2000 and 2007, regional production patterns and intra-African regional trade arrangements and outcomes changed little in either absolute or relative terms. For example, Yang and Gupta (2007) show that since the mid-1990s, intra-African trade has stagnated at about 10 percent of total African trade; and Akpan (2014) shows that trade between countries in the Economic Community of West African States (ECOWAS), for example, remained the same in 2003–2007 for many countries, including Nigeria, Ghana, and Côte d'Ivoire, which together account for about 80 percent of the total agricultural gross domestic product in ECOWAS. Beyond this assumption, reflecting the specific data limitations of this study, is the broader recognition of two ubiquitous challenges in conforming and applying spatial data: the "ecological fallacy" (Robinson 1950) and the modifiable areal unit problem (MAUP) (Openshaw 1984).

The ecological fallacy concerns the statistical limitations of conferring properties of a population on individuals from that population. Two of our underlying spatial datasets—cropland and population distribution—are constructed by taking statistics reported for administrative units (districts and census tracts, respectively) and conferring measures of spatial intensity derived at that scale (the area share of cropland and population density, respectively)

TABLE 3.1 Spatial data used in exploring the spatial patterns of partial productivity in crop production in Africa south of the Sahara

Dataset	Input data components	Construction
Primary datasets		
Farming systems ~2000 (Dixon, Gulliver, and Gibbon 2001)	Crop and livestock statistics, agroecological zone, population density, land use, and urban grids (5 arc-minute)	Expert-based classification logic applied to multiple map overlays, so as to best represent best understanding of the known spatial distribution of dominant farming systems
Population density (CIESIN et al. 2011)	Population census data at district/ward level (109,172 subnational reporting units) (5 arc-minute)	Population density assumed constant across all grid cells within each census tract boundary
Cropland distribution 2000 (Ramankutty et al. 2008)	Subnational land-use statistics (242 subnational reporting units) (5 arc-minute)	Cropland area intensity assumed constant across all grid cells within each statistical reporting unit
Crop distribution SPAM 2000 (You, Wood, and Wood-Sichra 2009)	Subnational crop, land-use, and price statistics (2,520 subnational reporting units); rainfed/irrigated cropland extents, population density, and crop suitability by agroecological zone grids (5 arc-minute)	Spatially distributed prior probabilities estimated from independent grid-scale variables, followed by entropy-based optimization of grid cell allocation across multiple crops
Underlying datasets		
Agroecological zones (Fischer et al. 2012)	Major climate, water availability (length of growing period), and thermal conditions, and elevation grids (5 arc-minute)	Expert-based classification of zones with similar agricultural production potential
Elevation (Jarvis et al. 2008)	Satellite-observed (stereoscopic) imagery by grid cell (90 m)	Digital elevation model with advanced "hole-filling" algorithm
Climate stations (multiple sources: WHO, FAO)	Rainfall, temperature, radiation, sunshine (point data, spatially interpolated to 5 arc-minute grid cells)	Spatial interpolation (such as *kriging*) of station values and derived grid cell values of evapotranspiration and length of growing period based on elevation and absolute location

Source: Authors' representation based on stated references.

Note: Regardless of date of publication, all datasets except climate stations represent conditions around the year 2000. Climate station means typically represent the period 1960–1990.

to each of the grid cells contained within the administrative reporting units. Clearly, it is unlikely that the density of population or cropland in each grid cell within an administrative district is uniform, but the degree of imprecision is shaped by the intrinsic spatial patterns of the two variables, and by the extent to which we attempt to use additional knowledge we may have on the factors shaping those patterns. In the case of population, we assumed homogeneous population density within each of the ~90-km^2 grid cells of an administrative unit, but we consider those areas to be sufficiently extensive, and the rural population to be sufficiently dispersed, to believe this to be a plausible hypothesis in the absence of other data.

In the case of crop distribution, however, we derived grid cell–specific area estimates from administrative area summary statistics through an analytical approach informed by a number of additional independent grid-scale variables. (For example, the share of district maize-harvested area attributed to each grid cell within the district was assumed to be conditioned by the potential suitability of each grid cell to grow maize based on independently derived grid cell–specific estimates of climate, soil, and terrain conditions.) While recognizing the potential pitfalls associated with the ecological fallacy, we believe our hypotheses and analytical approach have minimized its effects in the context of the geographical scope and goals of this study. A more recent study of the ecological fallacy has suggested that there are similar shortcomings in the inverse case of overreliance on just the properties of individuals in inferring the characteristics of the population, and that balanced interpretation of subpopulation properties should best encompass relevant individual and population properties (Subramanian et al. 2009).

The MAUP can be thought of as a spatial representation of the ecological fallacy "in which conclusions based on data aggregated to a particular set of districts may change if one aggregates the same underlying data to a different set of districts" (Waller and Gotway 2004, 104). Thus, regardless of the scale, the statistical properties of spatial variables are conditioned by the distinct and arbitrary boundaries for which they are variously reported.

While the deconstruction of administrative unit cropland and population summary statistics to grid cell estimates renders them vulnerable to ecological fallacy, and the aggregation of grid cell–specific land and labor productivity estimates into farming system classes opens up potential MAUP limitations, we maintain such effects are likely mitigated by the scale of spatial aggregation at which we report the productivity results, relative to the spatial scale underlying the primary statistical data used in generating the gridded databases used in the analysis. Thus, although the gridded cropland, population, and crop production spatial databases were constructed using primary input data from 242,109, 172, and 2,520 subnational statistical reporting units, respectively, our partial productivity results are cited only at the aggregated scale of 14 farming system classes.[3] While each of those reporting units differs widely in its areal extent (appendix Table 3A.2), it is clear that the average productivity results for the larger farming systems in the continent, such as root crops

3 The number of different farming systems encountered in each of the three subregions reported is actually fewer than 14: West Africa—9, East Africa—13, and southern Africa—12 (ignoring minor systems with a subregion extent of less than 10,000 ha).

in West Africa, encompass a significant degree of spatially disaggregated variation in the underlying primary statistical data.

Regionally important farming systems

To derive richer insights into the spatial variation in productivity across Africa than is possible based on national totals and means alone, we exchange the productivity reporting unit from country (geopolitical) boundaries to farming system extents (Figure 3.1). We define farming systems as geographical areas or sets of noncontiguous geographical areas that, largely through similarities of biophysical endowments, demographics, and built infrastructure (roads and irrigation), support similar patterns of agricultural livelihood choices. Figure 3.1 clearly reveals how individual countries can encompass several major farming system zones. It is also intuitive that levels of productivity between systems, such as agropastoral systems and highland-perennial systems, are likely to differ widely. (See Dixon, Gulliver, and Gibbon 2001 for a complete description of each farming system type.) We briefly examine some key attributes of these systems (rainfall, population, market accessibility, and cropland distribution), and then present our assessment of the spatial patterns of partial productivity (specifically, land and labor productivity in crop production) that those attributes play a large part in conditioning.

With regard to Africa south of the Sahara (SSA) cropland areas, Figure 3.1 and the appendix tables highlight how just 4 of the 14 systems reported dominate, representing almost 60 percent of the entire SSA cropland extent. The root crop system is the most extensive (37 Mha), spanning the more humid parts of the Guinea savanna in West Africa, as well as southern central Africa and the border zones between Tanzania and Mozambique, and accounting for almost 20 percent of the SSA cropland area. Next in extent are the cereal-root and the agropastoral millet/sorghum systems, which each accounts for around 15 percent of SSA cropland (30.2 Mha and 28.5 Mha, respectively). These two systems dominate the east–west expanse of cropland running from West Africa to Ethiopia, where the more humid cereal-root systems lie to the south of the drier agropastoral areas, and otherwise appear widely in western and southern Angola, western Zambia, and central Mozambique. Since this band lies south of the equator, however, the drier agropastoral systems lie to the south of the cereal-root systems. The fourth major crop-based system is the maize mixed system (23.5 Mha) that dominates the eastern region and parts of southern Africa.

FIGURE 3.1 Major farming systems of Africa

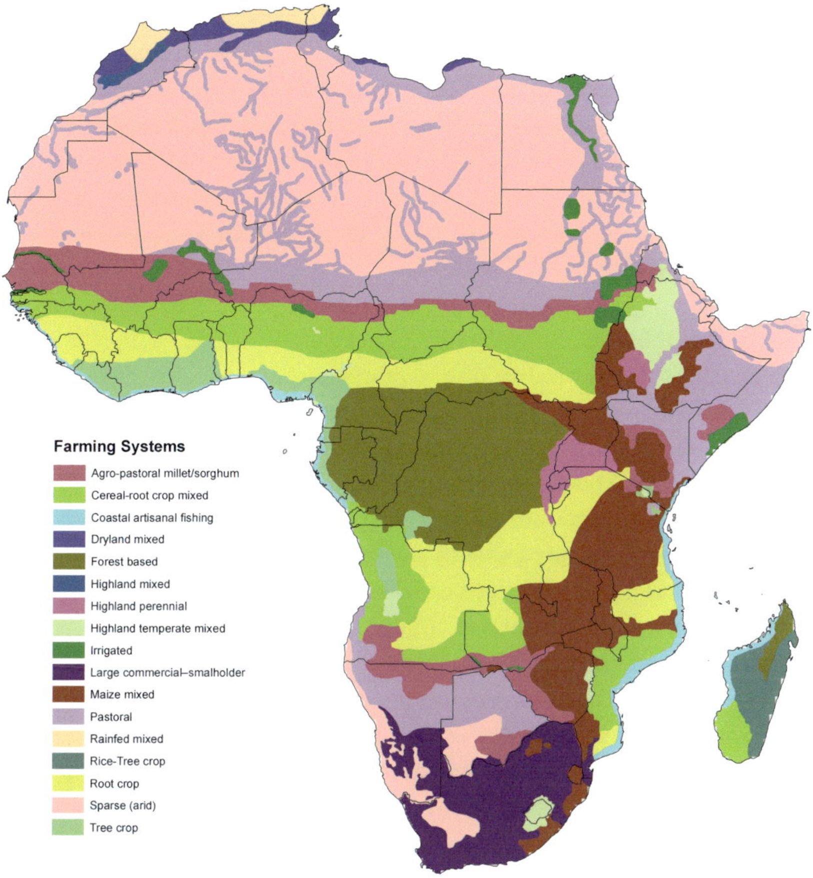

Source: Authors' illustration based on Dixon, Gulliver, and Gibbon (2001).

Other notable systems are the large-scale irrigated systems principally sustained by the Niger in West Africa, the Nile in Sudan, and the Juba and Shebelle rivers in southern Somalia. Elsewhere, formal irrigation infrastructure is limited. The large commercial and smallholder farming system[4] (13.2

4 The large commercial and smallholder farming system represents those geographic areas where both large-scale commercial and smallholder enterprises coexist. Typically, these enterprises specialize in producing a small number of dominant commercial crops, such as maize, wheat, and sugarcane. Nearly one-half of the continent's wheat is produced in such dualistic farming systems.

Mha) is particularly noteworthy in its spatial concentration, encompassing South Africa and southern Namibia and small areas of Mozambique. The two highland-based systems, while accounting for just more than 6 percent of total cropland area (12.7 Mha), are notable both for sustaining high levels of population and, as we shall see, for generally higher levels of productivity. Elsewhere, pastoral systems (21.7 Mha) dominate in the drier regions and forest- and tree-based systems (18.8 Mha) in the more humid areas. The coastal artisanal fishing[5] systems are diverse, with livelihoods based on fishing, crop production, multistoried tree crop gardens, and opportunities for lowland cultivation of rice.

Other data issues

Data on national crop production and prices were obtained from the FAOSTAT database (FAO 2012), and the number of economically active agricultural workers was obtained from the International Labor Organization (ILO 2010). We acknowledge legitimate concerns about the reliability of using such government-reported data, when as highlighted in Jerven (2013), national statistical systems are often poorly resourced and may be exposed to political sway. Furthermore, using "economically active in agriculture" as a proxy for agricultural workers in crop production, for example, is likely to introduce bias, where a high share of labor is used in livestock keeping or in postharvest processing. However, our hypothesis is that those shares may be fairly consistent among crop-based farming systems.

By combining different independent sources and types of data, several of which are based on observed measures of outcomes rather than self-reported data, we believe we have reduced some potential data pitfalls. Such a data assimilation or data fusion approach could be used more extensively to improve the reliability of international databases, such as FAOSTAT and ILO. This could include greater harmonization and integration of routine data-reporting systems with nationally representative, production-, labor-accounting-, and welfare-focused household surveys, such as the Living Standards and Measurement Survey, and the associated Integrated Survey on Agriculture and National Living Standards Survey, as well as other national agricultural censuses and sample surveys.

5 The coastal artisanal fishing system is defined as a narrow buffer area around coastlines where small-scale fishing livelihoods are known to be widespread. Those areas also encompass coastal zone crop production.

Spatial Measures of Partial Factor Productivity

Assessment of spatially disaggregated measures of partial factor productivity relied on assembling three primary spatial data sources drawn or generated from those described in Table 3.1, limited to the extent of SSA:[6] total value of crop production, cropland area, and number of agricultural workers. (See Box 3.1 for key issues regarding use of especially land productivity as a measure of partial factor productivity.) The 2005–2007 gridded value of production dataset was constructed by first estimating the average national value of production for each crop over the 2005–2007 period using FAOSTAT national production and price time series data (FAO 2012), and then distributing those national totals across grid cells in proportion to the share of 2000 national production of each crop in each grid cell (extracted from the International Food Policy Research Institute's Spatial Production Allocation Model [SPAM] 2000 dataset).[7] Gridded cropland data (cropland area share of each grid cell) were constructed similarly, using national cropland statistics and the 2000 gridded cropland database (Ramankutty et al. 2008). Agricultural workforce per grid cell was estimated by distributing the national 2005–2007 estimates of population economically active in agriculture into each grid cell in proportion to the 2000 share of national rural population in each grid cell (obtained from CIESIN et al.'s [2011] Global Rural-Urban Mapping Project). Once these gridded data layers were constructed, partial productivity measures were derived directly. First, aggregate production values were computed for each subregional farming system by summing the individual grid cell values within each farming system. The corresponding average land and labor productivity measures were obtained from those production value aggregates through division by aggregates of cropland area and agricultural workers, respectively, for the same subregional farming systems. (See appendix Tables 3A.1–3A.3 for more detailed presentation of these results.)

West Africa is both the largest and the most extensively cropped region of SSA, accounting for about half of the total cropland area but generating close to 60 percent of the SSA total value of crop production. A significant share of West Africa's production value can be traced to the regional preference for high-yielding, high-value yams, as well as traditional cash crops (for example,

---

6 Of the 291,892 SSA grid cells, 289,103 (99 percent) include land within which 164,761 (56 percent) include cropland.

7 The SPAM crops are wheat, rice, maize, barley, millet, sorghum, potatoes, sweet potatoes, cassava, bananas and plantains, soybeans, beans, oilseeds and pulses, sugarcane, sugar beets, coffee, cotton, other fiber crops, groundnuts, and other oilseeds. The SPAM approach is fully documented in You, Wood, and Wood-Sichra (2009).

BOX 3.1 Land productivity: An appropriate denominator?

While conceptual, definitional, and measurement issues surround metrics of land and labor as production inputs, the question of appropriate measures of land variables has received particular attention in the literature on agricultural productivity and intensification (Boserup 1965; Pingali and Binswanger 1988; Headey and Jayne 2014). The work of Boserup (1965) on the linkages between population pressure, technical change, and agricultural productivity drew attention to the issue of an appropriate measure of land area when assessing population density as a driver of agricultural intensification. Boserup concluded that the most relevant metric of population pressure in that context was neither total land area, which may include extensive areas unsuited to agriculture (such as deserts or shallow soils), nor currently planted cropland areas that omit fallow or other suitable lands that could be brought into production, but rather total land net of land area unsuited to agricultural production. Pingali and Binswanger (1988) went further by controlling for spatial variation in the quality of land from an agricultural production perspective, normalizing land-area measures on the basis of soil and climate conditions to produce agroclimatically adjusted population densities.

Similarly, appropriate land-area metrics are needed for assessing land productivity with respect to crop production. Two basic options are readily accessible: total harvested area of annual and perennial crops and total cropland. These options are different, since FAOSTAT cropland estimates include "the sum of temporary and permanent crop areas, temporary meadows, kitchen gardens and temporarily fallow (<5 years)" (FAO 2012). If a measure of productivity in terms of the land actually used for crop production in a given cropping year is most relevant, then total harvested area is a more appropriate productivity denominator. If deriving a measure of the average return on land available to farmers for crop cultivation is more important, then cropland area is a more logical choice. Cropland area may also be considered a better choice, since it implicitly introduces some accounting for land quality. Poorer-quality land in low-input systems needs to be fallowed for longer periods. Thus, while a smallholder in Africa may own 2 hectares of land, in any given season at least half of that land may need to be fallowed, implying that the cropland area-based land productivity measure will be about half that of a harvested area-based metric, but will better reflect the returns to the farm household's cropland holdings. Harvested area is also a less attractive option in the case of multiple growing seasons or sequential cropping within the annual cycle, in which case the physical land area will be double counted. This study followed the established practice of using total cropland as the land productivity denominator, not controlling for variations in land quality.

cocoa and cotton), in addition to cereals and other root crops. Within West Africa the root crop, cereal-root crop, and agropastoral systems together provide almost 70 percent of the cropland area and almost 60 percent of the value, dominated by yam and cassava in the wetter zones and millet and sorghum in the drier areas; the more humid tree crop area, in which cocoa production is widespread, occupies about 11 percent of West Africa's cropland but generates about 20 percent of the value of production.

In eastern and central Africa (ECA), the spatial concentration of production extents and values is even more pronounced in the highland areas (the highland perennial and the highland temperate mixed systems), accounting for around 8 percent of the land area, 15 percent of the cropland, and 30 percent of the ECA value of production. The other dominant ECA system is the maize mixed system, extending south from Ethiopia along the Rift Valley into Tanzania (and continuing, in southern Africa, down into Zimbabwe and Mozambique). The maize mixed system occupies just less than a quarter of the ECA cropland, and delivers a similar share of the value of production.

Southern Africa is dominated by two systems: the southern extension of the maize mixed system, which contributes about 30 percent of the value of production, and the large commercial and smallholder system, tracing its root to colonial and post-colonial commercial farm tenure practices in South Africa, which contributes around 40 percent of the southern Africa crop production value.

Partial productivity results are provided in two formats. Figure 3.2 shows indicative maps of grid cell–scale estimates of land and labor productivity, noting that grid cell data and variable construction issues described above suggest this representation should be treated as broadly illustrative only. The aggregated subregional farming system estimates, the main results of the analysis, are presented in Table 3.2 and Table 3.3.

Spatial patterns of land productivity in crop production

Overall, the regional land productivity levels show some consistency, with a progression from eastern ($555/ha), through southern ($604/ha), to western ($671/ha) Africa (Table 3.2).[8] The per-hectare patterns of land productivity in West Africa show an expected progressive increase from the semiarid agropastoral-millet/sorghum systems of the Sahel ($337/ha), through the higher-rainfall cereal-root crop system ($613/ha) and root crop system ($1,070/ha), to the subhumid and humid coastal artisanal fishing system ($1,125/ha). In

8 All currency is in US dollars, unless specifically noted as "international dollars."

TABLE 3.2 Land productivity: average value of annual crop production ($) per hectare cropland by subregion and farming system, 2005–2007

Farming system	Eastern and central Africa	Western Africa	Southern Africa	Africa south of the Sahara
Agropastoral-millet/sorghum	289	337	465	340
Cereal-root crop mixed	372	613	437	572
Coastal artisanal fishing	688	1,125	357	870
Forest based	523	839	1,315	575
Highland perennial	822	n.a.	n.a.	822
Highland temperate mixed	530	1,103	368	547
Irrigated	268	440	439	344
Large commercial and smallholder	n.a.	n.a.	850	850
Maize mixed	592	721	563	582
Pastoral	418	240	660	326
Rice-tree crop	853	n.a.	n.a.	853
Root crop	658	1,070	544	945
Sparse (arid)	246	735	545	278
Tree crop	710	1,108	1,064	1,093
Not labeled	625	949	778	878
SSA average	**555**	**671**	**604**	**624**

Source: Authors' calculations based on SPAM crop distribution (HarvestChoice 2014); farming systems (Dixon, Gulliver, and Gibbon 2001); FAO crop prices (FAO 2012); and cropland distribution (Ramankutty et al. 2008).

Notes: n.a. = not applicable; Not labeled = areas made of grid cells that do not have a farming system because of differences in the delineation of water and land interface (such as coastlines and lake areas) between data layers.

the 10 percent of the humid tree crop systems, land productivity is assessed at $1,108/ha. The higher productivity in more humid systems reflects not only higher yields, but also higher-value cash crops, especially cocoa and rice and, likely, higher levels of market accessibility. While the pastoral systems produce only $240/ha of crop production, these areas are, by definition, primarily livestock-oriented livelihood zones. The progression of land productivity values in West Africa from $240/ha in the semiarid marginal croplands that fringe the Sahel to $1,108/ha in the most humid coastal areas represents an almost fivefold range, and reflects a striking pattern of alignment between the gradients of rainfall and productivity. Given the lower rainfall variability observed in more humid zones, the higher returns to land in those zones are also likely to be more stable from year to year.

In contrast, in the semiarid pastoral systems crop production is not only less suited but also more erratic from year to year. A surprising finding is the

FIGURE 3.2 Land and labor productivity of crop production in Africa south of the Sahara (circa 2006)

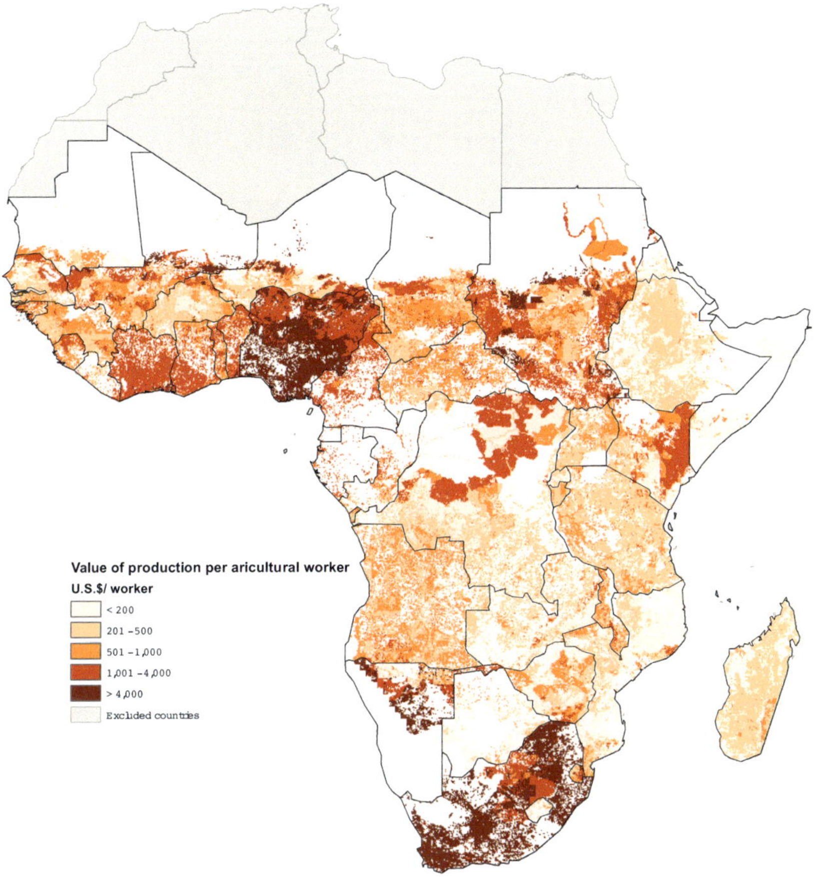

(continued)

FIGURE 3.2 (continued)

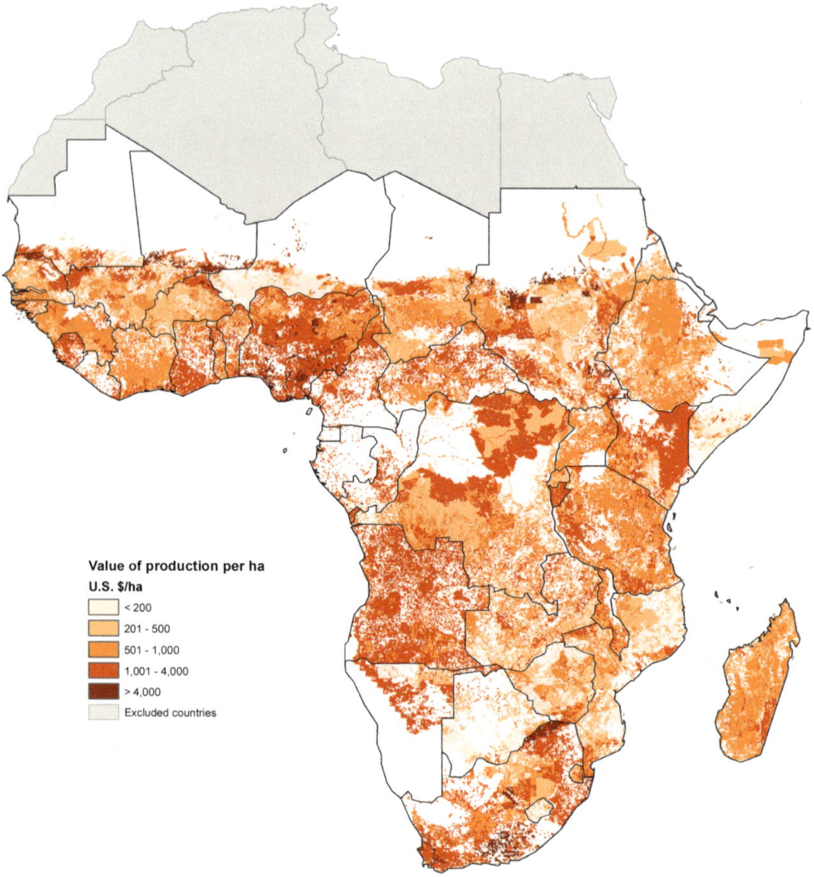

Sources: Authors' calculations and illustration based on SPAM crop distribution (HarvestChoice 2014); FAO crop prices and agricultural labor (FAO 2012); cropland distribution (Ramankutty et al. 2008); and rural population distribution (CIESIN et al. 2011).

modest value of land productivity in the formal irrigated systems ($440/ha) that occupy just more than 2 percent of the region's cropland, primarily in the semiarid Niger basin (Office du Niger). Differences in resolution across multiple data layers when examining these small geographic areas are suspected of producing less reliable results.

Land productivity patterns in eastern Africa also vary significantly by system. The highest land productivity of the major systems is around $820/ha, assessed for the high-population-density, high-market-access highland perennial systems of Ethiopia, Uganda, Rwanda, and Burundi (bananas, plantains, enset, coffee, cassava, sweet potatoes, beans, cereals, livestock). Land productivity in the more remote, less densely (but still highly) populated, and less humid highland temperate mixed system (wheat, barley, teff, peas, lentils, broad beans, potatoes, livestock) is significantly lower at $530/ha. The largest system, maize mixed, is estimated to provide returns to land of just under $600/ha, less than the $660/ha estimated for the root crop system (cassava, legumes) that enjoys around 15 percent higher annual rainfall but only about one-third of the population density. Although small in extent, tree crop and rice-tree crop systems (the latter only distinguished in Madagascar) provide high returns to land of $710/ha and $853/ha, respectively.

In southern Africa, by far the predominant system, the large commercial and smallholder system, also provides the highest land productivity ($850/ha) outside the small, humid forest-based systems ($1,315/ha) and tree crop ($1,064/ha) systems. The second-largest system, maize mixed, shows similar land productivity levels ($563/ha) as in eastern Africa, while the return to land through cropping in pastoral systems is significantly higher ($660/ha). The extensive large commercial and smallholder system is associated with by far the largest return to land of all the major southern African systems ($1,010/ha), compared with the maize mixed ($670/ha), cereal-root crop ($603/ha), root crop ($650/ha), and agropastoral ($720/ha) systems. These findings suggest that, by virtue of a commercial focus encompassing use of fertilizer inputs over many years, the soils of the large commercial and smallholder system remain more fertile than those in other rainfed, cereal-based systems.

Spatial patterns of labor productivity in crop production

Details of labor productivity are presented in Figure 3.2 and Table 3.3. There are striking differences in labor productivity across both subregions and farming systems. Perhaps the most notable individual estimate is that of labor productivity in the large commercial and smallholder systems of southern Africa ($3,620 per worker)—some sevenfold larger than the regional average ($544 per worker). This system comprises a mix of scattered smallholders among large-scale commercial operations that are, generally, highly mechanized. All other systems with high levels of labor productivity are found in western Africa, of which two—root crop and tree crop—systems predominate (representing about 40 percent of West Africa's cropland). Tree crop systems in West Africa ($1,626 per worker) include many cash crops—cocoa, coffee, oil palm, rubber, and yams—that are high value and, in the case of perennials, require less intensive labor inputs than annual crops. Root crop systems

TABLE 3.3 Labor productivity: average value of annual crop production ($) per agricultural worker, 2005–2007

Farming system	Eastern and central Africa	Western Africa	Southern Africa	Africa south of the Sahara
Agropastoral-millet/sorghum	235	580	264	461
Cereal-root crop mixed	360	985	215	699
Coastal artisanal fishing	300	1,534	175	696
Forest based	235	576	512	273
Highland perennial	381	n.a.	n.a.	381
Highland temperate mixed	206	1,974	296	234
Irrigated	246	644	187	374
Large commercial and smallholder	n.a.	n.a.	3,620	3,620
Maize mixed	269	489	388	300
Pastoral	305	610	277	382
Rice-tree crop	371	n.a.	n.a.	371
Root crop	312	1,588	247	867
Sparse (arid)	337	799	619	373
Tree crop	315	1,626	415	1,473
Not labeled	240	967	504	680
SSA Average	**287**	**1,084**	**461**	**544**

Sources: Authors' calculations based on SPAM crop distribution (HarvestChoice 2014); farming systems (Dixon, Gulliver, and Gibbon 2001); FAO crop prices and agricultural labor (FAO 2012); and rural population distribution (CIESIN et al. 2011).

Notes: n.a. = not applicable; Not labeled = areas made of grid cells that do not have a farming system because of differences in the delineation of water and land interface (such as coastlines and lake areas) between data layers.

($1,588 per worker) are characterized by crops with high yields and, in the case of yams, high value.

In contrast, labor productivity in eastern Africa is remarkably uniform and low, ranging from $235 to $380 per worker, with the highland perennial and rice-tree crop systems having the highest productivity ($381 and $371 per worker, respectively). Comparing the labor productivity of the major cereal-based farming systems, there is typically a two- to threefold higher productivity in West Africa compared with East Africa, though the productivity relativities among systems in each region are fairly consistent with expectations. These results are consistent with the findings of Block (2010—Figure 3.2) and Wiebe, Soule, and Schimmelpfennig (2001—Table 4.1). While the results of Block (2010) and Wiebe at al. (2001) include both crop and livestock sectors, they also show consistently higher labor productivity at the subregional scale for West Africa versus East Africa (Block 2010) and for a range of individual countries in West versus East Africa (Wiebe et al. 2001). Several potential structural and data-related factors underpin these regional differences. From a production structure perspective, there is a greater prevalence of informal irrigation in West Africa in "inland valley" production systems that are characterized as rainfed and not irrigated farming systems, but that boost yields and hence production values. There is also a greater prevalence of higher-value cash crops in the West African production mix (for example, rice, cocoa, and cotton), as is illustrated in higher land productivity values for the farming systems in which these crops predominate. These results also suggest *relativities* fairly consistent with expectations among systems.

Potential productivity effects of climate change

One of the major uncertainties with regard to the future trajectory of agricultural productivity in Africa is the likely impact of climate change (IPCC 2007). Several studies (Kurukulasuriya et al. 2006; Seo et al. 2008; Nelson et al. 2010) provide strong evidence that predicted changes in temperature and rainfall caused by global warming may, overall, impose serious constraints to agricultural growth in Africa, but that the changes are likely to have different effects in different locations. The findings of Seo et al. (2008) have the most direct relevance for this study,[9] suggesting that the impacts of climate change will vary across different agroecological zones (AEZs): farms in the savanna areas are seen as the most vulnerable to higher temperatures and reduced precipitation, while those in the subhumid or humid forest could gain even from a severe climate change. More specifically, Seo et al. (2008) find that households in the cereal-root crop mixed, dryland mixed, agropastoral, and pastoral farming systems (that is, those most equivalent to the savanna AEZs of Seo et al. 2008) are likely to be the most vulnerable to climate change. However, because climate warming is likely to increase livestock income while reducing crop income (Seo et al. 2008; Nelson et al. 2010), climate change may have a zero net effect on the total agricultural income of households engaging in both crop and livestock production in these systems, depending on the relative importance of the two subsectors in their livelihoods. Households engaging solely or mostly in crop production stand to lose the most, while those engaging solely or mostly in livestock stand to gain the most. Households

9 Seo et al. (2008) use a set of AEZs based on the same climate variable used in defining the farming systems depicted in Figure 3.1, so there is strong correspondence between the two schemas. We report here the AEZ results of Seo et al. (2008) summarized into the most equivalent farming systems.

in the forest-based and tree-crop farming systems (most of the subhumid or humid forest AEZs) are predicted to gain even from severe climate change.

The Seo et al. (2008) study of potential climate change impacts in the 16 AEZs in Africa relied on regressions of net farm revenues on climate variables and other socioeconomic factors at the grid cell level using data from 9,597 household surveys and climate stations in 10 countries in 2008.[10] Seo et al.'s (2008) work highlighted weaknesses in the parameterization of regional economic models, such as that used by Nelson et al. (2010), and reinforced the findings of this chapter, with greater empirical granularity, that considerable heterogeneity exists even within AEZs and farming systems that attempt to delineate regionally significant areas of broadly similar conditions under which agriculture may be practiced. This implies that examination of specific agricultural development issues, such as the impacts of climate change on the factors influencing the yield of crops or the prevalence and severity of pests and diseases, would be best served by more spatially disaggregated time-series analysis on a broader range of socioeconomic variables that affect agricultural productivity and output. Disaggregating AEZs into finer spatial system-based characteristics (as discussed in Chapter 4 and called agricultural productivity zones) can, therefore, help support more localized assessment of the potential impacts of climate change. Similar advances are also required in the regional specificity of climate change projections compatible with projected global scenarios, especially as they relate to assessing the changing patterns of extreme events, such as droughts and floods for which historical local data are available to support local validation and calibration.

Conclusions and Implications

We have highlighted the spatial dimensions of agriculture in SSA using a reporting unit—the farming system—that increases our ability to discriminate spatial patterns of (1) major cropping systems and their productivity and (2) a number of underlying conditioning factors. We have also developed indicative grid-scale representations of partial productivity that appear to reflect a spatial concordance among the location and (although not demonstrated by these results) the evolution of agricultural potential, agricultural production, population density, infrastructure, and market access, and, ultimately, patterns of agricultural productivity. Clearly, many other policy,

10 The 10 countries are Burkina Faso, Cameroon, Egypt, Ethiopia, Ghana, Kenya, Niger, Senegal, South Africa, and Zambia.

cultural, socioeconomic, and environmental factors come into play in determining farm-level productivity across individual districts and within individual plots. However, there is no escaping the overarching observation that geographic context plays a significant role in conditioning both the baselines and the likely trajectory of productivity growth possibilities. While that insight alone is intuitive, this and other studies reported in this volume point to our increasing ability to explore and account for the location-specific determinants and outcomes of agricultural development. Our maintained hypothesis is that improving our capacity to readily and routinely examine the patterns and processes of agricultural production at increasingly higher levels of spatial (and temporal) resolution can improve our ability to set and achieve more realistic agricultural development goals.

Setting aside findings for the least important systems in each region, our results show land productivity values for crop production as low as \$240–\$290/ha (agropastoral-millet/sorghum in East Africa and pastoral in West Africa) and as high as \$1,125/ha in the humid coastal systems of West Africa, where cash crops are widespread. Land productivity values are in the ranges of \$290–\$820/ha in East Africa (agropastoral to highland perennial), \$240–\$1,125/ha in West Africa (pastoral to coastal), and \$440–\$850/ha in southern Africa (cereal-root to large commercial and smallholder). With typical holdings of 1–3 hectares and 5–8 family members per farm household, it is easy to understand both why rural poverty below the \$1.25/person/day international poverty line is so prevalent and persistent, and why raising land productivity, and doing so in a sustainable manner, remains such a fundamental development goal for Africa.

With respect to labor productivity (and noting our comments on shortcomings in the absolute values of our agricultural labor proxy), we see regional labor productivity estimates spanning a much broader range, from \$206/worker in East Africa (highland temperate mixed) to the singularly high \$3,620/worker in the large commercial and smallholder systems, primarily located in southern Africa, where large commercial operators are highly mechanized in comparison with the rest of SSA. Labor productivity ranges are \$206–\$380/worker in East Africa (highland temperate mixed to highland perennial), \$580–\$1,626/worker in West Africa (agropastoral to tree crop), and \$247–\$3,620/worker in southern Africa (root crop to large commercial and smallholder). Again, the pervasive and persistent poverty consequence of low-productivity agriculture is apparent, with a significant share of agricultural workers in SSA generating only a gross crop value of around \$1 for each day of back-breaking drudgery.

One message of relevance to national practitioners and decisionmakers from even this qualitative review of aggregate evidence is that local factors shaping agricultural productivity and opportunities for growth can be increasingly recognized and accounted for in the targeting and formulation of agriculture-related policies, investments, and interventions aimed at farmers and their service providers. It is no longer necessary, efficient, or justifiable to make blanket provisions, regulations, and investments that do not adequately account for important spatial variation in development constraints or that do not better capitalize on more local opportunities, since many other policy, cultural, socioeconomic, and environmental factors also shape agricultural productivity and the search for viable and efficient options for bringing about substantial productivity increases in the region.

Other insights are particularly relevant to the role of subregional and international institutions. The spatial patterns of land and labor productivity highlight at least two opportunities for cross-border collaboration: learning and, where appropriate, collaboration and coordinated action. First, examining the panels of Figure 3.2, it is apparent that distinct, contiguous geographical clusters exist throughout Africa, where suitable conditions for and the practice of agriculture are concentrated; however, these clusters are divided artificially by national boundaries. Clear examples are the Great Lakes region of East Africa, the northern savannas of Nigeria and its near neighbors, and the southern more humid coastal belt spanning much of West Africa. In such contexts, promoting institutions and mechanisms that address cluster-specific challenges of improving productivity in a coherent manner and that respect, but are not constrained by, the presence of national boundaries, is likely a high-payoff strategy. Second, reflecting on Figure 3.1, is the value of applying a consistent production system framework across Africa as a means of better revealing and assessing the potential scope for regionwide technology and knowledge spillover. There is likely much to gain from understanding the challenges faced and solutions found (or emerging) at different locations within individual farming systems that occur repeatedly across Africa (for example, agropastoral-millet/sorghum, cereal-root crop, maize mixed, and root crop). A productivity-enhancing technology developed in the root crop areas of Mozambique may perfectly address a problem faced in the root crop areas of Côte d'Ivoire, presuming the mechanisms were in place for that knowledge to be documented, discovered, and acted upon across the entire geographic span of each unique farming system.

While neither detailed nor exhaustive, the data and insights provided in this chapter do point to the enduring validity of long-established development theories (von Thuenen 1826; Ricardo 1891; Boserup 1965). We clearly see evidence of larger returns to land and labor in areas of comparative rainfall advantage, larger returns in the more market-accessible systems, and suggestions of higher returns to land (if not to labor) in areas of high population density (in the East Africa highland perennial systems, for example), where pressure on natural resources is known to have led to improved management practices (Machakos being the storied example in this region). However, there are early indications and still emerging projections about the potential for climate change to play a significant role in redefining the spatial pattern of both challenges to but also opportunities for accelerating productivity growth in the region.

Appendix for Chapter 3

TABLE 3.A1 Distribution of value of crop production by farming system ($ millions), 2005–2007

Farming system	Eastern and central Africa	Western Africa	Southern Africa	Africa south of the Sahara
Agropastoral-millet/sorghum	8,133	65,471	7,474	81,078
Cereal-root crop mixed	9,955	148,846	14,988	173,789
Coastal artisanal fishing	2,934	29,570	4,133	36,637
Forest based	23,298	7,141	125	30,564
Highland perennial	57,589	n.a.	n.a.	57,589
Highland temperate mixed	29,064	3,956	2,182	35,202
Irrigated	5,928	7,593	74	13,595
Large commercial and smallholder	—	n.a.	52,428	52,428
Maize mixed	74,437	6	39,029	113,472
Pastoral	28,108	22,669	3,764	54,541
Rice-tree crop	13,282	n.a.	n.a.	13,282
Root crop	32,677	161,246	5,994	199,917
Sparse (arid)	2,745	165	521	3,431
Tree crop	2,971	119,987	172	123,130
Not labeled	820	6,452	1,161	8,433
Total	**291,941**	**573,101**	**132,043**	**997,085**

Sources: Authors' calculations based on SPAM crop distribution (HarvestChoice 2014); farming systems (Dixon, Gulliver, and Gibbon 2001); and FAO crop prices (FAO 2012).

Notes: — = data not available; n.a. = not applicable; Not labeled = areas made of grid cells that do not have a farming system because of differences in the delineation of water and land interface (such as coastlines and lake areas) between data layers.

TABLE 3A.2 Distribution of cropland area by farming system (1,000 hectares), 2005

Farming system	Eastern and central Africa	Western Africa	Southern Africa	Africa south of the Sahara
Agropastoral-millet/sorghum	5,594	21,008	1,926	28,527
Cereal-root crop mixed	4,778	21,657	3,808	30,242
Coastal artisanal fishing	350	2,243	364	2,957
Forest based	2,910	1,503	159	4,572
Highland perennial	5,317	n.a.	n.a.	5,317
Highland temperate mixed	6,101	434	868	7,402
Irrigated	1,879	2,333	39	4,251
Large commercial smallholder	n.a.	n.a.	13,219	13,219
Maize mixed	13,823	1	9,636	23,460
Pastoral	9,010	11,719	976	21,705
Rice-tree crop	1,825	n.a.	n.a.	1,825
Root crop	8,920	25,222	3,317	37,459
Sparse (arid)	161	6	178	344
Tree crop	263	11,930	165	12,358
Not labeled	163	604	268	1,034
Total	**61,092**	**98,659**	**34,924**	**194,675**

Sources: Authors' calculations based on SPAM crop distribution (HarvestChoice 2014); farming systems (Dixon, Gulliver, and Gibbon 2001); and cropland distribution (Ramankutty et al. 2008).

Note: n.a. = not applicable; Not labeled = areas made of grid cells that do not have a farming system because of differences in the delineation of water and land interface (such as coastlines and lake areas) between data layers.

TABLE 3A.3 Distribution of rural population headcount by farming system (number), 2005

Farming system	Eastern and central Africa	West Africa	Southern Africa	Africa south of the Sahara
Agropastoral-millet/sorghum	5,387,031	32,808,336	3,143,550	41,338,917
Cereal-root crop mixed	9,301,065	48,709,206	13,472,937	71,483,208
Coastal artisanal fishing	3,391,375	10,257,584	3,651,007	17,299,966
Forest based	29,966,512	4,136,853	129,907	34,233,272
Highland perennial	40,217,290	n.a.	n.a.	40,217,290
Highland temperate mixed	34,974,185	536,657	4,033,144	39,543,986
Irrigated	3,559,508	4,767,946	75,607	8,403,061
Large commercial and smallholder	n.a.	n.a.	13,124,074	13,124,074
Maize mixed	59,566,865	1,064	27,801,226	87,369,155
Pastoral	23,294,528	8,316,146	789,492	32,400,166
Rice-tree crop	8,967,891	n.a.	n.a.	8,967,891
Root crop	14,715,031	29,322,488	3,036,864	47,074,383
Sparse (arid)	4,881,068	416,379	100,120	5,397,567
Tree crop	2,521,193	32,975,758	831,987	36,328,938
Not labeled	1,877,891	3,580,151	882,511	6,340,553
Average	**242,621,433**	**175,828,568**	**71,072,426**	**489,522,427**

Sources: Authors' calculations based on SPAM crop distribution (HarvestChoice 2014); farming systems (Dixon, Gulliver, and Gibbon 2001); and rural population distribution (CIESIN et al. 2011).

Notes: n.a. = not applicable; Not labeled = areas made of grid cells that do not have a farming system because of differences in the delineation of water and land interface (such as coastlines and lake areas) between data layers.

References

Akpan, U. 2014. "Impact of Regional Road Infrastructure Improvement on Intra-Regional Trade in ECOWAS." *African Development Review* 26 (1): 64–76.

Benin, S., A. Nin-Pratt, S. Wood, and Z. Guo. 2011. *Trends and Spatial Patterns in Agricultural Productivity in Africa, 1961–2010.* ReSAKSS Annual Trends and Outlook Report 2011. Washington, DC: International Food Policy Research Institute.

Block, S. 2010. *The Decline and Rise of Agricultural Productivity in Sub-Saharan Africa since 1961.* NBER Working Paper 16481. Cambridge, MA: National Bureau of Economic Research.

Boserup, E. 1965. *The Conditions of Economic Growth.* London: Allen and Unwin.

CIESIN (Center for International Earth Science Information Network, Columbia University), International Food Policy Research Institute, World Bank, and Centro Internacional de Agricultura Tropical. 2011. *Global Rural-Urban Mapping Project, Version 1 (GRUMPv1): Population Density Grid.* Palisades, NY: National Aeronautics and Space Administration Socioeconomic Data and Applications Center. Accessed March 2012. http://sedac.ciesin.columbia.edu/data/set/grump-v1-population-density.

Denninger, K., and D. Byerlee. 2011. *Rising Global Interest in Farmland: Can It Yield Sustainable and Equitable Benefits?* Washington, DC: World Bank.

Dixon, J., A. Gulliver, and D. Gibbon. 2001. *Farming Systems and Poverty: Improving Farmers' Livelihoods in a Changing World.* Rome: Food and Agriculture Organization of the United Nations; Washington, DC: World Bank.

Dorosh, P., H. Wang, L. You, and E. Schmidt. 2010. *Crop Production and Road Connectivity in Sub-Saharan Africa: A Spatial Analysis.* Policy Research Working Paper 5385. Washington, DC: World Bank.

ESA (European Space Agency). 2012. *The Earth Observation Handbook: 2012 Special Edition for Rio+20.* Noordwijk, Netherlands: ESA Communications.

FAO (Food and Agriculture Organization of the United Nations). 2012. FAOSTAT database. "Production and Resource Domains." Accessed March. http://faostat.fao.org/default.aspx?lang=en.

FAO (Food and Agriculture Organization of the United Nations) and World Bank. 2009. *Awakening Africa's Sleeping Giant: Prospects for Commercial Agriculture in the Guinea Savannah Zone and Beyond.* Rome: FAO; Washington, DC: World Bank.

Fischer, G., F. O. Nachtergaele, H. van Velthuizen, L. Verhelst, and D. Wiberg. 2012. "Global Agro-Ecological Zones (GAEZ v3.0)—Model Documentation." Laxenburg, Austria: International Institute for Applied Systems Analysis; Rome: Food and Agriculture Organization of the United Nations.

HarvestChoice. 2014. "Crop Production: SPAM." Washington, DC: International Food Policy Research Institute; St. Paul: University of Minnesota. http://harvestchoice.org/tools/crop-production-spam–0.

Headey, D. D., and T. S. Jayne. 2014. "Adaptation to Land Constraints: Is Africa Different?" *Food Policy* 48: 18–33.

ILO (International Labor Organization). 2010. ILOSTAT database of labor statistics. Accessed March. http://laborsta.ilo.org/.

IPCC (Intergovernmental Panel on Climate Change). 2007. *State of the Science.* Contribution of Working Group I to the Fourth Assessment Report of the Intergovernmental Panel on Climate Change. Cambridge, UK: Cambridge University Press.

Jarvis, A., H. I. Reuter, A. Nelson, and E. Guevara. 2008. CGIAR-CSI SRTM 90m database. "Hole-filled SRTM for the Globe. Version 4." http://srtm.csi.cgiar.org.

Jerven, M. 2013. *Poor Numbers: How We Are Misled by African Development Statistics and What to Do about It.* Ithaca, NY: Cornell University Press.

Kurukulasuriya, P., R. Mendelsohn, R. Hassan, J. Benhin, M. Diop, H. M. Eid, K.Y. Fosu, et al. 2006. "Will African Agriculture Survive Climate Change?" *World Bank Economic Review* 20 (3): 367–388.

Nelson, G. C., M. W. Rosegrant, A. Palazzo, I. Gray, C. Ingersoll, R. Robertson, S. Tokgoz, et al. 2010. *Food Security, Farming, and Climate Change to 2050: Scenarios, Results, Policy Options.* Washington, DC: International Food Policy Research Institute.

Openshaw, S. 1984. *The Modifiable Areal Unit Problem.* Norwich, UK: Geobooks.

Pingali, P., and H. Binswanger. 1988. "Population Density and Farming Systems—The Changing Locus of Innovations and Technical Change." In *Population, Food and Rural Development,* edited by Ronald Lee. Oxford, UK: Clarendon Press.

Ramankutty, N., A. T. Evan, C. Monfreda, and J. A. Foley. 2008. "Farming the Planet: 1. Geographic Distribution of Global Agricultural Lands in the Year 2000." *Global Biogeochemical Cycles* 22 (1003): 1–19. http://trane.evsc.virginia.edu/Publications_files/2007GB002952.pdf.

Ricardo, D. 1891. *Principles of Political Economy and Taxation.* London: G. Bell.

Robinson, W. S. 1950. "Ecological Correlations and the Behavior of Individuals." *American Sociological Review* 15: 351–357.

Seo, S. N., R. Mendelsohn, A. Dinar, R. Hassan, and P. Kurukulasuriya. 2008. *A Ricardian Analysis of the Distribution of Climate Change Impacts on Agriculture across Agro-Ecological Zones in Africa.* Policy Research Working Paper 4599. Washington, DC. World Bank.

Stock, R. 2012. *Africa South of the Sahara: A Geographical Interpretation,* 3rd ed. New York: The Guildford Press.

Subramanian, S. V., K. Jones, A. Kaddour, and N. Krieger. 2009. "Revisiting Robinson: The Perils of Individualistic and Ecological Fallacy." *International Journal of Epidemiology* 38: 342–360.

Thornton, P. K., P. G. Jones, G. Alagarswamy, and J. Andersen. 2009. "Spatial Variation of Crop Yield Response to Climate Change in East Africa." *Global Environmental Change* 19 (1): 54–65.

von Thuenen, J. 1826. 1966. *Isolated State,* translated by C. M. Wartenberg. New York: Pergamon Press.

Waller, L. A., and C. A. Gotway. 2004. *Applied Spatial Statistics for Public Health Data.* New York: John Wiley and Sons.

Wiebe, K., M. Soule, and D. Schimmelpfennig. 2001. "Agricultural Productivity for Sustainable Food Security." In *Agricultural Investment and Productivity in Developing Countries,* edited by L. Zepeda. FAO Economic and Social Development Paper 148. Rome: Food and Agriculture Organization of the United Nations.

Wood, S., and J. R. Anderson. 2009. "Strategic Priorities for Agricultural Development in Eastern and Central Africa: A Review of the Institutional Context and Methodological Approach for Undertaking a Quantitative, Subregional Assessment." In *Prioritizing Agricultural Research for Development: Experiences and Lessons,* edited by D. Raitzer and G. Norton. Wallingford, UK: CABI.

Wood, S., K. Sebastian, and J. Chamberlin. 2003. "Land Quality, Agricultural Productivity and Food Security: A Spatial Perspective." In *Land Quality, Agricultural Productivity, and Food Security: Biophysical Processes and Economic Choices at Local, Regional and Global Levels,* edited by K. Wiebe. Northampton, MA: Edward Elgar.

Yang, Y., and S. Gupta. 2007. "Regional Trade Arrangements in Africa: Past Performance and the Way Forward." *African Development Review* 19 (3): 399–431.

You, L., S. Wood, and U. Wood-Sichra. 2009. "Generating Plausible Crop Distribution Maps for Sub-Saharan Africa Using a Spatially Disaggregated Data Fusion and Optimization Approach." *Agricultural Systems* 99 (2–3): 126–140.

Chapter 4

TYPOLOGY OF AGRICULTURAL PRODUCTIVITY ZONES

Bingxin Yu and Zhe Guo

Introduction

The preceding chapter illustrated remarkable spatial heterogeneity in agricultural productivity across Africa at the system level, characterized by the inherent variations of climate, land suitability for agriculture, human and animal populations, expanding transportation networks, and other rural infrastructure. To help identify specific policies and investments to increase productivity in different locations, the analysis in Chapter 3 can be complemented by closer examination of the production agents within those systems. Within a given farming system, production agents also are heterogeneous, in terms of not only the resource constraints they face (Byerlee, Harrington, and Winkelmann 1982; Chambers and Jiggins 1987), but also how those resource constraints affect technology adoption and technological change that may result in productivity growth and possibly poverty reduction (Feder, Just, and Zilberman 1985; Feder and Umali 1993). The coevolution of these characteristics contributes to a dynamic context in which the nature and performance of agriculture manifest themselves through different spatial patterns of production portfolio and productivity. Identifying spatially common themes in these different portfolios can offer opportunities for developing and implementing strategies that cut across national boundaries and agroecological conditions.

Based on the farming systems defined by Dixon, Gulliver, and Gibbon (2001) and the development domains approach (Pender, Place, and Ehui 1999; Wood et al. 1999), this chapter characterizes these common themes via spatially similar local production units or "agricultural productivity zones" (APZs). The APZs are nested within farming systems, national borders, and agricultural production conditions (biophysical and socioeconomic), providing a finer system-based measurement. Each APZ is defined as a geographical region, or a set of noncontiguous geographic areas, exhibiting broadly homogeneous characteristics with regard to the potential productivity of agricultural production agents. In sum, a typology analysis is used to classify APZs within and across national borders based on similarities of production agents in resource bases, economic activities, demographics, and market access.

The assumption is that production agents in a specific APZ are more likely to follow a similar technology adoption pathway, leading to similar productivity and growth outcomes. As such, the resulting typologies can be a useful tool for designing a tailored agricultural development strategy and policy intervention at the subnational level, according to both local absolute and comparative advantages.[1] By revealing similarities beyond national borders and farming systems, this typology analysis can help regional organizations and national governments to pool resources to identify and pursue common solutions that any one country alone may find to be beyond its capacity.

Data

Because we use several of the spatial datasets described in Chapter 3 (including the farming systems (Dixon, Gulliver, and Gibbon 2001), the Spatial Production Allocation Model (SPAM; You, Wood, and Wood-Sichra 2009), and land area (Fischer et al. 2012), the issues discussed and how they are dealt with are the same in this chapter. Therefore, we will not repeat them here. We focus only on the specifics as well as the additional datasets used, including the normalized difference vegetation index (NDVI), which is a biophysical measurement of vegetation and agriculture land production potentials; those on the socioeconomic attributes, such as distribution of rural and urban populations; and market access.

Farming systems

In their study of 15 farming systems, Dixon, Gulliver, and Gibbon (2001) summarize similarities in biophysical endowments, demographics, farm practices, infrastructure, and livelihood choices across different parts of the continent. Each of the 15 farming systems covers tens of millions of hectares that hide the heterogeneous pattern of agricultural production and productivity within a country (Conradie, Piesse, and Thirtle 2009a, 2009b). The major systems include the highland, tree-root crop, maize mixed, and pastoral systems.

1 Absolute advantage in the production of a particular agricultural commodity is defined by the cost of producing it in different areas, which in the framework used here is influenced by the biophysical factors that the different areas have for producing the commodity. The comparative advantage is defined by the opportunity cost of producing it in different areas, which is influenced by the ability that the different areas have for trading the commodity.

The highland perennial farming system exists in Ethiopia, Uganda, Rwanda, and Burundi in areas with favorable natural resources and climate. The dominant crops are perennial crops, such as bananas/plantains and coffee; root crops (cassava and sweet potatoes); beans; and cereals. The highland temperate mixed farming system is similar to the highland perennial farming in its biophysical and socioeconomic characteristics. Major crops are wheat, barley, teff, pulses, and potatoes. Both of the highland systems have a considerable livestock component for mechanization, fuel, and savings. The tree-root crop system is mainly located in central and West Africa in countries like Côte d'Ivoire, Democratic Republic of the Congo, Ghana, and Nigeria, and in part of Tanzania. The other dominant system, maize mixed, stretches between eastern and southern Africa, whereas the cereal-root crop system is in two belts—one in western Africa and the other in central and southern Africa.

Normalized difference vegetation index

Derived from remote sensing of satellite imageries of moderate resolution, NDVI is used to measure and monitor plant growth (vigor), vegetation cover, and biomass production based on multispectral satellite data. Having been in use since the 1980s, NDVI is one of the vegetation indexes widely used to identify vegetated areas and detect their conditions according to plant cover in multispectral, remote-sensing data (Running et al. 1995; Doraiswamy et al. 2005). NDVI summarizes the effect of soil characteristics, rainfall, temperature, length of growing period, and irrigation (Dixon, Gulliver, and Gibbon 2001; Sanchez, Palm, and Buol 2003; Hijmans et al. 2005; NASA 2011; Fischer et al. 2012). This effect derives from differences in which different plants and soils, for example, absorb and reflect different light rays across the electromagnetic spectrum that are not visible to the human eye. For example, live vegetation strongly absorbs visible light for the use of photosynthesis, but strongly reflects near-infrared light, while bare soils reflect most of the lights in both bands.

Vegetation indexes are typically measured as ratios or linear combinations of light-reflectance ratios of red, green, and infrared spectral bands. As such, they are more robust measures than using light reflectance of individual bands to vegetation parameters, such as biomass and percentage of vegetation cover, thus facilitating the classification and monitoring of agricultural crops (Asrar et al. 1984). Because the NDVI measure could be affected by such factors as cloud cover that may block the solar rays and some soil characteristics that may reflect solar rays in a manner similar to vegetation, the measure is less sensitive when applied in areas with less cloud cover or higher-vegetation biomass levels.

NDVI is among the indexes most widely used by many researchers because of its stability in capturing vegetation growth status and conditions, and vegetation phenology. NDVI series are often used to monitor agricultural productivity, natural resources, and food security, because they provide repeated observations of the same location, allowing frequent updating of the vegetation status. For example, the NDVI time series have been applied in several African countries to successfully produce early warning of potential food production problems.

NDVI is calculated as:

$$\text{NDVI} = \frac{(NIR - VIS)}{(NIR + VIS)}$$

where *NIR* is the spectral reflectance measurement in the near-infrared region, and *VIS* is the visible (red) region. NDVI falls in the range between –1 and 1, but is always positive in Earth surface observations. Generally speaking, a high positive NDVI (0.5–1.0) is associated with areas under dense vegetation, exposed soil is characterized by a small positive value (0.1–0.2), and open water has a value close to zero. In this study, NDVI is calculated using monthly NDVI observations from 2000 to 2008 at the pixel level, and then is classified into three levels: low, medium, and high.

Market access

Improved market access is widely regarded as necessary to support agricultural and rural development (Calderón 2009; Dorosh et al. 2010), as access to markets and infrastructure is critical for determining the comparative advantage of a given location. Many parts of Africa are characterized by low road densities and poor conditions, long distances, and inadequate infrastructure, which add to travel times and transportation costs and, therefore, limit opportunities for farmers to participate in markets. Poor market access can negatively impact farm production, by limiting access to critical agricultural inputs, such as fertilizer, pesticides, and seed. Compared with urban households and those with easy access to markets, rural farm households with poor market access typically rely on their own production for most of their calorie intake. Inadequate market access, therefore, puts these households at greater risk of food insecurity. The more accessible markets are, the greater the population's ability to improve its economic performance and maintain food security.

We use travel time to the nearest major market (defined as a city or town with a population greater than 50,000) as a proxy for market access and infrastructure development. This is a type of a cost–distance function, where the "cost" is in hours to the nearest market center for each location (1 × 1–kilometer [km] pixel). First, market centers and their size were determined using population estimates from Global Rural Urban Mapping Project (GRUMP) data for the year 2000 (CIESIN et al. 2011). Travel time was estimated based on the combination of global spatial data layers, including road and river networks, assessed in terms of their "friction" or kilometers per hour (km/hr) travel time, and adjusted based on a number of input variables, including road location, road type, elevation, slope, country boundaries, bodies of water, coastline, and land cover. Each input variable was converted to a value representing the time it takes to travel 1 km. In the case of road type, for example, paved roads were given a value of 60 km/hr, while gravel roads were given a value of 15 km/hr. Bodies of water, land cover, slope, country boundaries, and elevation were also used to modify the speed of travel. For example, steeper areas were assigned slower speeds, and time delays were factored into travel that crossed borders. This allows us to estimate accessibility under different local topological conditions. Data from You and Guo (2011) are also used.

Production and other variables

Several spatial datasets are used to measure values of agricultural output, factors of production, and prices. The specific variables and the datasets include

- *Production information on 20 major crops*: Area (in hectares) and output (in tons) of wheat, rice, maize, barley, millet, sorghum, potatoes, sweet potatoes, cassava, bananas, soybeans, beans, other pulses, sugarcane, coffee, cotton, other fibers, groundnuts, other oil crops, and other crops. Spatial data are taken from SPAM (You, Wood, and Wood-Sichra 2009), and each pixel is measured at 5 arc-minute (about 10-km) grid-cell resolution.
- *Production information on two other major crops:* Area (in hectares) and production (in tons) of cocoa and tobacco. Spatial data on these crops are generated by the authors' own calculation, based on national statistics as reported by FAO (2011) to match those in the SPAM dataset.
- *Production information on five types of livestock:* Number of live animals of cattle, sheep, goats, chickens, and pigs. Spatial data are taken from Gridded Livestock of the World (Wint and Robinson 2007), and each pixel is measured at 5 arc-minute (about 10-km) grid-cell resolution.
- *International prices of crops and livestock:* Obtained from FAO (2011), measured in 2004–2006 average of constant dollars per ton to avoid substantial year-to-year price fluctuation.[2]

---

2 All currency is in US dollars, unless specifically noted as "international dollars."

- *Agricultural value of production:* Obtained from FAO (2011), measured at 2004–2006 constant thousand dollars per ton.
- *Land indicator:* Total and crop land area in hectares. Data are taken from Fischer et al. (2012), and each pixel is measured at 5 arc-minute (about 10-km) grid-cell resolution. The share of crop land in total land area is calculated at the pixel level.
- *Demographic characteristics:* Total, rural, and urban human populations. Data are from the joint project of CIESIN et al. (2011), and each pixel is measured at 5 arc-minute (about 10-km) grid-cell resolution. Population density is calculated by dividing total population by land area, measured in the number of people per square kilometer (km^2) in each pixel.

It is recognized that these data could misrepresent the current state of the economies for historical and political reasons, as pointed out by Jerven (2013). By combining different independent sources and types of data, several of which are based on observed measures of outcomes, such as the NDVI and land indicator, rather than self-reported data, such as the FAOSTAT database, we believe we have reduced some potential data pitfalls. For example, NDVI and land area obtained from remote sensing are generally viewed as reliable and subject only to the small differences of vegetation types on the ground, which can be greatly improved when combined with farming systems to identify dominant agricultural activities. Regarding the demographic indicators, CIESIN et al. (2011) follow procedures to ensure that data disseminated are of reasonable quality, and population estimates are comparable to national population statistics reported by the United Nations (UN 2011).

Data manipulation and descriptive statistics

Given the different sources, ranges, and measurements of the indicators of productivity correlates, the data were first compiled and harmonized at the pixel level. This involved several steps, including simplification of farming systems and determination of analytic units.

NEW FARMING SYSTEMS

Some of the closely related farming systems in terms of agroecological and socioeconomic conditions were combined, reducing the total number from 14 to 10 (Table 4.1). For example, agropastoral-millet/sorghum, pastoral, and sparse were grouped to form a new farming system called "pastoral-agropastoral." These are areas that have a harsh agricultural environment, limited cultivated land, low population density, and high dependence on

TABLE 4.1 Comparison of simplified and FAO-defined farming systems

Simplified farming system	Farming system*
Tree-root crop	Tree crop
	Root crop
Forest based	Forest based
Highlands	Highland perennial
	Highland temperate mixed
Cereal-root crop	Cereal-root crop mixed
Maize mixed	Maize mixed
Pastoral-agropastoral	Agropastoral-millet/sorghum
	Pastoral
	Sparse (arid)
Irrigated	Irrigated
Rice-tree crop	Rice-tree crop
Coastal	Coastal artisanal fishing
Large commercial and smallholder	Large commercial and smallholder

Source: Authors' representation based on Dixon, Gulliver, and Gibbon (2001).
Note: FAO = Food and Agriculture Organization of the United Nations. * Defined by Dixon, Gulliver, and Gibbon (2001)

livestock, and are vulnerable to drought. The pastoral-agropastoral system stretches from the arid and semiarid zone of the Sahel and the Horn of Africa to the western part of southern Africa. Similarly, highland perennial and highland temperate mixed farming systems were combined into one system labeled "highlands," and tree crop and root crop farming systems were grouped as "tree-root crop." These aggregations also help to avoid having a small number of pixels that will be difficult to work with statistically in the relatively small areas, such as those under the highland temperate mixed, tree crop, and agropastoral-millet/sorghum farming systems defined in Dixon, Gulliver, and Gibbon (2001).

AGRICULTURAL PRODUCTIVITY ZONES

Next, we generated the APZs, which derive from an overlay of the 10 new farming systems and the four-level NDVI. As such, a typical APZ covers an area that is larger than a pixel (which is about 100 km^2) but less than the average area of a farming system (about 1.7 million km^2). The final APZs are aggregates of common pixels within a country border and a farming system, with an average area of 5,381 km^2. The spatial distribution of APZs is illustrated in Figure 4.1. Compared with the initial farming systems of Dixon,

Gulliver, and Gibbon (2001) (Figure 3.1 in Chapter 3), the APZs provide a finer distinction of potential production systems across the continent.

The APZs highlight considerable variations of biophysical conditions and livelihoods within both country borders and agroecological zones. Looking at the distribution of APZs across countries, for example, 543 APZs are distributed in the 43 countries represented in the data for Africa south of the Sahara (SSA), and there is substantial variation across the countries, with greater diversity in Tanzania, Nigeria, Kenya, and Sudan (each boasts more than 24 APZs), whereas Djibouti and Equatorial Guinea exhibit little diversity (appendix Table 4A.2).

As one would expect, countries that exhibit greater heterogeneity in farming systems (Figure 3.1) and NDVI (Figure 4.2a and appendix Table 4A.1) will have a higher number of APZs, whereas those with more homogeneous farming systems and NDVI will have a smaller number of APZs. One noticeable feature is the concentration of areas with high NDVI in central and western Africa, especially north of the equator, as well as along the coast of Madagascar. The lowest NDVI areas are located mainly along the Sahara Desert in the north and the Namib Desert in the southwest of the continent, as well as the arid area in the region of the Horn of Africa.

On average, southern African countries enjoy more crop-friendly climate and have above-average NDVI levels. There are some observable high correlations between some farming systems and NDVI. For example, the highest NDVI areas are associated with the forest- and tree-based farming systems. High NDVI is also associated with the tree-root crop and maize mixed farming systems, as well as with the highland systems of Ethiopia, Kenya, and Uganda. Because of the relatively lower rainfall in the pastoral-agropastoral system, NDVI is also low there. It is important that there are large commercial and smallholder systems in South Africa with generally low NDVI. Although countries with smaller land areas tend to have fewer APZs, it is not always the case, as APZs capture heterogeneity in two indicators—farming system and NDVI. Hence, large Sahel countries like Niger, Chad, and Mauritania have less than a dozen APZs (with each APZ covering large areas of more than 50,000 km^2) because of low spatial variation in the two indicators. On the other hand, Tanzania, for example, has 29 APZs because it has eight farming systems and all four classes of NDVI.

Next, the other spatial data on productivity correlates (agricultural production, population density, and market access) (Figure 4.2b) are overlaid on the APZs to analyze the relative importance or incidence of the characteristics in each APZ. The results are summarized in Table 4.2. Before looking at the results, we first look at the spatial patterns of the productivity correlates.

FIGURE 4.1 Distribution of agricultural productivity zones

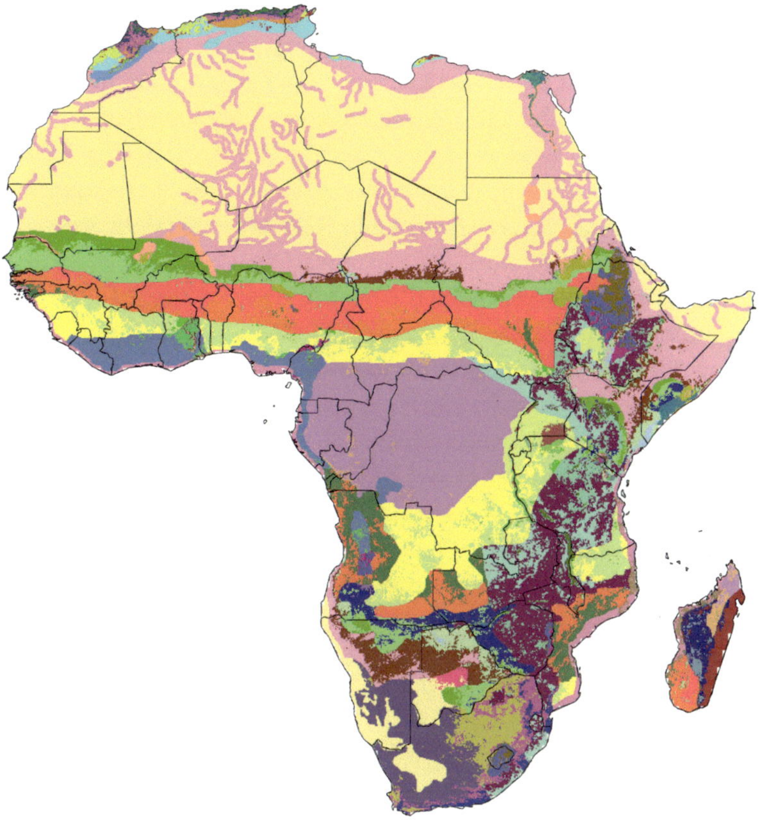

Source: Authors' illustration based on typology analysis.

Notes: There are 40 unique agricultural productivity zones (APZs), which makes it difficult to provide a key or legend.

SPATIAL PATTERNS OF THE PRODUCTIVITY CORRELATES

On average, cropland accounts for around 10 percent of the total land area in SSA (FAO 2011). Half of SSA's cropland is located in West Africa (appendix Table 4A.3), with more than 40–50 percent of land dedicated to agricultural production. There are also pockets of high cropland allocation in East African countries like Ethiopia, Rwanda, and Uganda, and central and northeast of southern Africa (Figure 4.2b). The top three farming systems, measured by cropland size, are root crop, cereal-root crop mixed, and agropastoral systems (appendix Table 4A.3). Maize mixed and pastoral systems are also important, with each accounting for more than 10 percent of total cropland area in the subcontinent.

With respect to market access, the average travel time to the nearest city with population above 50,000 is 12.7 hours for all of SSA taken together (Figure 4.2c and appendix Table 4A.4). Road access is more advanced in the highlands and coastal areas, while forest-based and pastoral systems face the biggest challenge in accessing markets. East Africa has the best market access, with 9.3 hours of average travel time, while West Africa has the worst—almost double East Africa's travel time. Average travel time is below 6.0 hours in four farming systems: highland perennial, irrigated, large commercial and smallholder, and tree crop systems. On the other hand, it usually takes more than 10 hours to visit the nearest city in the forest-based, pastoral, and root crop farming systems, and more than 30 hours to visit the sparse (arid) area.

Eastern Africa is home to about half of SSA's total population. It has a high rural population density of about 26.4 people per km^2, which is about one-third higher than that of western Africa, and more than twice as high as that of southern Africa. What is more revealing is land pressure for agricultural production, which is measured as the density of rural population per hectare (ha) of cropland. In eastern Africa, for example, each ha of cropland supports 4.0 rural people, compared with 1.8 in western Africa, and 2.0 in southern Africa. The highest rural population densities are found in the highland systems of eastern Africa (110–125 people/km^2) and the coastal systems of western Africa (80 people/km^2). Population is more dispersed in the pastoral system, with extremely low rural population densities found in the large commercial and smallholder system (about 11 people/km^2), compared with the average range of 20–50 people/km^2 for most of the other crop-based systems.

Looking at the interaction across the productivity correlates, it is logical to expect some coevolution over time. For example, it is generally expected that croplands adjacent to markets could be more productive because of easier access to a wide range of services, lower transaction costs on purchased inputs, and higher effective farmgate prices for outputs. The spatial correspondence between the productivity correlates is clearly revealed by comparing across the panels in Figures 4.2a–4.2d and appendix Tables 4A.1 and 4A.3–4A.5. For example, the intensity of cropland is closely associated with NDVI, population density, and market access in West Africa. In East and southern Africa, high population density overlaps with greater market access; hence, there is a high reliance on crop production, despite less-than-optimal agroecological conditions. In central Africa, high NDVI does not translate into crop-oriented agriculture, partly because of low market access and rural population density.

High NDVI, high population density, and high market access characterize most of the highland systems, which are dominated by small landholdings but intensive use of cropland. High cropland intensity and high concentrations of people are found in the savanna areas in West Africa, including northern Nigeria, southern Niger, and Burkina Faso, and in the belt running west from Cameroon, across the southern and coastal regions of Nigeria, Benin, Togo, Ghana, and Côte d'Ivoire. The most remote farming systems—forest-based and pastoral systems in the Horn of Africa, Sahel, and southwestern Africa—are also where population densities and cropping intensities are low. Regarding the other two top systems, the tree-root crop system has high population density and medium access to markets, whereas the maize mixed system has mostly a medium humid climate, moderately high population density, and low market access.

APZS AND PRODUCTIVITY CORRELATES

Table 4.2 summarizes all of the productivity correlates and APZs across the 10 new farming systems, based on aggregating pixel-level data within each farming system. Regarding the APZs, 141 of the total 543 are found in the pastoral-agropastoral farming system, followed by the tree-root crop system, with 94 APZs. The least diversity is found in the rice-tree crop system, with only 4 APZs.

Looking at the distribution of the population, most households (78.6 percent) manage or are sustained by five farming systems: tree-root crop (19.0 percent of the total population of SSA), pastoral-agropastoral (16.1 percent), maize mixed (15.7 percent), highlands (14.2 percent), and cereal-root crop (13.6 percent). These five systems account for 87 percent of the total cropland and more than 80 percent of the total agricultural output.

Specific commodities produced also differ across farming systems. For example, rice is mainly produced in the tree-root crop, cereal-root crop, and rice-tree crop farming systems, while tobacco is produced almost exclusively in the maize mixed system. The pastoral-agropastoral system is the major producer of sorghum, millet, soybeans, and groundnuts. Regarding livestock and cattle, for example, although 27 percent of cattle production comes from the pastoral-agropastoral system, the highlands, cereal-root crop, and maize mixed farming systems together account for more than half of total cattle production in SSA.

FIGURE 4.2 Spatial patterns of key factors influencing agricultural production and productivity at the system level

(a) Annual average (NDVI)

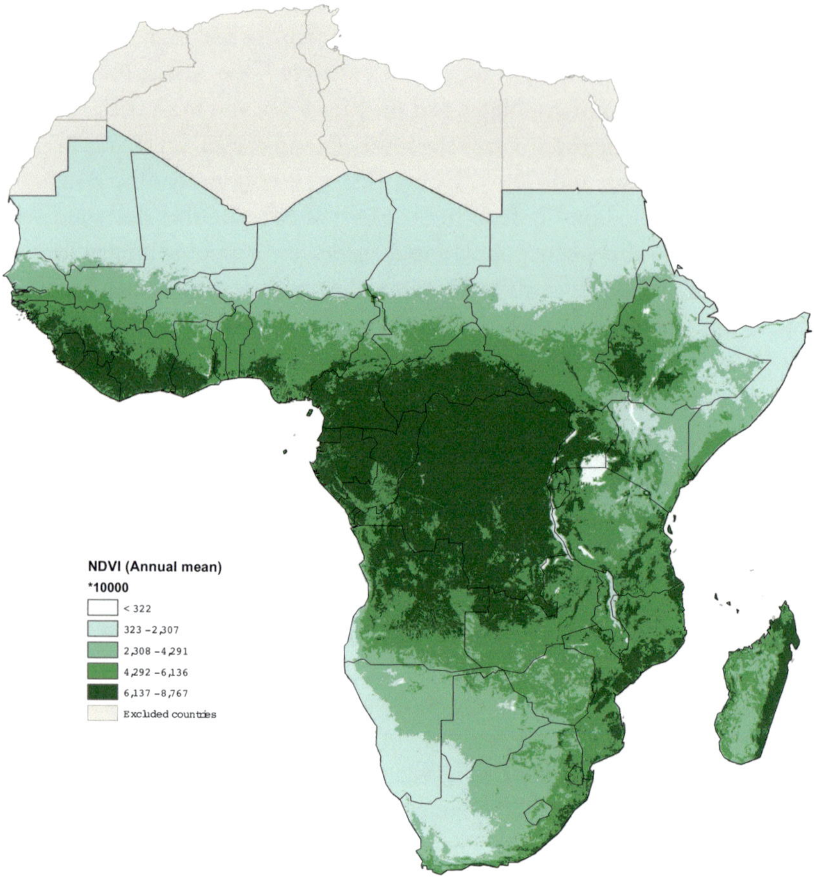

(b) Cropland intensity (c. 2000)

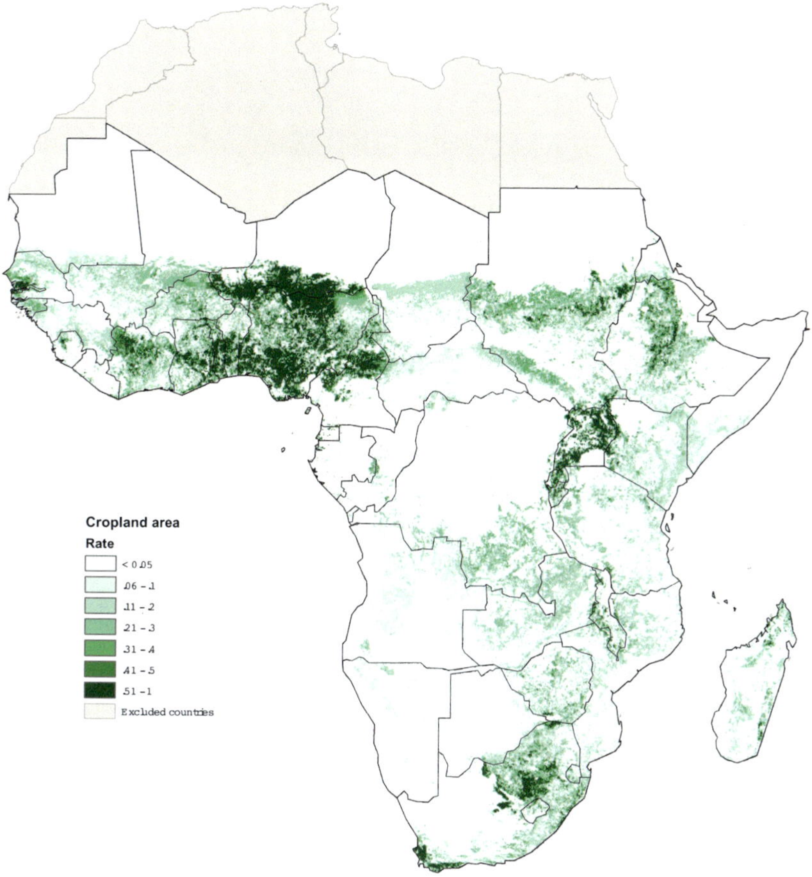

(continued)

FIGURE 4.2 (continued)

(c) Travel time to markets (>50,000 population)

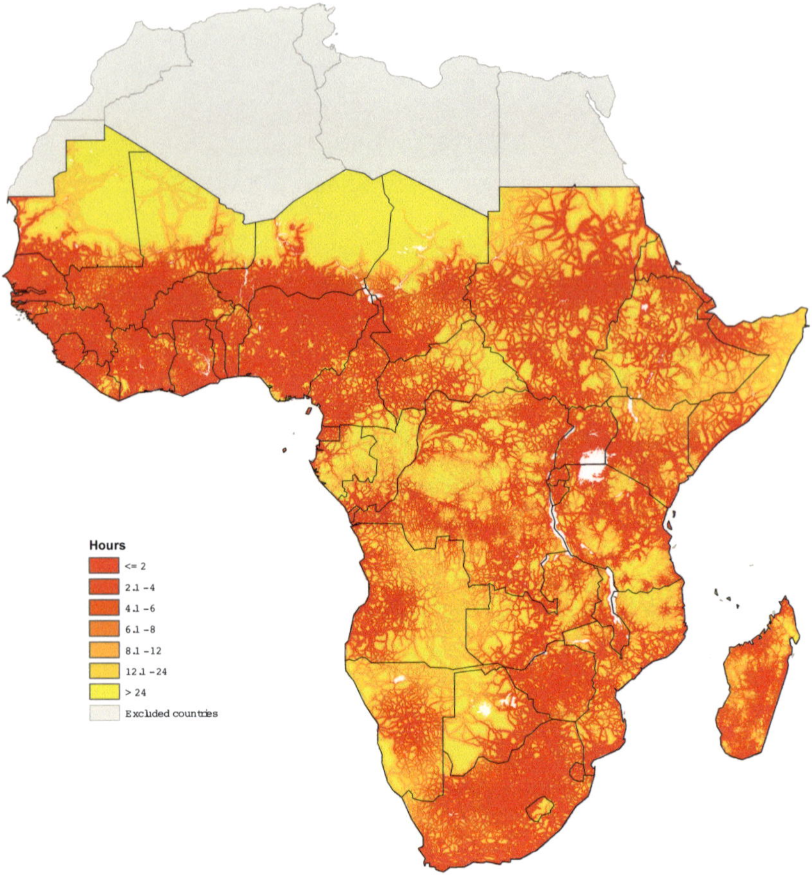

(d) Rural population density (c. 2000)

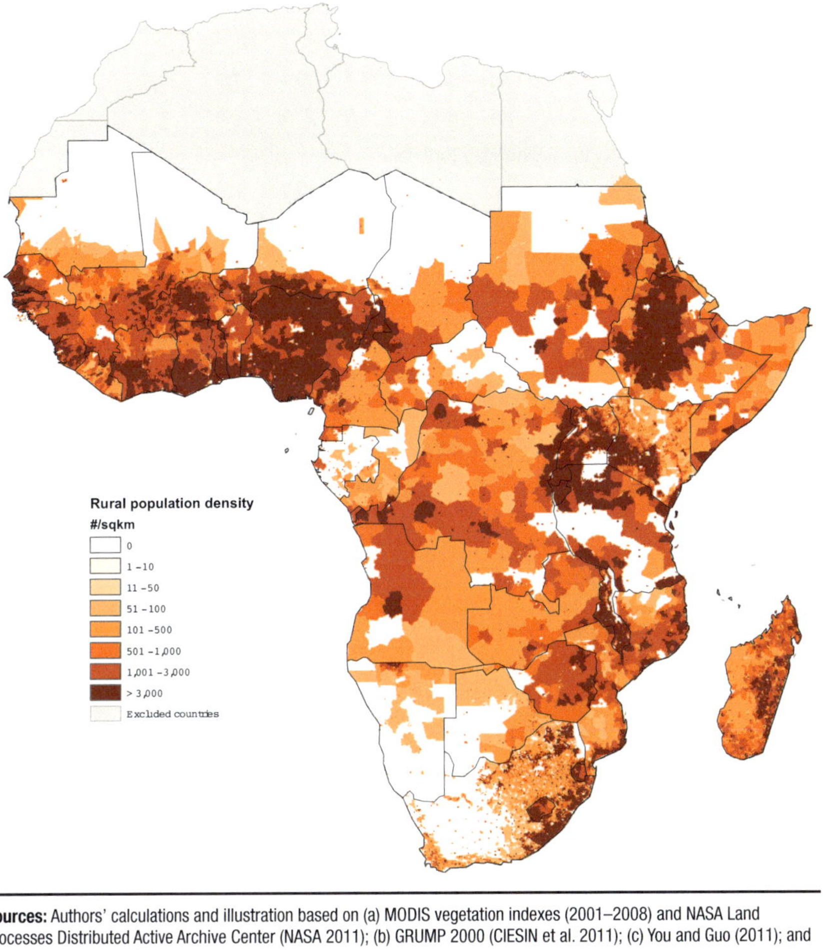

Sources: Authors' calculations and illustration based on (a) MODIS vegetation indexes (2001–2008) and NASA Land Processes Distributed Active Archive Center (NASA 2011); (b) GRUMP 2000 (CIESIN et al. 2011); (c) You and Guo (2011); and (d) Ramankutty et al. 2008.

Note: NDVI = normalized difference vegetation index.

TABLE 4.2 Share in Africa south of the Sahara and average by farming systems

Indicator	Tree-root crop	Forest based	Highlands	Cereal-root crop	Maize mixed	Pastoral-agropastoral	Irrigated	Rice-tree crop	Coastal	Large comm. & smallholder
Number of APZs	94	19	56	76	53	141	28	4	49	19
Share of total in farming system (%)										
Population	19.1	7.0	14.2	13.6	15.7	16.1	1.9	1.8	5.2	5.4
Crop area	20.9	3.3	7.4	19.9	11.5	27.2	2.9	0.9	2.3	3.8
Share of total value of production in farming system (%)										
Agriculture	23.9	3.2	10.6	18.1	14.0	18.2	2.4	1.3	2.5	5.8
Crop	28.2	3.7	9.6	17.6	13.7	14.9	1.9	1.2	2.9	6.2
Wheat	0.4	0.1	20.2	0.3	15.9	15.1	2.6	0.1	0.0	45.3
Rice	33.7	5.6	1.7	22.8	5.2	8.0	3.0	13.7	6.1	0.0
Maize	14.5	2.4	8.2	10.7	24.5	5.5	0.6	0.1	1.5	31.9
Sorghum/millet	6.6	0.2	4.0	27.8	6.0	46.7	7.8	0.0	0.4	0.6
Potato	4.7	3.3	23.0	7.3	49.5	9.9	1.1	0.5	0.7	0.1
Sweet potato	49.3	1.2	7.2	22.9	8.3	4.0	0.2	0.3	6.4	0.3
Cassava	48.3	11.9	5.9	12.3	10.5	3.4	0.1	1.3	6.3	0.0
Banana	22.9	6.4	44.8	6.1	12.2	3.8	0.2	0.6	2.6	0.4
Soybean	10.8	0.6	4.9	13.5	16.4	39.2	2.4	0.0	0.3	11.9
Pulses	8.4	1.3	14.9	34.0	13.2	24.3	1.6	0.4	1.3	0.7
Sugarcane	7.0	4.4	4.0	5.1	30.3	8.7	2.8	2.6	1.1	34.0
Coffee	43.2	4.2	16.7	3.8	19.9	6.3	0.0	4.4	1.5	0.0
Cotton	26.5	0.5	1.6	32.2	13.2	17.9	3.1	0.6	1.6	2.8
Groundnut	17.0	4.0	1.4	27.2	7.2	35.0	4.9	0.2	1.2	2.0
Cocoa	100.0	0.0	0.0	0.0	0.0	0.0	0.0	0.0	0.0	0.0
Tobacco	0.0	0.0	0.0	0.0	100.0	0.0	0.0	0.0	0.0	0.0
Livestock	11.1	1.9	13.3	19.4	14.8	28.4	3.9	1.3	1.1	4.8
Cattle	9.4	1.6	15.5	19.2	16.5	27.4	3.6	1.6	0.8	4.4
Sheep/goat	13.3	1.5	8.4	18.3	10.1	36.1	5.5	0.1	1.3	5.5
Chicken/pig	23.9	7.9	7.2	26.9	15.1	5.9	0.1	3.1	3.8	6.2
Average										
NDVI	High	High	Medium	Medium	Medium	Medium	Medium	Medium	High	Medium
Pop. density	1.3	0.5	1.0	0.7	0.8	0.3	0.5	0.2	5.7	0.5
Market access	6.6	10.2	5.1	6.3	7.2	9.4	4.8	8.1	4.5	6.0

Source: Authors' calculations based on SPAM results.

Notes: The shares are based on output values, and each row adds up to 100 percent. NDVI = normalized difference vegetation index.

Next, we present the methodology or typology analysis used to classify APZs according to similarities in the different productivity correlates.

Methodology for Typology Analysis

The typology analysis involves statistical and econometric methods—namely, spatial and cluster techniques—to group the APZs according to similarities in the various productivity correlates, including commodity value shares, NDVI, market access, urbanization, and population density in each farming system. We use the *k*-median cluster analysis (Everitt et al. 2011), which involves the following steps:

1. Assume a set *x* of n observations (APZs) in a *d*-dimensional space, where all variables are standardized to zero means and unit standard deviations to prevent any distortions caused by different measurement units of the variables.
2. Assume that the APZs can be grouped into an initial predetermined set of k separate clusters, and each APZ is assigned to one of the clusters only. Therefore, the first cluster contains the 1st—(1+*k*)th, (1+2*k*)th, … —APZs; and the second cluster contains the 2nd—(2+*k*)th, (2+2*k*)th, … —APZs; and so on.
3. For cluster *k*, the median m_k is computed by taking the sum of the absolute values of the differences in each dimension. This approach minimizes error over all clusters, because the cluster median is the point with the smallest sum of the distances from each observation in the cluster to the nearest median. In mathematical terms, define distance of observation *i* to the median point of cluster *k* as $d_{ik} = || x_i - m_k ||$, where || is the sum of the absolute values of all dimensions.
4. Calculate the sum of distance $\Sigma_k \Sigma_i d_{ik}$ for all observations *i*.
5. Assign each observation to the nearest cluster with the closest median.
6. Repeat step 3.
7. Stop when there is no change in the cluster median, or no observations change groups. That is, the *k* clusters formed from the data have the minimal sum of distance.

The number of *k* clusters is specified by the user in the iterative process, and there is no preset optimal number of clusters. The first goal is to find *k* clusters, such that the clusters are the most compact and distinct. Instead of predetermining a fixed number of clusters, this method computes and compares several k-median solutions with different numbers of *k* clusters for each farming system. The optimal number of distinct clusters is determined by a combination of the largest value of the Calinski–Harabasz pseudo-F index and the Duda–Hart Je(2)/Je(1) index and by the smallest value of the Duda–Hart pseudo-T-squared value (Milligan and Cooper 1985). In addition, we plot several statistics against the number of clusters to visually identify the solution, which is observed by a kink point in the curve. The statistics reported include the within sum of squares (WSS), the logarithm of WSS, the η^2 coefficient, and the proportional reduction of errors coefficient (Makles 2012).

Based on the resulting clusters, the common factors shared by APZs in the same cluster are used to identify typologies to summarize the comparative advantages and constraints in each unique cluster. Because typologies are not confined within country borders, they allow flexibility in capturing the binding factors across multiple countries. The final typology is reported by dominant crop and livestock production pattern. Countries can be included in more than one typology.

Results of Typology of APZs

The fundamental results of the typology analysis for identifying the optimum number of typologies or clusters are presented in Table 4.3 and Figure 4.3 using the tree-root crop farming system as an example. There are 94 APZs within the tree-root crop system. The Calinski–Harabasz pseudo-F index and the Duda–Hart index reported in Table 4.3 together indicate that the 94 APZs can be grouped into three distinct typologies or clusters. As presented in the methodology section, the Calinski–Harabasz pseudo-F index shows the kink at 5 clusters, indicating the maximum that is sufficient, and then the Duda–Hart indexes confirm that 3 clusters are sufficient, although these nuances (and the kink) do not appear clearly in Figure 4.3.

Similar analyses were conducted for each of the farming systems, which produced similar results. Because of space considerations, we do not report them. However, we have summarized the optimum number of typologies of APZs for each of the 10 farming systems in Table 4.4. The forest-based, cereal-root crop, and coastal farming systems also had three typologies of APZs each. The highlands and maize mixed farming systems had four typologies each, the pastoral-agropastoral and irrigated systems had five each, and the large commercial and smallholder system had two. Because the rice-tree crop

TABLE 4.3 Summary statistics of the cluster analysis for the tree-root crop farming system

Number of clusters or typologies	Calinski–Harabasz pseudo-F	Duda–Hart	
		Je(2)/Je(1)	Pseudo-T-squared
1	—	0.9282	7.12
2	7.12	0.9605	3.70
3	5.50	0.9612	3.59
4	4.96	0.7949	22.70
5	10.22	0.0000	—
6	8.67	0.9005	9.06
7	9.43	0.9708	2.34
8	8.55	0.9468	4.33
9	8.33	0.4013	5.97
10	7.80	0.9575	3.33
11	7.57	0.8658	11.31
12	8.86	0.1266	13.80
13	8.53	0.9295	4.70
14	8.65	0.4370	11.59
15	9.05	0.8894	7.34

Source: Authors' calculation based on cluster analysis.
Note: — = data not available.

farming system exists only in Madagascar, the cluster analysis was not done for that system.

In the upcoming sections, we present the typologies for each of the farming systems. In general, the typologies may be interpreted as subsystems of agricultural production activities, which we use interchangeably in describing the results. In accompanying tables, we report details of the typologies according to value of agricultural output (for specific commodities and in aggregate); characteristics of the productivity correlates (NDVI, cropland area, population, population density, and market access); and countries in which they are dominant.

Tree-root crop farming system

Table 4.5 describes the three typologies of APZs or subsystems within the tree-root crop farming system. They are labeled "roots+tubers," "cocoa+-cassava+banana," and "livestock," following the major constituent types of crop and livestock commodities produced. The roots+tubers subsystem

FIGURE 4.3 Plots of the cluster analysis for the tree-root crop farming system

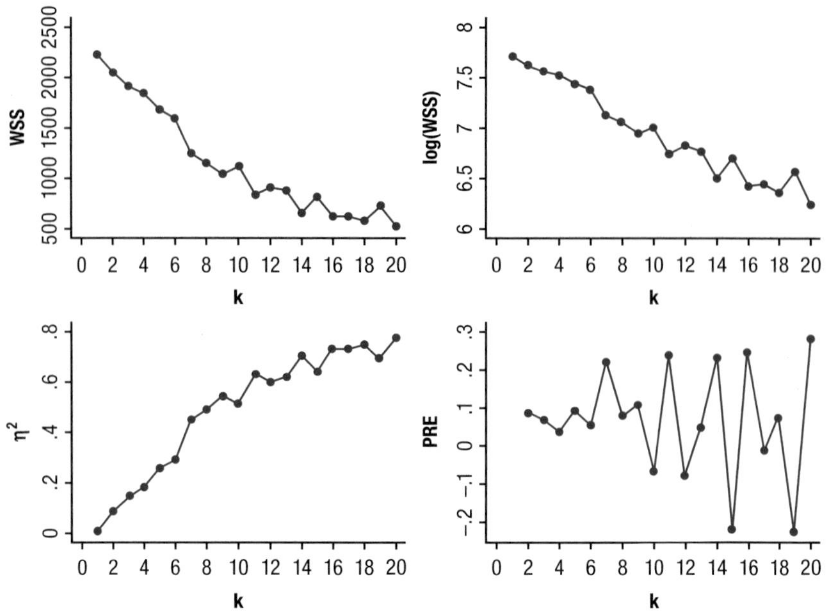

Source: Authors' calculation and illustration based on cluster analysis.

Notes: PRE = proportional reduction of errors; WSS = within sum of squares.

TABLE 4.4 Number of APZs and typologies of APZs by farming system

Farming system	Number of APZs	Number of clusters or typologies
Tree-root crop	94	3
Forest based	19	3
Highlands	56	4
Cereal-root crop	76	3
Maize mixed	53	4
Pastoral-agropastoral	141	5
Irrigated	28	5
Rice-tree crop	4	1
Coastal	49	3
Large commercial & smallholder	19	2

Source: Authors' calculation and illustration based on cluster analysis.

Notes: The rice-tree crop farming system exists only in Madagascar, so no cluster analysis was done for that system. APZs = agricultural productivity zones.

TABLE 4.5 Description of the typologies of APZs in the tree-root crop farming system

	Typology or subsystem		
Indicator	roots+tubers	cocoa+cassava+banana	livestock
Share in the farming system (%)			
Population	85.3	14.0	0.7
Crop area	75.9	23.9	0.1
Output value	74.9	24.7	0.3
Value share in the subsystem (%)			
Wheat	0.0	0.0	0.0
Rice	6.7	1.4	0.1
Maize	5.4	3.4	1.0
Barley	0.0	0.0	0.0
Sorghum/millet	2.9	0.3	7.7
Potato	0.3	0.3	0.0
Sweet potato	20.1	7.4	0.2
Cassava	22.9	14.9	0.5
Banana	4.6	11.4	0.2
Soybean	0.3	0.0	0.0
Pulses	2.2	0.7	0.4
Sugarcane	0.6	0.4	0.8
Coffee	2.9	1.0	0.0
Cotton	2.5	0.3	0.0
Groundnut	5.5	0.5	1.8
Cocoa	2.7	51.0	0.0
Tobacco	0.0	0.0	0.0
Cattle	8.5	1.2	52.6
Chicken/pig	1.5	0.5	0.1
Sheep/goat	3.9	1.4	34.4
Other	6.3	3.8	0.2
Total	100.0	100.0	100.0
*Average**			
Population density	Medium	High	Low
NDVI	High	High	High
Market access	Low	Medium	Medium
Dominant in country*	Central African Republic, Liberia,	Guinea, Sierra Leone	Ghana

Source: Authors' calculation based on cluster analysis.

Notes: *Details of agricultural output for the different matrix combinations of the productivity correlates (population density, NDVI, and market access) and dominating countries in each subsystem are presented in appendix Table 4A.6. APZs = agricultural productivity zones; NDVI = normalized difference vegetation index.

dominates and accounts for more than three-quarters of the total agricultural production, crop area, and population of the farming system. Whereas this subsystem can be found in many West African countries, the Central African Republic, and Sudan, it is dominant in the Central African Republic, Liberia, Guinea, and Sierra Leone. This subsystem typically is characterized by high NDVI, medium population density, and low market access.

The cocoa+cassava+banana subsystem is the next most important, and accounts for one half of the cocoa produced. Ghana, Côte d'Ivoire, Cameroon, and Togo account for the bulk of the agricultural production within the subsystem. Notably, the subsystem contributes almost three-quarters of the value of agricultural crop production in Ghana, accounts for about a quarter of the total crop area and agricultural production in the tree-root crop farming system, and has high population density and medium access to markets.

The livestock subsystem focuses on production of cattle, sheep, and goats. It is very marginal in the tree-root crop farming system, as it accounts for less than 1 percent of the system's agricultural production, population, and cropland. The subsystem has low population density and medium market access. Overall, the tree-root crop system is essential for food security and foreign exchange earnings, because nearly half of SSA's roots and tuber crops and the bulk of its cocoa are produced in two of its subsystems.

Forest-based farming system

The forest-based farming system also has three distinct subsystems: "cassava+banana," "cattle+rice+cassava," and "banana+cassava." All of these subsystems register high agricultural potential or NDVI and low market access (Table 4.6). From the labels, it is clear that cassava is an important crop in the entire farming system, although it is more dominant in the "cassava+banana" subsystem than in the other two, which also provides further insight on the nomenclature of the subsystems.

More than 80 percent of the total population lives in the "cassava+banana" subsystem and depends heavily on cassava and banana production, with more than half of the value of agricultural output being derived from cassava. This subsystem is mostly found in central Africa, such as in the Democratic Republic of the Congo and the Republic of the Congo, where the subsystem contributes 61 and 88 percent of national agricultural production, respectively (appendix Table 4A.6). The "cattle+rice+cassava" subsystem, which is mainly found in Madagascar, accounts for less than 10 percent of the total population of the forest-based system, 15 percent of the total cropland, and 24 percent of the total output value. Banana plays an essential role in the

TABLE 4.6 Description of the typologies of APZs in the forest-based farming system

	Typology or subsystem		
Indicator	cassava+banana	cattle+rice+cassava	banana+cassava
Share in the farming system (%)			
Population	81.7	8.9	9.4
Crop area	77.8	15.8	6.3
Output value	65.8	23.7	10.5
Value share in the subsystem (%)			
Wheat	0.0	0.1	0.0
Rice	3.8	17.4	0.0
Maize	7.5	2.3	3.1
Barley	0.0	0.0	0.0
Sorghum/millet	0.5	0.2	0.1
Potato	0.1	6.3	0.1
Sweet potato	1.4	5.6	6.8
Cassava	51.2	10.2	18.8
Banana	13.1	4.0	32.3
Soybean	0.1	0.0	0.1
Pulses	1.8	3.1	1.2
Sugarcane	2.9	1.6	4.4
Coffee	1.3	3.2	1.3
Cotton	0.3	0.3	0.3
Groundnut	8.3	5.7	5.4
Cocoa	0.0	0.0	0.0
Tobacco	0.0	0.0	0.0
Cattle	1.4	30.4	6.6
Chicken/pig	1.6	4.4	10.0
Sheep/goat	2.0	2.7	8.2
Other	2.7	2.7	1.2
Total	100.0	100.0	100.0
*Average**			
Population density	Medium	Medium	Low
NDVI	High	High	High
Market access	Low	Low	Low
Dominant in country*	Republic of the Congo, Democratic Republic of the Congo		Equatorial Guinea

Source: Authors' calculation based on cluster analysis.

Notes: *Details of agricultural output for the different matrix combinations of the productivity correlates (population density, NDVI, and market access) and dominating countries in each subsystem are presented in appendix Table 4A.6. APZs = agricultural productivity zones; NDVI = normalized difference vegetation index.

"cassava+banana" subsystem, yielding about one-third of the subsystem's total agricultural production.

Highlands farming system

Livestock plays an important role in the highlands farming system, contributing more than half of agricultural output in three of the four distinct subsystems identified (Table 4.7). The "banana+roots" and "cattle+maize" subsystems dominate, and together account for 88 percent of the total population, 92 percent of the total cropland, and 91 percent of the total value of output. The banana+roots subsystem is mostly found in several East African countries, and is the most important subsystem in Burundi, Rwanda, and Uganda. This subsystem provides nearly half of the banana production in the subcontinent. Regarding the cattle+maize subsystem, more than half of its total agricultural output originates from cattle, predominantly found in the Horn of Africa, Ethiopia, Eritrea, Zimbabwe, and Lesotho. Whereas all the subsystems are characterized by high population density, their agricultural potential and market access vary considerably, which makes them conducive to the type of mixed crop and livestock production systems found there.

Cereal-root crop farming system

Results of the typology of the APZs in the cereal-root crop farming system are shown in Table 4.8. The major subsystem, "cassava+coarse grain+groundnuts," accounts for slightly more than half of the total population, two-thirds of the total cropland, and 60 percent of total output value. The subsystem can be found mostly in central and western Africa, including the Central African Republic, Benin, Ghana, Guinea, Guinea-Bissau, and Nigeria. It is also important in Mozambique and Malawi.

Regarding the other two subsystems identified, "cattle+pulses+coarse grain" and "cattle+cassava," livestock is important and accounts for 15–49 percent of the total agricultural output value produced there. The cattle+ pulses+coarse grain subsystem is more diversified, and includes cattle, pulses, sorghum, millet, and groundnuts. It can be found in parts of the Sahelian countries, such as Sudan, Chad, Mali, Burkina Faso, Cameroon, and The Gambia. The cattle+cassava subsystem is found in parts of southern Africa, including Angola, Zambia, Madagascar, and Sudan. The subsystems here can be described generally as having medium population density, with medium-to-high agricultural potential, and low-to-medium market access. With poor road infrastructure, the average travel time to the nearest major market is about 4.3 hours.

TABLE 4.7 Description of the typologies of APZs in the highlands farming system

	Typology or subsystem			
Indicator	banana+roots	cattle+maize	cattle+maize+ sheep	livestock+ cassava
Share in the farming system (%)				
Population	44.6	43.8	11.3	0.3
Crop area	59.7	32.6	7.7	0.0
Output value	60.1	31.2	8.6	0.0
Value share in the subsystem (%)				
Wheat	0.1	7.9	2.9	0.6
Rice	1.0	0.0	0.1	1.5
Maize	2.9	9.5	17.7	0.9
Barley	0.0	3.2	0.7	0.0
Sorghum/millet	1.8	5.8	2.8	1.1
Potato	4.4	1.7	3.2	0.7
Sweet potato	8.4	1.2	2.1	3.3
Cassava	9.5	0.0	0.5	9.0
Banana	46.2	0.0	0.7	0.9
Soybean	0.3	0.1	0.2	0.1
Pulses	8.5	6.6	3.3	5.9
Sugarcane	0.8	0.4	1.7	0.1
Coffee	2.4	1.5	2.8	1.8
Cotton	0.1	0.2	1.3	0.1
Groundnut	1.1	0.1	1.5	4.7
Cocoa	0.0	0.0	0.0	0.0
Tobacco	0.0	0.0	0.0	0.0
Cattle	8.5	52.2	48.2	10.5
Chicken/pig	1.2	0.1	1.1	41.5
Sheep/goat	2.3	8.9	8.6	17.3
Other	0.6	0.7	0.5	0.0
Total	100.0	100.0	100.0	100.0
*Average**				
Population density	High	High	High	Medium
NDVI	High	Medium	High	Low
Market access	Medium	Low	Low	High
Dominant in country*	Rwanda, Burundi, Uganda	Ethiopia, Lesotho		

Source: Authors' calculation based on cluster analysis.

Notes: *Details of agricultural output for the different matrix combinations of the productivity correlates (population density, NDVI, and market access) and dominating countries in each subsystem are presented in appendix Table 4A.6. APZs = agricultural productivity zones; NDVI = normalized difference vegetation index.

TABLE 4.8 **Description of the typologies of APZs in the cereal-root crop farming system**

	Typology or subsystem		
Indicator	cassava+coarse grain+ groundnuts	cattle+pulses+ coarse grain	cattle+ cassava
Share in the farming system (%)			
Population	52.9	29.7	17.3
Crop area	67.7	26.8	5.5
Output value	60.1	30.9	9.0
Value share in the subsystem (%)			
Wheat	0.0	0.0	0.1
Rice	6.4	2.3	3.5
Maize	5.9	2.7	3.9
Barley	0.0	0.0	0.1
Sorghum/millet	14.8	11.5	3.6
Potato	0.9	0.2	0.5
Sweet potato	16.4	0.9	2.6
Cassava	8.5	0.6	19.6
Banana	3.0	0.1	4.1
Soybean	0.6	0.1	0.0
Pulses	5.5	20.5	3.2
Sugarcane	0.5	0.4	1.5
Coffee	0.1	0.4	1.2
Cotton	1.9	6.1	1.0
Groundnut	9.7	9.7	2.6
Cocoa	0.0	0.0	0.0
Tobacco	0.0	0.0	0.0
Cattle	9.1	31.1	38.3
Chicken/pig	1.9	2.2	1.2
Sheep/goat	3.9	9.9	9.3
Other	11.1	1.4	3.7
Total	100.0	100.0	100.0
*Average**			
Population density	Medium	Medium	Medium
NDVI	Medium	Medium	High
Market access	Medium	Medium	Low
Dominant in country*	Guinea-Bissau	Burkina Faso	

Source: Authors' calculation based on cluster analysis.

Notes: *Details of agricultural output for the different matrix combinations of the productivity correlates (population density, NDVI, and market access) and dominating countries in each subsystem are presented in appendix Table 4A.6. APZs = agricultural productivity zones; NDVI = normalized difference vegetation index.

Maize mixed farming system

The maize mixed farming system spans eastern and southern Africa. About half of SSA's total maize output is produced within this system. Although maize is identified with the farming system—and is, hence, called "maize mixed"—it accounts for less than 30 percent of the farming system's total agricultural output value, indicating the importance of other commodities in the system (Table 4.9).

Four typologies of APZs are identified, with the "cattle+maize" subsystem dominating. It accounts for 68 percent of the total population in the system, 64 percent of the total cropland, and 61 percent of the total value of output, and is characterized by medium population density, low market access, but favorable climate and soils. Countries along the eastern and southern coastlines, including Tanzania, Kenya, Zimbabwe, Ethiopia, and Uganda, are home to this subsystem, which is dominant in Tanzania and Zimbabwe (Table 4A.6).

The "roots+maize+tobacco" subsystem dominates agricultural production in Malawi, and is also key to the livelihoods of the 42 million people in Zimbabwe and Kenya. Overall, the subsystem supports 16 percent of the total population within the system, and accounts for nearly a quarter of the cropland and value of output. It is characterized by high population density, medium agricultural potential, and medium market access.

Whereas the third subsystem focuses on the cultivation of tobacco and maize, mainly in parts of Malawi, Mozambique, and Zambia, the fourth subsystem specializes in the production of sugarcane and cattle, and is mainly found in areas of medium agricultural potential in South Africa and Swaziland. In general, the potential of the four subsystems is constrained by their low-to-medium market access.

Pastoral-agropastoral farming system

The five subsystems identified in the pastoral-agropastoral system are characterized mostly by low population density, low NDVI, and low market access, except in a couple of cases that have medium population density and in one case medium NDVI (Table 4.10).

The "coarse grain+cattle+groundnuts" subsystem accounts for 62 percent of the total population, 80 percent of the cropland, and two-thirds of the agricultural output. It combines cattle with the production of drought-resistant sorghum, millet, and groundnuts. About half of SSA's total coarse grain production comes from this subsystem, which is widely distributed in the Sahel zone parallel to the equator in such countries as Niger, Mali, and Senegal.

TABLE 4.9 Description of the typologies of APZs in the maize mixed farming system

Indicator	Typology or subsystem			
	cattle+ maize	roots+maize+ tobacco	tobacco+ maize	sugarcane+ cattle
Share in the farming system (%)				
Population	68.0	16.3	10.0	5.7
Crop area	64.2	22.2	11.2	2.3
Output value	61.5	23.4	9.8	5.3
Value share in the subsystem (%)				
Wheat	1.9	1.1	0.6	2.7
Rice	2.1	0.4	0.7	0.1
Maize	12.5	19.5	14.9	6.9
Barley	0.2	0.2	0.0	0.0
Sorghum/millet	5.1	0.8	1.7	1.2
Potato	1.7	17.7	3.7	0.2
Sweet potato	4.2	7.3	5.5	0.4
Cassava	8.9	7.3	6.3	0.5
Banana	7.6	4.2	0.4	0.5
Soybean	0.8	0.2	0.5	0.7
Pulses	5.3	5.7	4.3	0.6
Sugarcane	3.0	1.4	4.3	35.8
Coffee	2.5	1.5	0.4	0.0
Cotton	1.6	0.8	3.9	1.4
Groundnut	3.1	3.0	4.0	1.8
Cocoa	0.0	0.0	0.0	0.0
Tobacco	5.6	13.2	33.8	6.0
Cattle	25.3	10.5	9.8	28.9
Chicken/pig	1.5	0.9	2.2	0.9
Sheep/goat	4.6	3.7	2.5	9.1
Other	2.4	0.6	0.5	2.2
Total	100.0	100.0	100.0	100.0
*Average**				
Population density	Medium	High	Medium	Medium
NDVI	High	Medium	High	Medium
Market access	Low	Medium	Low	Medium
Dominant in country*	Tanzania, Zimbabwe	Malawi	Zambia	Swaziland

Source: Authors' calculation based on cluster analysis.

Notes: *Details of agricultural output for the different matrix combinations of the productivity correlates (population density, NDVI, and market access) and dominating countries in each subsystem are presented in appendix Table 4A.6. APZs = agricultural productivity zones; NDVI = normalized difference vegetation index.

TABLE 4.10 Description of the typologies of APZs in the pastoral-agropastoral farming system

	Typology or subsystem				
Indicator	coarse grain+ cattle+ groundnuts	cattle dominated	cattle+ cassava+ maize	livestock	small ruminants
Share in the farming system (%)					
Population	61.7	17.8	14.0	4.5	2.0
Crop area	80.5	8.8	8.4	2.2	0.1
Output value	67.0	15.2	12.6	3.9	1.3
Value share in the subsystem (%)					
Wheat	1.0	0.5	3.1	0.3	1.6
Rice	2.2	0.5	0.5	1.5	0.6
Maize	1.0	1.2	12.0	1.7	0.7
Barley	0.0	0.1	0.8	0.0	0.0
Sorghum/millet	29.3	6.6	3.0	6.4	0.4
Potato	0.6	0.9	2.3	0.4	0.0
Sweet potato	1.9	0.5	2.8	2.1	0.0
Cassava	0.3	0.3	12.1	3.9	0.4
Banana	1.4	0.2	3.2	0.6	0.5
Soybean	1.6	0.0	0.4	0.0	0.1
Pulses	8.1	3.3	8.2	2.3	0.9
Sugarcane	0.8	1.3	1.8	0.7	0.7
Coffee	0.0	0.4	3.1	0.0	0.0
Cotton	2.0	1.1	1.0	1.4	0.0
Groundnut	15.5	4.2	2.3	7.7	0.1
Cocoa	0.0	0.0	0.0	0.0	0.0
Tobacco	0.0	0.0	0.0	0.0	0.0
Cattle	18.2	52.6	30.6	49.6	11.0
Chicken/pig	0.3	0.1	1.5	0.7	0.1
Sheep/goat	8.7	24.0	6.8	19.6	82.7
Other	7.2	2.1	4.5	0.9	0.0
Total	100.0	100.0	100.0	100.0	100.0
*Average**					
Population density	Medium	Low	Medium	Low	Low
NDVI	Low	Low	Medium	Low	Low
Market access	Low	Low	Low	Low	Low
Dominant in country*	Niger, Mali, Senegal	Namibia, Somalia, Djibouti, Mauritania	Angola	Botswana	

Source: Authors' calculation based on cluster analysis.

Notes: *Details of agricultural output for the different matrix combinations of the productivity correlates (population density, NDVI, and market access) and dominating countries in each subsystem are presented in appendix Table 4A.6. APZs = agricultural productivity zones; NDVI = normalized difference vegetation index.

TABLE 4.11 Description of the typologies of APZs in the irrigated farming system

	Typology or subsystem				
Indicator	coarse grain+ groundnuts+ livestock	livestock+ coarse grain	groundnuts+ coarse grain+ cattle	rice+ livestock	cattle
Share in the farming system (%)					
Population	56.5	31.7	4.4	7.3	0.0
Crop area	63.6	24.8	6.5	5.1	0.0
Output value	47.4	37.0	8.3	7.2	0.0
Value share in the subsystem (%)					
Wheat	2.5	0.9	0.0	0.5	0.1
Rice	2.0	0.8	4.2	45.0	0.0
Maize	1.4	3.3	1.8	0.2	2.2
Barley	0.0	0.0	0.0	0.0	0.3
Sorghum/millet	44.1	9.8	18.7	9.4	5.5
Potato	1.1	0.6	0.0	0.0	0.3
Sweet potato	1.4	0.1	0.0	0.0	0.4
Cassava	0.1	0.4	0.3	0.1	2.1
Banana	0.7	0.2	0.0	0.0	0.2
Soybean	1.1	0.0	0.0	0.0	0.2
Pulses	5.5	1.6	0.3	3.9	1.0
Sugarcane	2.5	1.7	0.5	7.1	0.0
Coffee	0.0	0.0	0.0	0.0	0.7
Cotton	2.0	3.4	0.2	0.1	0.0
Groundnut	12.3	3.8	59.3	0.3	0.4
Cocoa	0.0	0.0	0.0	0.0	0.0
Tobacco	0.0	0.0	0.0	0.0	0.0
Cattle	11.1	49.7	10.8	15.3	83.9
Chicken/pig	0.0	0.0	0.5	0.1	0.0
Sheep/goat	8.8	22.4	2.5	17.8	2.8
Other	3.5	1.2	0.7	0.1	0.0
Total	100.0	100.0	100.0	100.0	100.0
Average*					
Population density	High	Medium	High	Low	Low
NDVI	Low	Medium	Medium	Low	High
Market access	Medium	Medium	High	Medium	Low
Dominant in country*			The Gambia		

Source: Authors' calculation based on cluster analysis.

Notes: *Details of agricultural output for the different matrix combinations of the productivity correlates (population density, NDVI, and market access) and dominating countries in each subsystem are presented in appendix Table 4A.6. APZs = agricultural productivity zones; NDVI = normalized difference vegetation index.

In the "cattle dominated subsystem," more than half of the agricultural production comes from cattle. The "cattle+cassava+maize" subsystem, where cattle production is accompanied by growing cassava and maize, is mostly found in southern and eastern African countries, including Angola, Zambia, Zimbabwe, Ethiopia, and Kenya. The other two subsystems focus on livestock. However, whereas the "livestock" subsystem involves cattle, and sheep, and goats, the "small ruminants" subsystem is dominated by sheep and goats.

Irrigated farming system

Although the irrigated farming system is relatively small in terms of the share of SSA's total population or area, it is very diverse, with five distinct subsystems (Table 4.11) that involve large-scale irrigation schemes, such as those found in The Gambia, Senegal, Somalia, and Sudan. All of the five subsystems involve livestock.

The two major subsystems, "coarse grain+groundnuts+livestock" and "livestock+coarse grain," together account for 88 percent of the total population and cropland and 84 percent of the total output value of the farming system. Whereas coarse grain production dominates in the first subsystem, livestock dominates in the second (mostly in the Horn of Africa), with more than 70 percent of the agricultural output value being derived from livestock production. The third and fourth subsystems are also diversified, involving groundnuts, coarse grain, and cattle in the third subsystem, and rice and livestock in the fourth. The fifth subsystem, which is not diversified, focuses on cattle production.

Coastal farming system

The coastal farming system consists of three subsystems, each with a unique combination of commodities (Table 4.12). The "roots+tubers" subsystem, which dominates the system, depends on cassava and sweet potatoes. The "rice+cattle" subsystem involves a combination of rice and cattle production. The third subsystem is more diverse, with groundnut, sheep, coarse grain, and rice. It can be generally characterized as being densely populated, having medium-to-favorable agricultural potential, and having a relatively well-developed road network.

Large commercial and smallholder farming system

The large commercial and smallholder system, which exists mainly in southern Africa, is overwhelmingly dominated by one subsystem specialized in producing maize, wheat, sugarcane, and cattle (Table 4.13). Commonly observed

TABLE 4.12 Description of the typologies of APZs in the coastal farming system

	Typology or subsystem		
Indicator	roots+tubers	rice+cattle	groundnuts+sheep+ coarse grain+rice
Share in the farming system (%)			
Population	87.8	6.8	5.4
Crop area	87.7	11.7	0.7
Output value	83.5	15.9	0.6
Value share in the subsystem (%)			
Wheat	0.0	0.0	0.0
Rice	3.6	40.0	14.5
Maize	5.8	0.9	1.8
Barley	0.0	0.0	0.0
Sorghum/millet	1.3	0.4	16.7
Potato	0.2	1.7	0.1
Sweet potato	24.8	2.6	0.1
Cassava	30.2	7.4	1.0
Banana	7.7	2.5	0.1
Soybean	0.1	0.0	0.0
Pulses	2.8	2.8	2.7
Sugarcane	0.9	1.4	0.0
Coffee	0.5	2.3	0.0
Cotton	1.3	0.4	0.5
Groundnut	3.0	2.3	31.8
Cocoa	0.0	0.0	0.0
Tobacco	0.0	0.0	0.0
Cattle	1.8	25.3	6.7
Chicken/pig	1.7	3.3	0.5
Sheep/goat	3.4	1.4	21.3
Other	11.1	5.1	2.0
Total	100.0	100.0	100.0
*Average**			
Population density	High	Medium	High
NDVI	High	High	Medium
Market access	Medium	Medium	High

Source: Authors' calculation based on cluster analysis.

Note: *Details of agricultural output for the different matrix combinations of the productivity correlates (population density, NDVI, and market access) and dominating countries in each subsystem are presented in appendix Table 4A.6. APZs = agricultural productivity zones; NDVI = normalized difference vegetation index.

TABLE 4.13 Description of the typologies of APZs in the large commercial and smallholder farming system

	Typology or subsystem	
Indicator	maize+cattle+sugarcane+wheat	livestock
Share in the farming system (%)		
Population	99.3	0.7
Crop area	99.5	0.5
Output value	98.9	1.1
Value share in the subsystem (%)		
Wheat	11.2	0.3
Rice	0.0	0.0
Maize	44.2	0.1
Barley	0.4	0.0
Sorghum/millet	0.8	0.7
Potato	0.0	0.0
Sweet potato	0.4	0.0
Cassava	0.0	0.0
Banana	0.5	0.0
Soybean	1.1	0.0
Pulses	0.6	1.4
Sugarcane	12.1	0.4
Coffee	0.0	0.0
Cotton	0.8	3.5
Groundnut	2.1	0.1
Cocoa	0.0	0.0
Tobacco	0.0	0.0
Cattle	12.7	45.9
Chicken/pig	1.4	0.3
Sheep/goat	5.3	45.9
Other	6.5	1.3
Total	100.0	100.0
*Average**		
Population density	Medium	Low
NDVI	Low	Low
Market access	Medium	Low

Source: Authors' calculation based on cluster analysis.

Notes: *Details of agricultural output for the different matrix combinations of the productivity correlates (population density, NDVI, and market access) and dominating countries in each subsystem are presented in appendix Table 4A.6. APZs = agricultural productivity zones; NDVI = normalized difference vegetation index.

in South Africa, this subsystem employs about 40 million people and covers 6 million ha of land. Although nearly half of the continent's wheat output is produced in this subsystem, wheat accounts for only about 5 percent of the subsystem's total agricultural output.

Use of Typology within a Country

The typology of APZs can be reorganized by country. This is potentially useful for improving the productivity of an individual country within the broader cross-country context. The detailed results are summarized in appendix Tables 4A.7–4A.9 for major, minor, and marginal subsystems. Major subsystems contribute more than 20 percent of national agricultural output (Table 4A.7), minor subsystems contribute 5–20 percent (Table 4A.8), and marginal subsystems contribute less than 5 percent (Table A4.9). We use Ethiopia and Ghana to illustrate how the typology can be used within specific countries.

Typology of APZs in Ethiopia

For Ethiopia, nine subsystems are identified—one major, three minor, and five marginal (Table 4.14). Because livestock is important in all the subsystems (because it contributes more than half of agricultural production), and all the subsystems are characterized by low market access, key to enhancing agricultural development and improving rural livelihoods in Ethiopia will be the challenges and opportunities presented by differences in population density and agricultural potential and other productivity correlates across the subsystems.

For example, the cattle+maize subsystem is associated with the highlands farming system in areas with high population density and high or medium agricultural potential. The subsystem is also associated with the maize-mixed farming system in areas characterized by medium population density and medium agricultural potential. An immediate implication of this is that intensive cattle production systems, for example, may be more suitable for the highland system, where population density is high, whereas less intensive systems may be more suitable for the maize-mixed system, where population density is medium. Similarly, different breeds of cattle may be needed for the highland system, with high agricultural potential, compared with those in the maize-mixed system, with medium agricultural potential.

The same logic applies to the other systems and subsystems. For example, the pastoral-agropastoral system has three subsystems: cattle+cassava+maize, cattle dominated, and coarse grain+cattle+groundnuts. In the high

TABLE 4.14 Typology of APZs in Ethiopia

Farming system	APZ or subsystem	Population density	NDVI	Market access	Share in agricultural value (%) Maize	Coarse grain	Coffee	Cattle	Sheep	Share in national agriculture (%)
Major										
Highlands	cattle+maize	High	Medium	Low	9.7	8.9	1.5	52.3	8.6	55.6
Minor										
Maize mixed	cattle+maize	Medium	Medium	Low	11.5	12.7	7.5	45.4	7.6	17.2
Pastoral-agropastoral	cattle+cassava+maize	High	Medium	Low	10.5	5.8	7.6	46.6	7.8	12.3
Highlands	cattle+maize	High	High	Low	10.6	5.1	5.0	60.9	6.6	8.3
Marginal										
Pastoral-agropastoral	cattle dominated	Medium	Low	Low	4.1	2.9	4.9	52.5	17.7	2.7
Cereal-root crop	cattle+cassava	Medium	Medium	Low	6.9	6.2	0.1	64.0	7.6	2.2
Pastoral-agropastoral	coarse grain+ cattle+groundnuts	Medium	Medium	Low	6.3	33.5	2.0	46.0	7.9	1.5
Irrigated	coarse grain+ groundnuts+ livestock	Low	Low	Medium	9.7	22.6	1.4	49.3	7.4	0.2
Irrigated	cattle	Low	Medium	Low	2.2	6.3	0.8	86.2	3.0	0.0

Source: Authors' calculation based on typology and cluster analysis.
Notes: APZs = agricultural productivity zones; NDVI = normalized difference vegetation index.

population density and medium agricultural potential areas (cattle+cassava+maize), a strategy that promotes labor-intensive cereal production will likely have higher returns, compared with the medium population density and low agricultural potential areas (cattle dominant), where cereal production will not likely be profitable without irrigation or drought-resistant varieties (Pender, Place, and Ehui 2006). Because all the subsystems are characterized by low market access, improving the rural road network in most places will be beneficial, although improving roads that link high-producing areas to urban centers is more likely to generate the greatest benefits.

In general, therefore, having blanket agricultural development policies for the whole nation, instead of policies that target specific production environments, is not likely to be beneficial. Policies and programs differentiated by local demographic and biophysical conditions will be necessary to achieve maximal impact based on the typology developed in this analysis.

TABLE 4.15 Typology of APZs in Ghana

Farming system	APZ or subsystem	Population density	NDVI	Market access	Share in agricultural value (%): Maize	Coarse grain	Sweet potato	Cassava	Banana	Cocoa	Ground-nut	Share in national agriculture (%)
Major												
Tree-root crop	cocoa+ cassava+ banana	High	High	Medium	3.1	0.2	12.7	26.6	15.1	30.2	0.3	73.2
Minor												
Cereal-root crop	cassava+ coarse grain+ groundnuts	Medium	Medium	High	10.6	13.2	15.2	3.8	0.0	0.0	21.0	16.2
Tree-root crop	roots+ tubers	Medium	Medium	Medium	7.2	4.2	26.5	18.8	6.5	0.0	6.6	5.9
Marginal												
Coastal	roots+ tubers	High	Medium	High	6.7	0.1	9.0	42.3	17.5	0.0	0.0	4.7
Tree-root crop	livestock	Low	Low	High	0.0	0.0	0.0	0.0	0.0	0.0	0.0	0.0
Coastal	ground-nuts+ sheep+ coarse grain+rice	High	Low	High	0.0	0.0	0.0	0.0	0.0	0.0	0.0	0.0

Source: Authors' calculation based on typology and cluster analysis.

Notes: APZs = agricultural productivity zones; NDVI = normalized difference vegetation index.

Typology of APZs in Ghana

Table 4.15 shows the results for Ghana, which differ significantly from the results for Ethiopia. For example, Ghana has only six subsystems—one major, two minor, and three marginal. The major subsystem, cocoa+cassava+banana, dominates and accounts for about 73 percent of Ghana's total agriculture, while the marginal roots+tubers subsystem accounts for only 5 percent.

Similar to the Ethiopia case, agricultural development strategies depend on the challenges and opportunities presented by differences in the productivity correlates across the subsystems. For example, because labor is generally abundant, promoting labor-intensive, high-yielding technologies will be critical everywhere. To maximize the impact of those interventions, however, additionally improving market access in the cocoa+cassava+banana subsystem (which is characterized by high population density, medium NDVI, and high market access) and promoting larger-scale production and drought-resistant varieties in the cassava+coarse grain+groundnuts subsystem (which is characterized by medium population density, medium NDVI, and high market access) will be important.

The implications of the typology analysis in tailoring policies and programs to local demographic and biophysical conditions can be further demonstrated by comparing the systems and subsystems across countries. For example, the cereal-root crop farming system is common to Ghana and Ethiopia, which indicates that there may be benefits from cross-country learning and technology transfer. In Ethiopia, the cattle+cassava subsystem is characterized as having medium population density, medium NDVI, and low market access. The cassava+coarse grain+groundnuts subsystem in Ghana also has medium population density and medium NDVI, but has high market access. Aside from maize and coarse grains, which are produced in both subsystems, the focus is different. In Ethiopia, livestock dominates, whereas in Ghana sweet potatoes and groundnuts dominate. While strategies may differ from country to country, one country can learn from the other, such as learning how common productivity correlates within and across farming systems. Specific policies and programs must be tailored to local demographic and biophysical conditions.

Conclusions and Implications

To effectively raise agricultural productivity in different parts of Africa in a sustainable and inclusive manner, investment and policy interventions must take into account the considerable spatial diversity of the potentials and constraints that local farmers face. Having a typology of the possible pathways based on key agricultural productivity correlates can be useful for identifying specific investment and policy interventions. The cluster analysis conducted in this chapter helps to fill the knowledge gap by identifying APZs, which are defined as noncontiguous geographic areas that share similar biophysical, demographic, and socioeconomic conditions. Drawing from the literature on farming systems, agricultural technology adoption, and development domains, the analysis involved statistical and econometric methods (spatial and cluster techniques) using several spatial datasets with observations at the pixel level (mostly 5 arc-minute, or 10-km, grid-cell resolution).

The results show that the 15 farming systems identified in the continent encompass a considerable degree of heterogeneity in the potentials and constraints that local farmers face, which is reflected by the 543 resulting APZs. The greatest diversity is found in the pastoral-agropastoral farming system, with 141 APZs, followed by the tree-root crop system, with 94 APZs. The least diversity is found in the rice-tree crop system, with only 4 APZs.

The typology analysis shows commonalities across the APZs in different countries. For example, the tree-root crop, forest-based, cereal-root crop, and coastal farming systems had three typologies of APZs each; the highlands and maize mixed farming systems had four each; the pastoral-agropastoral and irrigated systems had five each; and the large commercial and smallholder system had two. The typologies were described in several aspects, including major agricultural activities, area covered, population density, market access, and countries where they are found.

Two countries, Ethiopia and Ghana, were analyzed in detail to demonstrate the utility of the typologies at the country level, as well as their implications for cross-country learning and technology transfer. The results clearly show that farmers in spatially similar localities undertake different agricultural production activities that are shaped by how the different factors mentioned above are exhibited in a locality. Some broad strategies and interventions to consider for improving agricultural productivity are also provided, noting that complementary production analysis within each system or APZ (as done in Chapter 5) is needed for obtaining sharper policy implications.

Although this typology analysis has many benefits, it also has some limitations. It uses a few selected indicators (or productivity correlates) to represent the biophysical, demographic, and socioeconomic factors that affect agricultural production and productivity. The analysis also presents a static analysis of the APZs to represent a long-term pattern of agricultural production and productivity, for which a time-series analysis of APZs will be needed.

Despite these limitations, the overall utility of the typology analysis cannot be overemphasized. By triangulating aggregate national-level data with pixel-level data and analysis, the typologies provide a missing link in identifying regionally consistent strategies and locally relevant policies and programs. With more spatial data on different productivity variables becoming available, both across countries and over time, tailoring policies and programs to local biophysical, demographic, and socioeconomic conditions should be imperative.

Appendix for Chapter 4

TABLE 4.A1 Average annual NDVI by farming system

Farming system	Eastern and central Africa	Western Africa	Southern Africa	Africa south of the Sahara
Agropastoral	0.299	0.263	0.407	0.312
Cereal-root crop mixed	0.450	0.446	0.583	0.489
Coastal	0.566	0.605	0.627	0.601
Forest based	0.763	0.700	0.668	0.740
Highland perennial	0.638	n.a.	n.a.	0.638
Highland temperate mixed	0.451	0.637	0.478	0.465
Irrigated	0.288	0.214	0.473	0.266
Large commercial and smallholder	n.a.	n.a.	0.316	0.316
Maize mixed	0.529	0.710	0.514	0.523
Pastoral	0.225	0.155	0.325	0.221
Rice-tree crop	0.545	n.a.	n.a.	0.545
Root crop	0.601	0.609	0.608	0.606
Sparse (arid)	0.119	0.104	0.170	0.116
Tree crop	0.626	0.634	0.659	0.635
Average	**0.299**	**0.263**	**0.407**	**0.312**

Source: Authors' calculation based on the product of NDVI indexes from NASA (2011).

Notes: n.a. = not applicable; NDVI = normalized difference vegetation index.

TABLE 4A.2 Number and size of agricultural productivity zones (APZs) by country

Country	Number of APZs	Total APZ size (km²)	Average APZ size (km²)
Angola	24	1,240,087	51,670
Benin	9	115,108	12,790
Botswana	14	578,741	41,339
Burkina Faso	8	275,063	34,383
Burundi	7	26,880	3,840
Cameroon	23	463,948	20,172
Central African Republic	9	620,270	68,919
Chad	13	1,259,542	96,888
Congo, R	9	341,100	37,900
Congo, DRC	20	2,327,139	116,357
Côte d'Ivoire	7	316,304	45,186
Djibouti	2	19,679	9,839
Equatorial Guinea	5	24,440	4,888
Eritrea	9	113,304	12,589
Ethiopia	21	1,131,650	53,888
Gabon	7	255,418	36,488
Gambia, The	5	8,402	1,680
Ghana	13	234,609	18,047
Guinea	10	241,743	24,174
Guinea-Bissau	5	26,857	5,371
Kenya	25	578,681	23,147
Lesotho	6	30,802	5,134
Liberia	5	92,159	18,432
Madagascar	15	567,498	37,833
Malawi	9	117,769	13,085
Mali	13	1,236,707	95,131
Mauritania	6	1,026,828	171,138
Mozambique	19	770,688	40,563
Namibia	14	815,022	58,216
Niger	11	1,169,758	106,342
Nigeria	25	899,113	35,965
Rwanda	6	25,372	4,229
Senegal	13	190,715	14,670
Sierra Leone	5	68,985	13,797
Somalia	15	614,756	40,984
South Africa	14	1,203,945	85,996

Country	Number of APZs	Total APZ size (km²)	Average APZ size (km²)
Sudan	25	2,478,093	99,124
Swaziland	5	17,400	3,480
Tanzania	30	934,170	31,139
Togo	11	57,147	5,195
Uganda	12	240,915	20,076
Zambia	19	751,390	39,547
Zimbabwe	20	390,612	19,531
Total	**543**	**23,898,806**	**44,013**

Source: Authors' calculation based on typology analysis.
Note: APZs = agricultural productivity zones.

TABLE 4A.3 Cropland area by farming system, in 1,000 hectares in 2005

Farming system	Eastern and central Africa	Western Africa	Southern Africa	Africa south of the Sahara
Agropastoral	5,594	21,008	1,926	28,527
Cereal-root crop mixed	4,778	21,657	3,808	30,242
Coastal	350	2,243	364	2,957
Forest based	2,910	1,503	159	4,572
Highland perennial	5,317	n.a.	n.a.	5,317
Highland temperate mixed	6,101	434	868	7,402
Irrigated	1,879	2,333	39	4,251
Large commercial and smallholder	n.a.	n.a.	13,219	13,219
Maize mixed	13,823	1	9,636	23,460
Pastoral	9,010	11,719	976	21,705
Rice-tree crop	1,825	n.a.	n.a.	1,825
Root crop	8,920	25,222	3,317	37,459
Sparse (arid)	161	6	178	344
Tree crop	263	11,930	165	12,358
Not labeled	163	604	268	1,034
Total	**61,092**	**98,659**	**34,924**	**194,675**

Source: Authors' calculation based on Ramankutty et al. (2008).
Notes: n.a. = not applicable; Not labeled = areas made of grid cells that do not have a farming system because of differences in the delineation of water and land interface (such as coastlines and lake areas) between data layers.

TABLE 4A.4 Travel time by farming system, in hours to cities with population greater than 50,000 inhabitants in 2005

Farming system	Eastern and central Africa	Western Africa	Southern Africa	Africa south of the Sahara
Agropastoral	7.3	4.7	10.1	6.8
Cereal-root crop mixed	8.2	5.4	11.4	7.9
Coastal	6.8	5.6	7.0	6.4
Forest based	11.2	14.2	9.4	12.2
Highland perennial	5.3	n.a.	n.a.	5.3
Highland temperate mixed	8.5	9.9	7.0	8.2
Irrigated	5.8	4.5	7.3	5.4
Large commercial and smallholder	n.a.	n.a.	5.7	5.7
Maize mixed	9.3	14.0	7.7	8.6
Pastoral	9.3	16.4	12.5	12.4
Rice-tree crop	7.3	n.a.	n.a.	7.3
Root crop	8.5	8.1	15.4	10.1
Sparse (arid)	11.5	41.1	15.5	30.1
Tree crop	6.6	5.4	8.7	5.8
Not labeled	7.5	12.2	4.9	8.6
Average	**9.3**	**17.9**	**10.1**	**12.7**

Source: Authors' calculation based on You and Guo (2011).

Notes: n.a. = not applicable; Not labeled = areas made of grid cells that do not have a farming system because of differences in the delineation of water and land interface (such as coastlines and lake areas) between data layers.

TABLE 4A.5 Rural population density by farming system, in people per km^2 in 2005

Farming system	Eastern and central Africa	Western Africa	Southern Africa	Africa south of the Sahara
Agropastoral	13.7	32.4	5.6	21.0
Cereal-root crop mixed	13.2	33.7	14.3	23.1
Coastal	33.5	80.6	31.4	50.2
Forest based	17.7	4.6	6.4	13.1
Highland perennial	125.6	n.a.	n.a.	125.6
Highland temperate mixed	110.2	30.0	39.6	90.5
Irrigated	16.8	39.4	9.7	24.7
Large commercial and smallholder	n.a.	n.a.	10.8	10.8
Maize mixed	42.9	0.1	26.5	35.7
Pastoral	15.0	7.3	1.2	9.6
Rice-tree crop	29.5	n.a.	n.a.	29.5
Root crop	15.6	26.1	4.1	16.8
Sparse (arid)	4.4	0.2	0.2	1.4
Tree crop	43.8	53.8	13.9	49.7
Not labeled	22.2	45.9	18.3	30.1
Average	**26.4**	**19.5**	**11.9**	**20.2**

Source: Authors' calculations based on GRUMP 2000 (CIESIN et al. 2011).

Notes: km^2 = square kilometer; n.a. = not applicable; Not labeled = areas made of grid cells that do not have a farming system because of differences in the delineation of water and land interface (such as coastlines and lake areas) between data layers.

TABLE 4A.6 Typology of major subsystems in Africa south of the Sahara

Farming system	Subsystem	Population density	NDVI	Market access	Country	Area (Mha)	Pop. (million)	Share in national agriculture (%)
Tree-root crop	cocoa+ cassava+ banana	Low	High	Medium	Equatorial Guinea	0.1	0.4	13.5
		Medium	High	Medium	Cameroon	2.9	14.6	37.5
		High	Medium	High	Côte d'Ivoire	6.1	14.3	38.4
		High	High	Medium	Ghana	5.0	17.2	73.2
		High	High	Medium	Togo	1.3	4.1	35.8
		High	High	High	Sierra Leone	0.4	3.3	11.7
	roots	Low	Medium	Low	Sudan	11.3	30.8	5.6
		Low	High	Low	Angola	2.0	12.2	11.1
		Low	High	Low	Central African Republic	0.7	3.7	54.8
		Low	High	Low	Gabon	0.2	0.9	21.3
		Low	High	Low	Mozambique	4.9	16.4	11.1
		Low	High	Low	Republic of the Congo	0.2	2.4	5.1
		Low	High	Low	Zambia	1.2	10.6	7.8
		Medium	Medium	Medium	Ghana	5.0	17.2	5.9
		Medium	High	Low	Democratic Republic of the Congo	5.9	50.8	19.4
		Medium	High	Low	Tanzania	6.5	32.9	38.8
		Medium	High	Medium	Cameroon	2.9	14.6	24.2
		Medium	High	Medium	Côte d'Ivoire	6.1	14.3	56.4
		Medium	High	Medium	Liberia	0.3	2.1	64.0
		Medium	High	High	Guinea	1.6	7.6	76.5
		Medium	High	High	Sierra Leone	0.4	3.3	80.5
		High	High	Medium	Burundi	1.0	6.3	19.3
		High	High	High	Benin	1.9	5.8	47.4
		High	High	High	Nigeria	39.6	107.9	36.0
		High	High	High	Togo	1.3	4.1	39.9
Forest based	banana+ cassava	Low	High	Low	Gabon	0.2	0.9	51.5
		Medium	High	Low	Cameroon	2.9	14.6	10.7
		Medium	High	Medium	Equatorial Guinea	0.1	0.4	54.5

Farming system	Subsystem	Population density	NDVI	Market access	Country	Area (Mha)	Pop. (million)	Share in national agriculture (%)
	cassava+ banana	Low	High	Low	Republic of the Congo	0.2	2.4	88.3
		Medium	High	Low	Democratic Republic of the Congo	5.9	50.8	61.4
	cattle+rice+ cassava	Low	High	Low	Central African Republic	0.7	3.7	39.3
		Medium	High	Low	Madagascar	2.4	14.8	20.9
Highlands	banana+ roots	High	High	Medium	Democratic Republic of the Congo	5.9	50.8	9.6
		High	High	Medium	Rwanda	1.4	7.8	99.6
		High	High	High	Burundi	1.0	6.3	80.6
		High	High	High	Kenya	3.9	29.8	8.8
		High	High	High	Uganda	6.2	23.1	57.0
	cattle+maize	High	Low	High	Eritrea	0.4	3.5	33.8
		High	Medium	Low	Ethiopia	6.5	63.0	55.6
		High	Medium	Low	Lesotho	0.2	2.0	91.4
		High	High	Low	Ethiopia	6.5	63.0	8.3
		High	High	Medium	Zimbabwe	2.7	12.6	5.0
Cereal-root crop	cassava+ coarse grain+ groundnuts	Low	Medium	Low	Central African Republic	0.7	3.7	5.9
		Medium	Medium	Medium	Benin	1.9	5.8	34.1
		Medium	Medium	High	Ghana	5.0	17.2	16.2
		Medium	High	Low	Mozambique	4.9	16.4	36.3
		Medium	High	Medium	Guinea	1.6	7.6	17.3
		Medium	High	Medium	Guinea-Bissau	0.2	0.7	75.3
		High	Medium	High	Nigeria	39.6	107.9	36.3
		High	High	High	Malawi	3.0	11.2	14.0
	cattle+ cassava	Low	High	Low	Angola	2.0	12.2	26.6
		Low	High	Low	Zambia	1.2	10.6	11.7
		Medium	Medium	Medium	Madagascar	2.4	14.8	13.6
		Medium	High	Low	Sudan	11.3	30.8	6.1

(continued)

TABLE 4A.6 (continued)

Farming system	Subsystem	Population density	NDVI	Market access	Country	Area (Mha)	Pop. (million)	Share in national agriculture (%)
		High	Low	High	Togo	1.3	4.1	13.2
	cattle+ pulses+ coarse grain	Low	Low	Low	Sudan	11.3	30.8	19.3
		Medium	Medium	Medium	Burkina Faso	4.2	11.5	76.0
		Medium	Medium	Medium	Chad	2.8	7.7	47.7
		Medium	Medium	Medium	Mali	3.4	11.2	27.6
		Medium	Medium	Medium	Senegal	2.4	6.3	18.8
		High	Medium	High	Cameroon	2.9	14.6	16.7
		High	Medium	High	The Gambia	0.2	0.6	33.9
Maize mixed	cattle+maize	Medium	Medium	Low	Tanzania	6.5	32.9	53.5
		Medium	Medium	Medium	Zimbabwe	2.7	12.6	69.4
		Medium	High	Low	Ethiopia	6.5	63.0	17.2
		High	High	Medium	Kenya	3.9	29.8	29.3
		High	High	Medium	Uganda	6.2	23.1	42.2
	roots+ maize+ tobacco	Medium	Low	Medium	Zimbabwe	2.7	12.6	10.7
		High	Medium	Medium	Kenya	3.9	29.8	35.1
		High	Medium	Medium	Malawi	3.0	11.2	74.8
	sugarcane+ cattle	High	Medium	Medium	South Africa	5.9	40.3	5.1
		High	High	Medium	Swaziland	0.1	0.9	98.9
	tobacco+ maize	Medium	High	Low	Malawi	3.0	11.2	11.1
		Medium	High	Low	Mozambique	4.9	16.4	31.1
		Medium	High	Low	Zambia	1.2	10.6	67.5
Pastoral-agropastoral	cattle dominated	Low	Low	Low	Kenya	3.9	29.8	6.3
		Low	Low	Low	Namibia	0.3	1.4	56.4
		Low	Low	Low	Somalia	1.0	7.3	51.6
		Low	Low	Medium	Botswana	0.2	1.6	35.1
		Low	Low	Medium	Djibouti	0.0	0.2	100.0
		Low	Low	Medium	Mauritania	0.3	1.1	74.2
		Medium	Low	Low	Eritrea	0.4	3.5	39.0
		Medium	Low	Medium	Sudan	11.3	30.8	18.1

Farming system	Subsystem	Population density	NDVI	Market access	Country	Area (Mha)	Pop. (million)	Share in national agriculture (%)
	cattle+ cassava+ maize	Low	Medium	Low	Angola	2.0	12.2	50.5
		Low	Medium	Low	Kenya	3.9	29.8	16.8
		Low	Medium	Medium	Zambia	1.2	10.6	9.8
		Medium	Medium	Medium	Zimbabwe	2.7	12.6	11.1
		High	Medium	Low	Ethiopia	6.5	63.0	12.3
	coarse grain+ cattle+ groundnuts	Low	Low	Low	Sudan	11.3	30.8	29.3
		Low	Low	Medium	Chad	2.8	7.7	32.9
		Low	Medium	Low	Namibia	0.3	1.4	7.4
		Medium	Low	Low	Niger	11.6	10.8	93.5
		Medium	Low	Medium	Burkina Faso	4.2	11.5	23.0
		Medium	Low	Medium	Eritrea	0.4	3.5	27.0
		Medium	Low	Medium	Mali	3.4	11.2	60.7
		Medium	Low	High	Senegal	2.4	6.3	70.4
		High	Low	High	Nigeria	39.6	107.9	20.3
	livestock	Low	Low	Low	Chad	2.8	7.7	18.2
		Low	Medium	Low	Botswana	0.2	1.6	57.7
		Low	Medium	Low	Namibia	0.3	1.4	13.4
		Medium	Low	Low	Angola	2.0	12.2	6.6
		Medium	High	Low	Somalia	1.0	7.3	11.3
	sheep	Low	Low	Low	Mauritania	0.3	1.1	10.6
		Low	Low	Low	Somalia	1.0	7.3	18.6
Irrigated	coarse grain+ groundnuts+ livestock	Medium	Low	Medium	Sudan	11.3	30.8	7.2
	groundnuts+ coarse grain+ cattle	High	Medium	High	The Gambia	0.2	0.6	64.6
		High	Medium	High	Senegal	2.4	6.3	5.6
	livestock+ coarse grain	Medium	Low	Medium	Sudan	11.3	30.8	9.9
		Medium	Medium	Medium	Somalia	1.0	7.3	18.4
	rice+ livestock	Medium	Low	High	Mauritania	0.3	1.1	11.0
Rice-tree crop	rice+cattle+ cassava	Medium	High	Medium	Madagascar	2.4	14.8	53.7

(continued)

TABLE 4A.6 (continued)

Farming system	Subsystem	Population density	NDVI	Market access	Country	Area (Mha)	Pop. (million)	Share in national agriculture (%)
Coastal	rice+cattle	Low	High	Medium	Madagascar	2.4	14.8	11.8
		Medium	High	High	Guinea-Bissau	0.2	0.7	24.5
		High	High	High	Sierra Leone	0.4	3.3	7.8
	roots	Low	High	Low	Gabon	0.2	0.9	27.2
		Medium	High	Medium	Liberia	0.3	2.1	36.0
		Medium	High	Medium	Mozambique	4.9	16.4	17.1
		Medium	High	High	Equatorial Guinea	0.1	0.4	32.0
		High	Medium	High	Togo	1.3	4.1	7.7
		High	High	High	Benin	1.9	5.8	18.5
Large commercial & smallholder	livestock	Low	Low	Low	Botswana	0.2	1.6	7.2
		Low	Low	Low	Namibia	0.3	1.4	15.5
	maize+ cattle+ sugarcane+ wheat	Medium	Low	Medium	South Africa	5.9	40.3	88.3
		High	Low	High	Lesotho	0.2	2.0	8.6

Source: Authors' calculation based on cluster analysis.
Note: NDVI = normalized difference vegetation index.

TABLE 4A.7 Typology of major subsystems (within systems) by country in Africa south of the Sahara

Country	Population density	NDVI	Market access		
			Low	Medium	High
Angola	Low	Medium	cattle+cassava+ maize (pastoral-agropastoral)		
	Medium	Medium	cattle+cassava (cereal-root crop)		
Benin	Medium	Medium		cassava+coarse grain+groundnuts (cereal-root crop)	
	High	High			roots (tree-root crop)
Botswana	Low	Low		cattle dominated (pastoral-agropastoral)	
	Low	Medium	livestock (pastoral-agropastoral)		

Country	Population density	NDVI	Market access: Low	Medium	High
Burkina Faso	Medium	Low		coarse grain+ cattle+groundnuts (pastoral-agropas-toral)	
	High	Low			cattle+pulses+coarse grain (cereal-root crop)
Burundi	High	High			banana+roots (high-lands)
Cameroon	Medium	Medium		roots (tree-root crop)	
	Medium	High		cocoa+cassava+ banana (tree-root crop)	
Central African Republic	Low	High	roots (tree-root crop)		
	Medium	High	cattle+rice+cassava (forest based)		
Chad	Medium	Low		cattle+pulses+ coarse grain (cere-al-root crop)	
	High	Low		coarse grain+ cattle+groundnuts (pastoral-agropas-toral)	
Côte d'Ivoire	Medium	High		roots (tree-root crop)	
	High	Medium			cocoa+cassava+ba-nana (tree-root crop)
Democratic Republic of the Congo	High	High	cassava+banana (forest based)		
Djibouti	Low	Low		cattle dominated (pastoral-agropas-toral)	
Equatorial Guinea	Medium	High		banana+cassava (forest based)	
	Medium	HIgh			roots (coastal)
Eritrea	Medium	Low		coarse grain+ cattle+groundnuts (pastoral-agropas-toral)	
	Medium	Low	cattle dominated (pastoral-agropas-toral)		

(continued)

TABLE 4A.7 (continued)

Country	Population density	NDVI	Market access: Low	Market access: Medium	Market access: High
	High	Low			cattle+maize (highlands)
Ethiopia	High	Medium	cattle+maize (highlands)		
Gabon	Low	High	roots (tree-root crop)		
	Low	High	banana+cassava (forest based)		
	Low	High	roots (coastal)		
Gambia, The	High	Medium			cattle+pulses+coarse grain (cereal-root crop)
	High	Medium			groundnuts+coarse grain+cattle (irrigated)
Ghana	High	High		cocoa+cassava+banana (tree-root crop)	
Guinea	Medium	Medium			roots (tree-root crop)
Guinea-Bissau	Medium	High		cassava+coarse grain+groundnuts (cereal-root crop)	
	Medium	High			rice+cattle (coastal)
Kenya	Medium	Low		roots+maize+tobacco (maize mixed)	
	High	High		cattle+maize (maize mixed)	
Lesotho	High	Low	cattle+maize (highlands)		
Liberia	Medium	High	roots (coastal)		
	Medium	High		roots (tree-root crop)	
Madagascar	Medium	Medium	cattle+rice+cassava (forest based)		
	Medium	Medium	rice+cattle+cassava (rice-tree crop)		
Malawi	High	Medium		roots+maize+tobacco (maize mixed)	
Mali	Medium	Low			cattle+pulses+coarse grain (cereal-root crop)

Country	Population density	NDVI	Market access: Low	Medium	High
	Medium	Low		coarse grain+ cattle+groundnuts (pastoral-agropastoral)	
Mauritania	Low	Low		cattle dominated (pastoral-agropastoral)	
Mozambique	Low	High	tobacco+maize (maize mixed)		
	Medium	High	cassava+coarse grain+groundnuts (cereal-root crop)		
Namibia	Low	Low	cattle dominated (pastoral-agropastoral)		
Niger	Medium	Low		coarse grain+ cattle+groundnuts (pastoral-agropastoral)	
Nigeria	High	Low			coarse grain+cattle+ groundnuts (pastoral-agropastoral)
	High	Medium		cassava+coarse grain+groundnuts (cereal-root crop)	
	High	Medium			roots (tree-root crop)
Republic of the Congo	Medium	High	cassava+banana (forest based)		
Rwanda	High	High		banana+roots (highlands)	
Senegal	Medium	Low			coarse grain+cattle+ groundnuts (pastoral-agropastoral)
Sierra Leone	High	High			roots (tree-root crop)
Somalia	Medium	Low	cattle dominated (pastoral-agropastoral)		
South Africa	Medium	Medium		maize+cattle+ sugarcane+wheat (large commercial & smallholder)	
Sudan	High	Medium		coarse grain+ cattle+groundnuts (pastoral-agropastoral)	

(continued)

TABLE 4A.7 (continued)

Country	Population density	NDVI	Market access: Low	Medium	High
Swaziland	High	Medium			sugarcane+cattle (maize mixed)
Tanzania	Medium	Medium	cattle+maize (maize mixed)		
	High	Medium		roots (tree-root crop)	
Togo	High	Medium			roots (tree-root crop)
	High	High			cocoa+cassava+banana (tree-root crop)
Uganda	High	Medium		cattle+maize (maize mixed)	
	High	High			banana+roots (highlands)
Zambia	Medium	High	tobacco+maize (maize mixed)		
Zimbabwe	Medium	Medium		cattle+maize (maize mixed)	

Source: Authors' calculation based on typology and cluster analysis.
Note: NDVI = normalized difference vegetation index.

TABLE 4A.8 Typology of minor subsystems by country in Africa south of the Sahara

Country	Population density	NDVI	Market access: Low	Medium	High
Angola	Low	High	roots (tree-root crop)		
Angola	Medium	Low	livestock (pastoral-agropastoral)		
Benin	High	High			roots (coastal)
Botswana	Medium	Low	livestock (large commercial & smallholder)		
Burundi	High	Medium		roots (tree-root crop)	
Cameroon	Medium	Low			cattle+pulses+coarse grain (cereal-root crop)
Cameroon	High	High		banana+cassava (forest based)	

Country	Population density	NDVI	Market access: Low	Market access: Medium	Market access: High
Central African Republic	Low	Medium	cassava+coarse grain+groundnuts (cereal-root crop)		
Chad	Low	Low	livestock (pastoral-agropastoral)		
Democratic Republic of the Congo	Medium	High	roots (tree-root crop)		
Democratic Republic of the Congo	High	High			banana+roots (highlands)
Equatorial Guinea	Low	High		cocoa+cassava+banana (tree-root crop)	
Ethiopia	Medium	Medium	cattle+maize (maize mixed)		
Ethiopia	High	Medium	cattle+cassava+maize (pastoral-agropastoral)		
Ghana	Medium	Medium			cassava+coarse grain+groundnuts (cereal-root crop)
Ghana	Medium	Medium		roots (tree-root crop)	
Guinea	Medium	High		cassava+coarse grain+groundnuts (cereal-root crop)	
Kenya	Low	Low	cattle dominated (pastoral-agropastoral)		
Kenya	Low	Medium	cattle+cassava+maize (pastoral-agropastoral)		
Kenya	High	High			banana+roots (highlands)
Lesotho	High	Low			maize+cattle+sugarcane+wheat (large commercial & smallholder)
Madagascar	Low	Medium		rice+cattle (coastal)	
Madagascar	Low	Medium		cattle+cassava (cereal-root crop)	
Malawi	Medium	High	tobacco+maize (maize mixed)		

(continued)

TABLE 4A.8 (continued)

Country	Population density	NDVI	Market access		
			Low	Medium	High
Malawi	High	Medium		cassava+coarse grain+groundnuts (cereal-root crop)	
Mauritania	Low	Low	sheep (pastoral-agropastoral)		
Mauritania	Medium	Low			rice+livestock (irrigated)
Mozambique	Low	Medium	roots (tree-root crop)		
Mozambique	Medium	High		roots (coastal)	
Namibia	Low	Low	livestock (large commercial & smallholder)		
Namibia	Low	Medium	coarse grain+cattle+groundnuts (pastoral-agropastoral)		
Namibia	Low	Medium	livestock (pastoral-agropastoral)		
Republic of the Congo	Low	High	roots (tree-root crop)		
Senegal	Medium	Low		cattle+pulses+coarse grain (cereal-root crop)	
Senegal	Medium	Medium			groundnuts+coarse grain+cattle (irrigated)
Sierra Leone	High	High			rice+cattle (coastal)
Sierra Leone	High	High			cocoa+cassava+banana (tree-root crop)
Somalia	Low	Low	sheep (pastoral-agropastoral)		
Somalia	Medium	High	livestock (pastoral-agropastoral)		
Somalia	High	Medium		livestock+coarse grain (irrigated)	
South Africa	High	Medium		sugarcane+cattle (maize mixed)	
Sudan	Low	Medium	roots (tree-root crop)		
Sudan	Medium	Low		livestock+coarse grain (irrigated)	

Country	Population density	NDVI	Market access Low	Medium	High
Sudan	Medium	Low		cattle+pulses+coarse grain (cereal-root crop)	
Sudan	Medium	Low		cattle dominated (pastoral-agropastoral)	
Sudan	Medium	High	cattle+cassava (cereal-root crop)		
Sudan	High	Low			coarse grain+groundnuts+livestock (irrigated)
Togo	High	Low			cattle+cassava (cereal-root crop)
Togo	High	Medium			roots (coastal)
Zambia	Low	Medium	roots (tree-root crop)		
Zambia	Low	Medium	cattle+cassava (cereal-root crop)		
Zambia	Low	Medium		cattle+cassava+maize (pastoral-agropastoral)	
Zimbabwe	Medium	Low		roots+maize+tobacco (maize mixed)	
Zimbabwe	Medium	Medium		cattle+maize (highlands)	
Zimbabwe	Medium	Medium		cattle+cassava+maize (pastoral-agropastoral)	

Source: Authors' calculation based on typology and cluster analysis.

Note: NDVI = normalized difference vegetation index.

TABLE 4A.9 Typology of marginal subsystems by country in Africa south of the Sahara

Country	Population density	NDVI	Market access Low	Medium	High
Angola	Low	Low	cattle dominated (pastoral-agropastoral)		
Angola	Low	Medium	cattle (irrigated)		
Angola	Low	High	cassava+banana (forest based)		

(continued)

TABLE 4A.9 (continued)

Country	Population density	NDVI	Market access		
			Low	Medium	High
Angola	Medium	Low			cattle+maize (highlands)
Angola	High	Low		cassava+coarse grain+groundnuts (cereal-root crop)	
Angola	High	High			coarse grain+cattle+ groundnuts (pastoral-agropastoral)
Angola	High	High			banana+roots (highlands)
Benin	High	Low			groundnuts+sheep+ coarse grain+rice (coastal)
Botswana	Low	Medium	cattle+maize (maize mixed)		
Burkina Faso	Medium	Medium		roots (tree-root crop)	
Burkina Faso	High	Medium			cattle+cassava+ maize (pastoral-agro-pastoral)
Burundi	Low	Low			livestock+cassava (highlands)
Burundi	High	Low			cattle+maize (highlands)
Cameroon	Low	Medium			cattle+cassava+ maize (pastoral-agro-pastoral)
Cameroon	Low	High	cassava+coarse grain+groundnuts (cereal-root crop)		
Cameroon	Medium	Low			cattle dominated (pastoral-agropastoral)
Cameroon	Medium	Medium		cattle+maize (highlands)	
Cameroon	Medium	High	banana+roots (highlands)		
Cameroon	High	Medium			coarse grain+cattle+ groundnuts (pastoral-agropastoral)
Cameroon	High	High			roots (coastal)
Central African Republic	Low	High	cattle+maize (maize mixed)		

Country	Population density	NDVI	Market access: Low	Medium	High
Chad	Low	High		cattle+cassava+ maize (pastoral-agropastoral)	
Chad	Medium	High		cassava+coarse grain+groundnuts (cereal-root crop)	
Chad	Medium	High		roots (tree-root crop)	
Côte d'Ivoire	Medium	Medium			cattle+cassava (cereal-root crop)
Côte d'Ivoire	High	Medium			groundnuts+sheep+ coarse grain+rice (coastal)
Côte d'Ivoire	High	High			roots (coastal)
Democratic Republic of the Congo	Low	Low			roots+maize+tobacco (maize mixed)
Democratic Republic of the Congo	Low	Low			livestock+cassava (highlands)
Democratic Republic of the Congo	Low	Low			livestock (tree-root crop)
Democratic Republic of the Congo	Medium	Low	cattle+rice+cassava (forest based)		
Democratic Republic of the Congo	Medium	Low			cattle+maize (high-lands)
Democratic Republic of the Congo	High	Medium			cattle+cassava (cereal-root crop)
Democratic Republic of the Congo	High	Medium			cattle+maize (maize mixed)
Equatorial Guinea	Low	Medium			groundnuts+sheep+ coarse grain+rice (coastal)
Eritrea	Medium	Low			livestock+coarse grain (irrigated)
Ethiopia	Low	Low		coarse grain+ groundnuts+live-stock (irrigated)	
Ethiopia	Low	Medium	cattle (irrigated)		

(continued)

TABLE 4A.9 (continued)

Country	Population density	NDVI	Market access: Low	Medium	High
Ethiopia	Medium	Low	cattle dominated (pastoral-agropastoral)		
Ethiopia	Medium	Medium	coarse grain+ cattle+groundnuts (pastoral-agropastoral)		
Ethiopia	Medium	Medium	cattle+cassava (cereal-root crop)		
Gabon	Low	Low	groundnuts+sheep+ coarse grain+rice (coastal)		
Gambia, The	High	Medium			groundnuts+sheep+ coarse grain+rice (coastal)
Ghana	Low	Low			livestock (tree-root crop)
Ghana	High	Low			groundnuts+sheep+ coarse grain+rice (coastal)
Ghana	High	Medium			roots (coastal)
Guinea	Medium	Low			cattle+pulses+coarse grain (cereal-root crop)
Guinea	Medium	High			cocoa+cassava+ banana (tree-root crop)
Guinea	High	Medium			rice+cattle (coastal)
Guinea-Bissau	Medium	Low		cattle+pulses+ coarse grain (cereal-root crop)	
Kenya	Medium	Low			livestock+cassava (highlands)
Kenya	Medium	Medium	roots (tree-root crop)		
Kenya	High	Low			cattle+maize (highlands)
Kenya	High	High			roots (coastal)
Liberia	Low	Low			groundnuts+sheep+ coarse grain+rice (coastal)
Malawi	Low	Low			cattle+maize (maize mixed)

Country	Population density	NDVI	Market access Low	Market access Medium	Market access High
Malawi	Medium	Low			cattle+pulses+coarse grain (cereal-root crop)
Mali	Low	Low	livestock (pastoral-agropastoral)		
Mali	Low	Low		rice+livestock (irrigated)	
Mali	Low	High		roots (tree-root crop)	
Mali	Medium	Low			coarse grain+ groundnuts+livestock (irrigated)
Mali	Medium	High		cattle+cassava (cereal-root crop)	
Mali	High	Medium			cattle+cassava+ maize (pastoral-agro-pastoral)
Mauritania	Medium	Low		livestock+coarse grain (irrigated)	
Mozambique	Low	Low			livestock (tree-root crop)
Mozambique	Low	Low			roots+maize+tobacco (maize mixed)
Mozambique	Low	Low		cattle+maize (maize mixed)	
Mozambique	High	Low			groundnuts+sheep+ coarse grain+rice (coastal)
Mozambique	High	Low			cattle+pulses+coarse grain (cereal-root crop)
Mozambique	High	High		banana+roots (highlands)	
Mozambique	High	High			maize+cattle+sug-arcane+wheat (large comm. & smallholder)
Namibia	Low	Low	sheep (pastoral-agropastoral)		
Namibia	Low	Medium	tobacco+maize (maize mixed)		
Namibia	Low	High	cattle+cassava+ maize (pastoral-agropastoral)		

(continued)

TABLE 4A.9 (continued)

Country	Population density	NDVI	Market access		
			Low	Medium	High
Niger	Low	Low	livestock (pastoral-agropastoral)		
Niger	Low	High		cattle+cassava+maize (pastoral-agropastoral)	
Niger	Medium	Low			coarse grain+groundnuts+livestock (irrigated)
Niger	Medium	Low		cattle+pulses+coarse grain (cereal-root crop)	
Nigeria	Low	High			cattle+cassava+maize (pastoral-agro-pastoral)
Nigeria	Medium	Medium	banana+roots (highlands)		
Nigeria	High	Low			coarse grain+groundnuts+livestock (irrigated)
Nigeria	High	Low			cattle+pulses+coarse grain (cereal-root crop)
Nigeria	High	Low			livestock (tree-root crop)
Nigeria	High	Low			groundnuts+coarse grain+cattle (irrigated)
Nigeria	High	High		roots (coastal)	
Republic of the Congo	Low	Low		cattle+rice+cassava (forest based)	
Republic of the Congo	Low	Medium		cocoa+cassava+banana (tree-root crop)	
Republic of the Congo	Medium	High		cattle+cassava (cereal-root crop)	
Republic of the Congo	High	High	roots (coastal)		
Rwanda	High	Low			livestock+cassava (highlands)
Rwanda	High	High	roots (tree-root crop)		
Senegal	Medium	Low			rice+livestock (irrigated)

Country	Population density	NDVI	Market access Low	Medium	High
Senegal	High	Medium			groundnuts+sheep+ coarse grain+rice (coastal)
Senegal	High	Medium			cattle+cassava+ maize (pastoral-agro-pastoral)
Senegal	High	High			cassava+coarse grain+groundnuts (cereal-root crop)
Somalia	Low	Medium	cattle+cassava+ maize (pastoral-agropastoral)		
South Africa	Low	Low		sheep (pastoral-agropastoral)	
South Africa	Medium	Low			tobacco+maize (maize mixed)
South Africa	Medium	Low		cattle+cassava+ maize (pastoral-agropastoral)	
South Africa	High	Medium		cattle+maize (highlands)	
South Africa	High	High		cattle+maize (maize mixed)	
Sudan	Low	Low	sugarcane+cattle (maize mixed)		
Sudan	Low	Medium		livestock (tree-root crop)	
Sudan	Low	Medium	cattle+maize (maize mixed)		
Sudan	Low	High	livestock (pastoral-agropastoral)		
Sudan	Medium	Medium		cattle+cassava+ maize (pastoral-agropastoral)	
Swaziland	Medium	High		maize+cattle+ sugarcane+wheat (large comm. & smallholder)	
Tanzania	Low	Low			livestock (tree-root crop)
Tanzania	Low	Low			livestock+cassava (highlands)

(continued)

TABLE 4A.9 (continued)

Country	Population density	NDVI	Market access: Low	Medium	High
Tanzania	Low	Low	cattle dominated (pastoral-agropastoral)		
Tanzania	Low	Medium	livestock (pastoral-agropastoral)		
Tanzania	Medium	Medium		cattle+cassava+ maize (pastoral-agropastoral)	
Tanzania	High	Medium		cattle+maize (highlands)	
Tanzania	High	High		banana+roots (highlands)	
Tanzania	High	High		roots (coastal)	
Togo	High	High			cassava+coarse grain+groundnuts (cereal-root crop)
Togo	High	High			groundnuts+sheep+ coarse grain+rice (coastal)
Uganda	Low	Medium	coarse grain+ cattle+ground-nuts (pastoral-agropastoral)		
Uganda	Medium	Low			livestock+cassava (highlands)
Uganda	Medium	Low			roots+maize+tobacco (maize mixed)
Zambia	Low	Low			livestock (tree-root crop)
Zambia	Low	High	cattle (irrigated)		
Zambia	Medium	Medium		livestock+coarse grain (irrigated)	
Zambia	High	Low		sugarcane+cattle (maize mixed)	
Zimbabwe	Low	Low	livestock (large commercial & smallholder)		
Zimbabwe	Low	Low		cattle dominated (pastoral-agropastoral)	
Zimbabwe	Medium	Low		tobacco+maize (maize mixed)	

Country	Population density	NDVI	Market access		
			Low	Medium	High
Zimbabwe	High	Low			livestock+coarse grain (irrigated)
Zimbabwe	High	High	cassava+coarse grain+groundnuts (cereal-root crop)		
Zimbabwe	High	High		sugarcane+cattle (maize mixed)	

Source: Authors' calculation based on typology and cluster analysis.

Note: NDVI = normalized difference vegetation index.

References

Asrar, G., M. Fuchs, E. T. Kanemasu, and I. L. Hatfield. 1984. "Estimating Absorbed Photosynthetic and Leaf Area Index from Spectral Reflectance in Wheat." *Agronomy Journal* 76: 300–306.

Byerlee, D., L. Harrington, and D. Winkelmann. 1982. "Farming Systems Research: Issues in Research Strategy and Technology Design." *American Journal of Agricultural Economics* 64: 897–904.

Calderón, C. 2009. *Infrastructure and Growth in Africa*. Policy Research Working Paper 4914. Washington, DC: World Bank.

Chambers, R., and J. Jiggins, 1987. "Agricultural Research for Resource-Poor Farmers." *Agricultural Administration and Extension* 27 (Part I): 35–52; 27 (Part II): 109–128.

CIESIN (Center for International Earth Science Information Network, Columbia University), International Food Policy Research Institute, World Bank, and Centro Internacional de Agricultura Tropical. 2011. *Global Rural-Urban Mapping Project, Version 1 (GRUMPv1): Population Density Grid*. Palisades, NY: National Aeronautics and Space Administration, Socioeconomic Data and Applications Center. http://sedac.ciesin.columbia.edu/data/set/grump-v1-population-density.

Conradie, B., J. Piesse, and C. Thirtle. 2009a. "District-level Total Factor Productivity in Agriculture: Western Cape Province, South Africa, 1952–2002." *Agricultural Economics* 40 (3): 265–280.

———. 2009b. "What Is the Appropriate Level of Aggregation for Productivity Indices? Comparing District, Regional and National Measures." *Agrekon* 48 (1): 9–20.

Dixon, J., A. Gulliver, and D. Gibbon. 2001. *Farming Systems and Poverty: Improving Farmers' Livelihoods in a Changing World*. Rome: Food and Agriculture Organization of the United Nations; Washington, DC: World Bank.

Doraiswamy, P., T. Sinclair, S. Hollinger, B. Akhmedov, A. Stern, and J. Prueger. 2005. "Application of MODIS Derived Parameters for Regional Crop Yield Assessment." *Remote Sensing of Environment* 97: 192–202.

Dorosh, P., H. Wang, L. You, and E. Schmidt. 2010. *Crop Production and Road Connectivity in Sub-Saharan Africa: A Spatial Analysis*. Policy Research Working Paper 5385. Washington, DC: World Bank.

Everitt, B. S., S. Landau, M. Leese, and D. Stahl. 2011. *Cluster Analysis*, 5th ed. Chichester, UK: Wiley.

FAO (Food and Agriculture Organization of the United Nations). 2011. FAOSTAT database. "Food and Agricultural Price Statistics." http://faostat.fao.org/site/351/default.aspx.

Feder, G., R. Just, and D. Zilberman. 1985. "Adoption of Agricultural Innovations in Developing Countries: A Survey." *Economic Development and Cultural Change* 30: 59–76.

Feder, G., and D. Umali. 1993. "The Adoption of Agricultural Innovations: A Review." *Technological Forecasting and Social Change* 43: 215–239.

Fischer, G., F. O. Nachtergaele, H. van Velthuizen, L. Verhelst, and D. Wiberg. 2012. "Global Agro-Ecological Zones (GAEZ v3.0)—Model Documentation." Laxenburg, Austria: International Institute for Applied Systems Analysis; Rome: Food and Agriculture Organization of the United Nations.

Hijmans, R. J., S. E. Cameron, J. L. Parra, P. G. Jones, and A. Jarvis. 2005. "Very High Resolution Interpolated Climate Surfaces for Global Land Areas." *International Journal of Climatology* 25: 1965–1978.

Jerven, M. 2013. *Poor Numbers: How We Are Misled by African Development Statistics and What to Do about It.* Ithaca, NY: Cornell University Press.

Makles, A. 2012. "Stata Tip 110: How to Get the Optimal K-means Cluster Solution." *The Stata Journal* 12 (2): 347–351.

Milligan, G. W., and M. C. Cooper. 1985. "An Examination of Procedures for Determining the Number of Clusters in a Dataset." *Psychometrika* 50: 159–179.

NASA (National Aeronautics and Space Administration). 2011. Land Processes Distributed Active Archive Center (LP DAAC). MODIS Vegetation Indices ver5. Sioux Falls, SD: United States Geological Survey, Earth Resources Observation and Science Center.

Pender, J., F. Place, and S. Ehui. 1999. *Strategies for Sustainable Agricultural Development in the East African Highlands.* Environment and Production Technology Department Working Paper 41. Washington, DC: International Food Policy Research Institute.

———, eds. 2006. *Strategies for Sustainable Land Management in the East African Highlands.* Washington, DC: International Food Policy Research Institute.

Ramankutty, N., A. T. Evan, C. Monfreda, and J. A. Foley. 2008. "Farming the Planet: 1. Geographic Distribution of Global Agricultural Lands in the Year 2000." *Global Biogeochemical Cycles* 22 (1003): 1–19. http://trane.evsc.virginia.edu/Publications_files/2007GB002952.pdf.

Running, S. W., T. R. Loveland, L. L. Pierce, R. R. Nemani, and E. R. Hunt. 1995. "A Remote Sensing Based Vegetation Classification Logic for Global Land Cover Analysis." *Remote Sensing of Environment* 51: 39–48.

Sanchez, P., C. Palm, and S. Buol. 2003. "Fertility Capability Soil Classification: A Tool to Help Assess Soil Quality in the Tropics." *Geoderma* 114 (3–4): 157–185.

UN (United Nations Statistics Division). 2011. "Demographic and Social Statistics." http://unstats.un.org/unsD/demographic/default.htm.

Wint, W., and T. Robinson. 2007. *Gridded Livestock of the World 2007.* Rome: Food and Agriculture Organization of the United Nations.

Wood, S., K. Sebastian, F. Nachtergaele, D. Nielsen, and A. Dai. 1999. *Spatial Aspects of the Design and Targeting of Agricultural Development Strategies.* Environment and Production Technology Division Discussion Paper 44. Washington, DC: International Food Policy Research Institute.

You, L., and Z. Guo. 2011. "Travel Time to Major Market Cities." Washington, DC: International Food Policy Research Institute, HarvestChoice. http://harvestchoice.org/labs/travel-time-major-market-cities.

You, L., S. Wood, and U. Wood-Sichra. 2009. "Generating Plausible Crop Distribution Maps for Sub-Saharan Africa Using a Spatially Disaggregated Data Fusion and Optimization Approach." *Agricultural Systems* 99 (2–3): 126–140.

Chapter 5

AGRICULTURAL INTENSIFICATION AND FERTILIZER USE

Alejandro Nin-Pratt

Introduction

A general question deriving from the preceding chapter is, what is the best approach for increasing agricultural productivity sustainably across and within different agricultural production zones? For example, is Asia's Green Revolution model of high-yielding cereal varieties, fertilizer, and intensive use of labor a viable approach for Africa? A common answer to this question is there is little incentive to adopt labor-intensive technologies like those promoted by the Asian Green Revolution, because population densities in Africa are low compared with those in many of the Asian countries. This implies that farmers in Africa will not find profitable technologies that save land (abundant) and use labor (scarce and relatively more expensive) more intensively (Binswanger and Pingali 1988). This does not mean that Africa is at a disadvantage compared with Asia regarding adopting high-yielding varieties or using chemical inputs like fertilizer, herbicides, and pesticides more intensively. What it actually means is that the path that many countries in Africa will follow to incorporate new technology and increase production and productivity in agriculture will be different from the path followed by Asian countries.

As Hayami and Ruttan (1970) made clear in their seminal paper comparing the evolution of agriculture in Japan and the United States, rapid growth in agricultural productivity depends on adapting agricultural technology to the factor proportions prevailing in the region. According to Hayami and Ruttan (1970), an important aspect of this adaptation is the ability of a country or region to generate innovations in agricultural technology biased toward saving the limiting factors. In the case of Asia, these innovations were primarily biological and chemical, with the purpose of saving land (the scarce factor) and using labor (the abundant factor) more intensively. In land-abundant labor-scarce countries, profitable technologies are normally biased toward saving scarce labor and bringing more land into production (for example, applying mechanical innovations and herbicides). Is this the case in Africa? Are

African economies mostly abundant in natural resources and scarce in labor? If they are, is this changing as a result of fast population growth observed in the past decades?

In this chapter we assess the evolution of agricultural intensification in Africa, and the changes in output composition and input use associated with different intensification patterns. The focus is on understanding the process of intensification in terms of the relationship between the relative abundance of factors of production (land, labor, and capital) and their relative prices; the role that population growth plays in this process; and how relative factor abundance and markets affect the demand for different technologies and the particular intensification paths followed by different countries.

Consistent with the preceding chapter and following Boserup (1965) and Ruthenberg (1980), "agricultural intensification" in this chapter is defined as the process of relative changes in the availability of land, labor, and capital driven by population growth and by higher returns to farming, which arise from improvements in market infrastructure and increases in farmgate prices. Therefore, intensification refers to the stock of available land, regardless of whether this land is under cultivation. Intensive use of this stock occurs, for example, when new land is placed under cultivation, when the length of the average fallow period is shortened, or when land under cultivation is used with increasing levels of inputs, labor, or capital per unit of land. Furthermore, whereas the intensification process may result in increased output per unit of land or yield, the two (that is, intensification and increasing yield) are not synonymous, which is important in distinguishing other uses of the term agricultural intensification in the agricultural and development literature.

This definition of intensification implies that a common use of the term that refers to the farm-level process of increasing inputs, labor, or capital per hectare of agricultural land for the purpose of increasing the value of output per hectare is only one of several different ways to increase intensification, and is contained in the more general concept used in this chapter (Tiffen, Mortimore, and Gichuki 1994; Carswell 1997). In the context of this other use of the term, intensification is "extensification," or the expansion of production into previously uncultivated areas, which may also require increased inputs, investments, and labor; however, the increased inputs, etc., do not result in higher output or input per unit of land. In sum, intensification as used in these other contexts is narrowed down to a technical or agronomic process without social or economic meaning.

According to our conceptual framework, we expect land-abundant countries to increase production by incorporating new land into agriculture, reducing fallows, and eventually increasing cropping intensity. These changes in general will result in increased labor productivity and little or no impact on land productivity. On the other extreme, labor-abundant countries that have reached their land frontier can increase production only by increasing land productivity through higher yields and/or increased cropping intensity, if they can still expand the area where they do multiple cropping.

National time-series data are used to calculate a measure of intensification based on the conceptual framework of Boserup (1965) and Ruthenberg (1980). This framework can be decomposed into different indicators, capturing the different possible paths a country can follow to increase intensification. These indicators are calculated for 40 countries in Africa south of the Sahara (SSA) covering the period between 1995 and 2011.The conceptual framework for the analysis is presented in the next section, where we look at the literature on the relationship between factor abundance, factor prices, and demand for new technology, and the role of population pressure and markets as drivers of innovation in agriculture. This is followed by the data and estimation technique, including the specific intensification index used. The results are then presented in terms of trends in intensification and pathways to increase intensification, including an analysis of fertilizer use and the role it played in the different intensification paths. Finally, the chapter summarizes the information and findings. and discusses the policy implications of the results.

Conceptual Framework and Literature Review

To conceptualize the relationship between factors of production, technology, their intensities of use, and the demand for different technologies under different population pressure scenarios, we draw on three related groups of the literature: (1) the induced innovation model, which posits the process by which technical and institutional changes are induced through the responses of farmers and other actors in the value chain to changes in relative resource endowments and prices of factors and outputs (Hayami and Ruttan 1970; Binswanger and Ruttan 1978); (2) the directed technological change, which distinguishes the different response paths (that is, price versus market-size effects) to improving the productivity of the relatively scarce or abundant factors (Acemoglu 1998, 2002, 2007); and (3) the Boserup framework of agricultural intensification, particularly the notion that technological change is induced by some critical or threshold population density (Boserup 1965; Ruthenberg 1980).

Technological change and technology adoption

The importance of relative resource endowments as determinants of technological change has been part of the economics of technological change since the 1970s, when Ruttan, Hayami, and Binswanger (Hayami and Ruttan 1970, 1985; Binswanger and Ruttan 1978) formulated a model of induced innovation, in which the development and application of new technology are endogenous to the economy. According to this model, the direction of technological change in agriculture is induced by changes (or differences) in relative resource endowments and factor prices. Because of the relatively high prices of less abundant resources, alternative agricultural technologies are invented to facilitate the substitution of relatively scarce (hence, expensive) factors for relatively abundant (hence, cheap) factors.

More recently, Acemoglu (1998, 2002, 2007), contributing to what is today known as the directed technological change literature, distinguishes the price and market-size effects on the incentives to improve the productivity of the relatively scarce or abundant factors. According to Acemoglu (2002), (1) when the price effect dominates the market size effect, it induces a change in technology biased toward improving the productivity of the relatively scarce factor (which is consistent with the induced innovation model); (2) when the market-size effect dominates, it induces a change in technology biased toward improving the productivity of the relatively abundant factors; and (3) economic and political institutions help shape the direction of technical change, considering the relative cost of different innovation. This result implies, for example, that whereas factor scarcity will lead to technological changes biased against that factor, it depends on whether the changes are gross substitutes or complements. For example, labor scarcity induces technological advances if the technology is strongly "labor substitute," whereas labor scarcity discourages technological advances if the technology is strongly "labor complementary" (Acemoglu 2009). Wage increases above the competitive equilibrium have similar effects on labor scarcity.

Because the productivity-increasing characteristics of the technologies are manifested only when farmers use them, the adoption patterns of different technologies will depend on their (expected) benefits under different factor scarcity regimes. The benefits of the technologies could be measured in terms of the reduction in the unit costs of production that result from adopting them (Binswanger 1986). The greater the reduction in unit costs, the greater the likelihood of adopting the technologies. The question that arises then is, how do different types of technologies contribute to unit cost reduction in land- or labor-scarce environments, for example? To address this, we adopt

Binswanger's (1986) classification of technologies according to how they use land, labor, and inputs as enhancing yield and saving labor.[1]

Yield-enhancing technologies are grouped further into three types: (1) input-using, such as fertilizers and pesticides; (2) stress-avoiding, based on genetic resistance or tolerance to pests, diseases, or water stress; and (3) husbandry, such as better land preparation or mechanical weeding. The group of labor-saving technologies includes the use of machines, draft animals, implements, and herbicides.

All three types of yield-enhancing technologies reduce the land area required to produce one unit of output, reducing not only land costs but also the cost of all inputs that are used proportional to the area saved. The difference between them is in how they use inputs in the remaining area. For example, the use of chemical inputs demands greater water, more human or mechanical power, etc., which together increase the total input costs substantially. In contrast, the use of drought-resistant varieties, for example, may raise the cost only marginally, and the extra cost of acquiring the new varieties is normally low. The impact on cost of using improved husbandry techniques may also be small, depending on how much complementary increased labor or machinery is required.

To be cost-efficient, the increase in the cost of using yield-enhancing technologies must be less than the cost of land and inputs saved from using less land. When land area can be easily expanded, yield-enhancing technologies are not likely to result in large savings in land costs. The major savings are likely to be the labor used in the different activities on the area saved. For these technologies to be adopted when land is abundant, the increased cash costs of inputs must be less than the value of labor savings alone. When land is abundant, the most likely yield-enhancing technologies to be adopted are those reducing environmental stress, as the extra cost of acquiring the technology is often low.

Labor-saving technologies do not usually reduce land area and have little, if any, effect on yields. For these technologies to be cost-efficient too, the value of labor savings needs to be larger than the extra machine or herbicide costs. A unique feature of these technologies is that their value rises with rising wages, and is not strongly dependent on land values or pre-existing technology levels (Binswanger 1986). The implications for their adoption in land-abundant regions are clear. Farmers demand labor-saving technologies in addition to

1 Binswanger's (1986) classification also includes quality-increasing innovations, which are not discussed here.

crops and varieties, which together enable them to produce more food or obtain a higher gross return for a lower labor input.

To understand the technological change process further, particularly relating to population growth and population pressure on land, the next two subsections review the literature on population-driven and market-driven technological change.

Population pressure and technological change

According to Boserup (1965), an agrarian community has a fixed territory and an array of discrete techniques that use land with different intensities consisting of five different categories: forest fallow, bush fallow, short fallow, annual cropping, and multiple cropping. Each successive category represents an intensification in the Boserupian sense of the use of land (Darity 1980). With low population densities, farmers cultivate the land for a few years, and then move on to another patch when fertility diminishes, leaving the land for several years to recover its natural fertility. With increasing population density for a particular production technique, output per capita declines and, when it depresses the average standard of living to some low point, more labor is allocated to bring new land under cultivation and/or shorten the fallow period, so that land is cultivated for longer periods. This process continues until annual cropping and later multiple cropping become standard. With more intensive use of land, the rate of natural replenishment is reduced, requiring a switch to new techniques of replenishment. With land becoming scarce, its value rises, and farmers find it cost-effective to use manure or chemical fertilizers to maintain soil fertility and use of low-cost irrigation can become profitable.

Thus, at the core of Boserup's model is the notion of technological change induced or impelled by a "critical" population density, which Turner and Fischer-Kowalskic (2010) claim offered a powerful set of ideas in opposition to the prevailing neo-Malthusian ideas. Boserup (1965) challenged Malthus' proposition that the relatively slow growth in the food ceiling served as the upper limit for the faster-paced potential growth in population. She reversed the causality, arguing that increases in population pressure trigger the development or the use of technologies and management strategies to increase production commensurate with demand and that, over the long run, this process transforms the physical and social landscapes.[2]

---

2 Turner and Fischer-Kowalskic (2010) assert that Boserup's (1965) thesis remains important today to the various subfields contributing to sustainable development. Its foundations have been tested and critiqued, generating a vast literature exploring the roles of environment, gender, social capital, household composition, tenure, off-farm employment opportunities, and

It is also important to consider that it is not inevitable that population pressure will lead to technological change, as Boserup (1965) acknowledged. Population pressure is a necessary but not a sufficient condition, since different communities may be faced with different technological elasticities because of differences in soils and climates, differences in the distribution of land between types of uses, and different external influences. All of these considerations could make communities with the same population characteristics emerge with different production techniques. Regardless, for Boserup (1965), population pressure must be present to precipitate a move toward more intensive uses of the land (Darity 1980).

Recent literature has relaxed assumptions imposed in Boserup's (1965) scheme, revealing the conditions leading to Boserupian, Malthusian, or other outcomes (Turner and Ali 1996; Place and Otsuka 2000; Gray and Kevane 2001; Reenberg 2001; Stone 2001; Malmberg and Tegenu 2006; Pascual and Barbier 2006; Demont et al. 2007; Tachibana, Nguyen, and Otskuka 2010). This literature assumes that population growth results in increasing hardship in meeting the prevailing standard of living, causing farmers to opt for more intensive agriculture or other paths not necessarily requiring intensification, as long as those paths allow them to maintain or improve their living standards.

For example, working with a survey of farms in northern Côte d'Ivoire, Demont et al. (2007) found that the Boserupian and Malthusian theses coexist, rather than compete. They observed that in a first stage, demographic pressure engenders Malthusian mechanisms (degradation of the environment and decline of profitability), leading to migration and, hence, Malthusian population control. They also showed that as long as the option to migrate is kept open, Malthusian population control will generally dominate Boserupian mechanisms of induced innovation. However, in the long run, it is expected that the saturation of sparsely populated regions will induce intensification and mechanization across farming systems. They also found that taking into account an urbanization level of 45 percent, the agrarian transition in Côte d'Ivoire will be induced not only by local demographic pressure, but also by the increase of urban food, feed, and fiber demand, and the development and expansion of marketing systems.

Another example is the case of Bangladesh, discussed in Turner and Ali (1996). Analyzing the evolution of agriculture from 1950 to 1986, these

state policies, among other factors, on agricultural intensification under different land pressures (for example, Brookfield 1972, 2001; Turner and Brush 1987; Morrison 1996; Turner and Ali 1996; Angelsen 1999; Dorsey 1999; Lambin, Rounsevell, and Geist 2000; Stone 2001; Carr 2004).

authors found complementary episodes of Boserupian and Malthusian response. Over the entire period, induced intensification proceeded in a Boserupian path marked by several thresholds, each of which had the potential to spin off into a Malthusian path. The first threshold was reached in the 1960s and was adverted by the adoption of high-yielding-variety technologies. The second threshold in the 1980s was overcome by a shift to crops with high market values, especially market gardening in more favorable locations. Yet another threshold was reached in the 1990s, when economic and policy barriers to irrigation technologies impeded production in food staples, and the poor state of transportation infrastructures inhibited most villages from moving into market gardening. Eventually, barriers to various technologies, such as low-lift pumps, were reduced, and their increased use throughout Bangladesh led to yet another spurt in land productivity through increased dry-season cultivation. Turner and Ali (1996) concluded that the discussion has thus moved beyond a simple Malthus–Boserup debate, demonstrating how both positions may be supported, depending on where in the intensification process the analysis is undertaken, or on the temporal scale of analysis employed. On the other hand, Turner and Ali (1996) indicate that the processes that divert intensification into the involution and stagnation paths are less developed conceptually, and that a better grasp of these processes is required for a fully developed theory of agricultural change among smallholders.

Market-driven technological change

Although population pressure is a focus of this chapter, it is not the only factor causing intensification, as farmers take steps toward more intensive uses of land for various reasons (Stone 2001). For example, market incentives can induce farmers to intensify land use in the absence of land shortage (Turner and Brush 1987; Netting, Stone, and Stone 1989). Even in low-density areas, farmers facing a growing demand arising from newly accessible markets, for example, will want to produce more, which will increase demand for land and spur more intensive use of land. Also, the density threshold at which there is significant demand for fertilizers can be quite low, provided other favorable conditions exist (Goldman and Smith 1995). This implies that natural resource-rich countries with low population densities, as many African countries, can take a market-driven intensification path that demands agricultural technologies with strong labor-saving components, rather than the land-saving technologies that were promoted in Asia under the Green Revolution. Binswanger (1986) reminds us that in Thailand, which has traditionally had an open land frontier, remarkable agricultural growth has come from area expansion, and fertilizer use levels and adoption of high-yielding varieties have been below those in other Asian countries.

Even if we accept that market-driven intensification in Africa could result in demand for labor-saving rather than land-saving technologies, it is still valid to assume that labor supply in rural areas will continue to grow, reducing labor costs in land-abundant countries and creating conditions for the adoption of labor-intensive technologies, at least in densely populated areas. In other words, labor-intensive technologies could still be promoted in natural resource-rich countries in densely populated areas where farms are small, incomes are low, and a high proportion of the rural poor live.

A first problem with this reasoning is that it does not consider the fact acknowledged by Boserup (1965) that population pressure is not a sufficient condition for intensification, as discussed above. For example, Goldman (1993) and Goldman and Smith (1995) argue that constraints to innovation could also appear in densely populated areas when there is little potential to increase farm sizes. If no land is available for expansion, the additional wealth that agricultural investment and new technology can generate is limited, and nonagricultural activities may then be preferable to investment in agriculture.

A second problem is that it assumes that densely populated areas in resource-rich countries behave like closed Boserupian economies, where population pressure increases labor supply and generates labor surpluses that farmers use to increase output through the introduction of land-saving, labor-intensive technologies, as the excess labor has no other employment opportunities, or the possibility to migrate is closed. Also, Schultz (1964) developed a critique of labor surpluses in agriculture as postulated by Lewis (1954), arguing that numerous case studies of the agriculture sector in less developed societies showed that when labor was withdrawn from the agrarian sector, "the output of the traditional sector falls." One of these studies by Hansen (1969), looking at agriculture in Egypt, a country with one of the highest population densities in the world, found that small farmers have a high level of employment because of the substantial opportunities for obtaining employment outside their own farm, both on other farms and outside agriculture. Hansen (1969) concludes that the active labor market observed in Egypt is difficult to reconcile with the idea of surplus labor and zero productivity of labor as a general phenomenon. If, in fact, a country with the population density of Egypt did not have rural labor surpluses, it is at least unlikely that resource-rich countries in SSA will conform to the surplus labor model.

High labor costs appear to be a structural characteristic of resource-rich economies as a consequence of a different agricultural intensification path when compared with that occurring in labor-abundant economies. One of the explanations for this persistence of high labor costs most commonly found in the literature relates to Dutch disease. This phenomenon arises when a strong upswing in the world price of the export commodity leads to increased purchasing power and increased demand for urban goods, real appreciation of the local currency, and an increase in the relative price of nontradable goods. The result of these changes is a shift of labor, pulled by the more attractive returns in the export commodity and in the nontraded goods and services, and a "push" of workers into urban areas.

Gollin, Jedwab, and Vollrath (2013) developed a model that formally explains urbanization without industrialization and the persistence of high labor costs, despite rapid population growth in Africa. One of the implications of natural resource rents is that natural resource-rich economies do not experience a stage of labor abundance with low labor costs in agriculture, as was observed in Asia. What is observed instead, as described by Gollin, Jedwab, and Vollrath (2013), is rapid urbanization, resulting in "consumption cities" that are made up primarily of workers in nontradable services, surrounded by rural areas with high population density. These densely populated rural areas either produce semi-subsistence agriculture while diversifying into nonfarm activities (services), or specialize in high-value crops. In addition, interspersed with these densely populated rural areas are vast areas of relatively low population density dedicated to the production of export crops and semi-commercial agriculture.

Multiple cropping and intensive use of chemical fertilizer associated with cereal production could be an option in densely populated areas if it can compete with production in sparsely populated areas, and if returns to family labor in this activity are higher than other farm and nonfarm activities that seem to be more attractive for smallholders. For example, in many countries, natural resources favor production of cassava and other noncereal staples that give higher marginal returns to labor than intensive cereal production. These developments stand in contrast to the Asian case of labor-abundant economies where, with labor shifts out of agriculture into industrial employment, we observe the typical substitution of industrial labor for agricultural labor, resulting in "production cities" that produce tradable goods (manufacturing).

Fertilizer use in Africa

It is reasonable to think that Africa needs some strategic stimulus like Asia's Green Revolution. As we have seen in Chapter 1 and from the evidence presented in this chapter so far, however, we know that the conditions of relative prices and factor scarcities that "induced" Asian agriculture toward adopting labor-intensive technologies are not present in many parts of Africa to induce similar intensification and modernization of agriculture. Despite the adoption of improved maize, wheat, and rice varieties in many parts of Africa since the early 1990s, with tangible evidence of increased food production and productivity where adoption has occurred (Maredia, Byerlee, and Pee 2000; Haggblade and Hazell 2010), it is clear that Africa needs its own agricultural "revolution," borrowing, of course, from existing technologies.

There is compelling argument for more intensive use of organic and chemical fertilizers in Africa, because the expansion of the agricultural frontier and the opening of less favorable soils for cultivation could lead to a disaster in the long run, given the difficulty of restoring tropical soils to productive capacity without nutrient replenishment (Morris et al. 2007). However, the estimates of improvements in soil fertility are substantial, in terms of increases in the amount of fertilizers needed to boost agricultural productivity growth, improve food security, and raise rural incomes.

Recognizing the importance of fertilizer from both organic and inorganic sources, the African Union heads of state and government in 2006 declared fertilizer as a strategic commodity for an African Green Revolution, and resolved to increase the level of use of fertilizer from the then average of 8 kilograms per hectare (kg/ha) to an average of at least 50 kg/ha by 2015 (AU 2006)—popularly known as the Abuja Declaration. It is thus not surprising to see the resurgence of input subsidy programs (ISPs) in Africa, which according to Jayne and Rashid (2013) "has arguably been the region's most important agricultural policy development in recent years." The evidence also shows that many of the ISPs have succeeded in temporarily increasing the use of fertilizer, which in turn is expected to increase output. However, the food production and food price responses to ISPs are generally significantly lower than commonly understood, because of the inefficient use of fertilizer that results from chronically late delivery of program fertilizer, nonresponsive soil conditions, poor management practices, and insufficient use of complementary inputs that are necessary to enable farmers to obtain higher rates of fertilizer use efficiency (Morris et al. 2007; Jayne and Rashid 2013).

Other issues with the ISPs, rather than fertilizer use, concern the development of the private sector and sustainability. For example, many of the initiatives have also been launched to remove fertilizer market distortions and promote private-sector participation in input markets; however, the evidence on these initiatives is mixed. On the one hand, private firms selected to distribute program fertilizer on behalf of government have benefited the most from such programs, often at the expense of firms that were excluded. On the other hand, the increased imports and supply of fertilizers have increased the number and sales of distributors and retailers in input markets.

The other problem is the sustainability of the programs without large infusions of external financial support that few African countries can afford. In 2011 alone, for example, 10 African countries spent roughly \$1.05 billion[3] on ISPs, which is equivalent to 29 percent of their public expenditures on agriculture on average (Jayne and Rashid 2013). In Malawi and Tanzania, the programs cost about 50 percent of the total public expenditures on agriculture. Such high expenditures on fertilizers alone have serious implications for spending on other agricultural public goods and investments—such as research, extension, infrastructure, and land and water management—that have been found to have high rates of return and longer-lasting benefits.

Overall, growth in the use of fertilizer has been slow in most African countries, because poor targeting of ISPs has crowded out commercial fertilizer, as the bulk of the subsidized fertilizer has been provided to farmers who would have purchased it regardless (Jayne et al. 2013). All of the evidence presented suggests that the current approach to promoting fertilizer use within the broader context of African agricultural development could result in, yet again, a frustrated attempt to replicating some of the lessons of the Asian Green Revolution for a sustainable African agricultural intensification.

Key issues for agricultural intensification in Africa

From the above literature review, two key relationships need to be understood well in the African context and considered in identifying the best approach for an African agricultural intensification. First is the relationship among factor scarcity or abundance, technology, input, and factor use intensity. Second is the demand for different technologies under different relative factor use intensities. A fundamental set of questions is, which factor is scarce or abundant, where is it scare or abundant, why is it scare or abundant, and how has the relative scarcity or abundance changed over time?

Despite rapid population growth, many parts of Africa are characterized by natural resource abundance, low population density, high labor cost, and low labor productivity (Binswanger and Pingali 1988; Woodhouse 2009; Gollin, Jedwab, and Vollrath 2013), suggesting that shifting cultivation may still be a viable system of farming in many of those areas. But the combination of increasing food deficits and rapidly growing urban demand, in addition to

3 All currency is in US dollars, unless specifically noted as "international dollars."

the socioeconomic and agroecological differences across African countries, also suggests that different intensification paths in different cereals and non-cereal staples in different areas will be critical. For instance, cassava production has expanded in different parts of Africa as a food security crop, replacing fallow. Generally, cassava can give reasonable yields in soils of low fertility, and is thought to require less labor per unit of output than most other major staples. Furthermore, expansion of cassava production in Africa appears to be leading to greater labor productivity in SSA (Hillocks 2002).

The remainder of this chapter analyzes these issues systematically, starting in the next section with a presentation of the specific measures of intensification used, followed by the data and estimation techniques. To differentiate the resulting intensification paths for different socioeconomic and agroecological conditions across different parts of Africa, the empirical analysis is applied to the maize mixed and highland temperate mixed farming systems.

Empirical Approach

Intensification indicators

Several indicators have been proposed to measure intensification in agricultural production. Some of the most commonly used indicators try to capture the intensity of land use by looking at the length of the cultivation and fallow periods. One of these indicators is the ratio R, calculated as the ratio of the length of the cultivation period to the total length of the cycle of land utilization, defined as the length of the cultivation period plus the length of the fallow period (Ruthenberg 1980). When R is less than 33, the corresponding system is classified as shifting cultivation or long-fallow agriculture. An R value between 33 and 66 is used to indicate a short-fallow, semi-permanent cultivation. When R is greater than 66, the system is classified as permanent cultivation with either single cropping or various degrees of multiple cropping. Cropping intensity measures the intensity of land use under cultivation as the ratio between gross and net cropped area, varying from 100 to 200 if there is complete double cropping.

Information to estimate these indicators is not always available, especially when comparing agriculture sectors across countries. On the other hand, these indicators present only a fragmented view of the process of intensification and its changes across time. For the purpose of this study, we propose an indicator to measure intensity of agricultural production at the sectoral level that can be decomposed into a set of other indicators reflecting the level of intensification reached by a particular country and the factors driving intensification.

Our overall intensification indicator (intensity index) is the ratio of total agricultural output and total stock of agricultural land in a country, including land under cultivation and also land not incorporated into production. This measure reflects the intensity of use of the available land in a country, and implies that intensification (1) could be increased by simply incorporating new land into production or by reducing the fallow period, or (2) could also result from a more intensive use of land under cultivation. Given the different nature of countries' production processes, we decompose this index into crop and livestock intensity indexes as follows:

$$AI = YT/TPA = CII \times LII = Yc/TPA + Ylv/TPA \qquad (5.1)$$

where AI is the agricultural intensity index; YT is total agricultural production; TPA is the total agricultural potential area or the total stock of agricultural land in the country; CII and LII are, respectively, the crop and livestock intensity indexes; and Yc and Ylv are crop and livestock outputs, respectively.

We further decompose CII as follows:

$$CII = CPA/TPA \times [Ar/CPA \times Ah/Ar \times Yc/Ah] \qquad (5.2)$$

where CPA is the stock of land suitable for crop production, so that CPA/TPA is a measure of the quality or potential of agriculture in the country and determines the contribution of crop intensification to overall agricultural intensification. The first term in the brackets is the ratio between arable land (Ar) and total land suitable for agriculture, which could be thought of as an indicator of land abundance. The second term in the brackets is the ratio of harvested land (Ah) and arable land (Ar), which is an indicator that could be used as a proxy for cropping intensity, as normally defined, the ratio of gross and net cropped area. The last term, Yc/Ah, reflects land productivity and measures crop output per hectare of harvested land.

We expect that in densely populated countries, Yc/Ah would contribute the largest share to crop intensification. On the other hand, we expect that crop production in sparsely populated countries would increase through a combination of more land being incorporated into crop production, and a more intensive use of that land (increasing double cropping, for example).

Finally, the livestock production intensity index has two components:

$$LII = Ylv/SK \times SK/TPA \qquad (3)$$

where SK is animal stock measured in tropical livestock units.[4] Comparing this index with the crop index, SK/TPA is the equivalent to land being incorporated into livestock production, while Ylv/SK, output per animal, is a productivity measure. Intensification in livestock production at low levels of population density is expected to occur through increases in SK/TPA, with no major changes in animal productivity. Increased animal productivity would require more inputs per animal, similarly to what is needed to increase yields in crop production.

Data and estimation

We use data for 40 SSA countries from the Food and Agriculture Organization of the United Nations (FAO 2013), which provides national time-series data from 1961 to 2011 for the total quantity of different agricultural input and output volumes measured in 2004–2006 US$. Total agricultural output is the sum of crop and livestock output. Inputs are labor, measured as the total economically active population in agriculture; fertilizer (metric tons of nitrogen, potash, and phosphates used measured in nutrient-equivalent terms); animal stock, which includes cattle, sheep, goats, pigs, and chickens aggregated as the total number of cow equivalents; and capital, which is estimated by FAO using physical data on livestock, tractors, irrigated land and land under permanent crops, etc., and the average prices for 1995.

We also use three measures of land: arable land, which is land under temporary agricultural crops (multiple-cropped areas are counted only once); harvested area, which is the area from which a crop is gathered; and temporary meadows for mowing or pasture, land under market and kitchen gardens, and land temporarily fallow (less than five years). The abandoned land resulting from shifting cultivation is not included in this category. If the crop under consideration is harvested more than once during the year as a consequence of successive cropping, the area is counted as many times as harvested. On the contrary, area harvested will be recorded only once in the case of successive gathering of the crop during the year from the same standing crops. Finally,

---

4 The tropical livestock unit is used for aggregating different types of livestock in mostly the tropics, and is based on the equivalent of one cow with an average weight of 250 kg. Other types of livestock are given a coefficient relative, which is their weight relative to that of the cow. Different terms are used in other areas, such as *livestock unit* in the United Kingdom, which is based on the dairy cow with an average weight of 600–650 kg; and *animal unit* in the United States, which is based on the beef cow with an average weight of 455 kg. As such, "cow equivalent" is the general term that is used. See Chilonda and Otte (2006), for example, for further discussion on different terms.

total agricultural land is total land being used in production, and is the sum of arable land and pasture land, which is land used permanently (five years or more) to grow herbaceous forage crops, either cultivated or growing wild (wild prairie or grazing land).

Data on potential agricultural land and land suitability were provided by the International Food Policy Research Institute's Spatial Production Allocation Model for all countries in SSA at the pixel level (discussed in Chapters 3 and 4), and were aggregated to be used at the country level in this study. Total agricultural land is identified according to topographic characteristics, length of growing period, volume of annual rain, etc., and is classified according to its suitability for agricultural production in six categories:

- S1 = Land very poorly suited for pasture and at best poorly suited for rainfed crops
- S2 = Land poorly suited for pasture and at best poorly suited for rainfed crops
- S3 = Land suited for pasture and at best poorly suited for rainfed crops
- S4 = Land suited for rainfed crops and pasture possible
- S5 = Land well suited for rainfed crops and pasture possible
- S6 = Prime land for rainfed crops and pasture possible

Land variables in equations (5.1)–(5.3) are defined as follows: potential agricultural area or the total stock of agricultural land is TPA = S1 + S2 + S3 + S4 + S5 + S6, and potential crop area is CPA = S4 + S5 + S6.

Information on fertilizer use and determinants was obtained from Heston, Summers, and Aten (2012); FAO (2013); and World Bank (2013).

Results

Agricultural intensification: present levels and trends

To shed light on the effect of population density on the intensity of output production and input use, we sort countries by their population density in 1995 (the beginning of the period covered in this study), and define four groups, each containing the same number of countries, with group 1 (G1) including countries with the lowest population densities and group 4 (G4) including those with the highest population densities.

Table 5.1 presents the values of population density and different measures of output per hectare. The first measure uses total available agricultural land; the second measure uses the same area, but adjusted by quality (number of hectares equivalent to land well suited for crop production); and the last measure employs actual agricultural land used in production. Estimates of output per hectare are presented for the 40 countries included in the analysis. We will refer to the quality-adjusted total available land as potential agricultural area (TPA_{adj}). The measure of population density used here is calculated as total rural population divided by TPA_{adj}.

The first thing to notice is that highest population pressure on agricultural land in SSA (G4) occurs in Burundi, Rwanda, Kenya, Uganda, Ethiopia (East Africa); Nigeria, The Gambia, Guinea-Bissau (West Africa); and Malawi and Swaziland (southern Africa). On the other end of the ranking of countries, G1 includes countries with forest-based agriculture (Republic of the Congo, Central African Republic, and Gabon); semiarid countries with very low population density (Botswana, Chad, and Namibia); and large southern countries with high agricultural potential (Angola, Mozambique, and Zambia). The intermediate groups (G2 and G3) are composed mostly of West African countries; southern African countries (South Africa, Zimbabwe, and Madagascar); and East African countries (Sudan, Tanzania, and Somalia). Within these two groups, Sierra Leone has the highest population density (0.68), while Sudan has the lowest population density (0.20).

According to the numbers in Table 5.1, and despite some expected variability in part explained by land quality, our measure of output per hectare of potential agricultural land is clearly related to population density, as shown in Table 5.2. Correlation values in the last two rows of the table show that the expected relationship between population density and production per hectare holds, and is highly significant for the measures using potential area. Average values of output per hectare seem to show evidence of the existence of population density thresholds for intensification, given the differences in output and input per hectare between G4 and all other groups. Output per hectare of potential agricultural area is three times larger in G4 ($175) than in G2 and G3, which show almost the same values ($61 and $67, respectively). There are also large differences between G1 ($17) and all other groups.

How does intensity in the use of inputs relate to population density? Table 5.3 shows values of population density, fertilizer, and capital per hectare of different measures of agricultural area. Values for intensity of input use show greater variability at similar levels of population density than output per hectare. Without considering South Africa, which is an outlier in this sample,

TABLE 5.1 Population density and output per hectare of agricultural area (average values for 1995–2000)

Quantile	Country	Population density	YT/TPA	YT/TPA_{adj}	YT/Ar
G1	Central African Republic	0.047	$12	$14	$122
	Namibia	0.068	$5	$19	$8
	Botswana	0.078	$3	$16	$6
	Gabon	0.104	$51	$68	$34
	Zambia	0.115	$11	$14	$32
	Angola	0.117	$11	$16	$16
	Congo, Rep.	0.127	$17	$20	$18
	Chad	0.139	$13	$29	$23
	Mozambique	0.183	$21	$23	$31
	Liberia	0.198	$26	$40	$95
G2	Sudan	0.202	$30	$56	$44
	Mali	0.239	$22	$63	$52
	Mauritania	0.257	$9	$65	$9
	Madagascar	0.263	$42	$63	$61
	Equatorial Guinea	0.294	$25	$34	$95
	Cameroon	0.297	$83	$96	$269
	Côte d'Ivoire	0.329	$160	$177	$235
	Benin	0.360	$150	$158	$540
	Congo, Dem. Rep.	0.364	$33	$39	$130
	Zimbabwe	0.373	$53	$86	$127
G3	Tanzania	0.412	$49	$62	$105
	Guinea	0.448	$42	$84	$73
	Burkina Faso	0.448	$61	$73	$152
	Niger	0.503	$21	$85	$36
	Senegal	0.515	$57	$98	$111
	Ghana	0.516	$173	$184	$264
	Somalia	0.606	$22	$187	$31
	Togo	0.611	$128	$148	$190
	South Africa	0.659	$82	$324	$95
	Sierra Leone	0.675	$36	$66	$89

Quantile	Country	Population density	YT/TPA	YT/TPA_{adj}	YT/Ar
G4	Guinea-Bissau	0.759	$77	$158	$106
	Gambia, The	0.980	$84	$130	$152
	Swaziland	1.005	$150	$324	$196
	Nigeria	1.065	$260	$341	$301
	Ethiopia	1.128	$22	$56	$67
	Uganda	1.181	$226	$267	$339
	Kenya	1.190	$71	$198	$138
	Malawi	1.463	$168	$222	$287
	Burundi	4.189	$294	$526	$321
	Rwanda	5.571	$400	$910	$521

Source: Author's calculations based on agricultural intensity index analysis.

Notes: Ar = arable land in hectares; TPA = total potential agricultural area in hectares; TPA_{adj} = TPA adjusted for quality: equivalent hectares of land well suited for rainfed crops (based on crop suitability index [CSI]: 50–80) and pasture possible (based on pasture suitability index [PSI] >0); population density = total rural population divided by TPA_{adj}; YT = total agricultural production in 2004–2006 US$ constant prices.

TABLE 5.2 Population density and output per hectare of different measures of agricultural area by quantile of population density and correlation values, 1995–2000

Indicators	Population density	YT/TPA	YT/TPA_{adj}	YT/Ar
Quantile				
G1	0.12	$17	$26	$39
G2	0.30	$61	$84	$156
G3	0.54	$67	$131	$114
G4	1.85	$175	$313	$243
Correlation between population density and averaged values, 1995–2000	1.00	0.985	0.99	0.872
P-value		0.015	0.010	0.128
Correlation between population density and pooled country-year values	1.00	0.81	0.88	0.65
P-value		0.000	0.000	0.000

Source: Author's calculations based on agricultural intensity index analysis.

Notes: Ar = arable land in hectares; TPA = total potential agricultural area in hectares; TPA_{adj} = TPA adjusted for quality: equivalent hectares of land well suited for rainfed crops (based on crop suitability index [CSI]: 50–80) and pasture possible (based on pasture suitability index [PSI] >0); population density = total rural population divided by TPA_{adj}; YT = total agricultural production in 2004–2006 US$ constant prices.

TABLE 5.3 Population density and inputs per hectare of different measures of agricultural area (average values for 1995–2000)

Quantile	Country	Population density	Fertilizer/ CPA	Capital/ CPA	Fertilizer/ Ar	Capital/ Ar	Fertilizer/ Ar_{adj}	Capital/ Ar_{adj}
G1	Central African Republic	0.047	0.010	0.012	0.222	0.253	0.380	0.433
	Namibia	0.068	0.017	0.048	0.203	0.584	0.455	1.308
	Botswana	0.078	0.748	0.057	13.244	0.983	32.479	2.411
	Gabon	0.104	0.121	0.136	0.617	0.695	1.227	1.382
	Zambia	0.115	0.926	0.036	15.782	0.609	28.666	1.106
	Angola	0.117	0.072	0.056	1.161	0.905	2.291	1.786
	Congo, Rep.	0.127	0.444	0.033	7.786	0.588	14.340	1.083
	Chad	0.139	0.390	0.036	3.731	0.345	6.476	0.599
	Mozambique	0.183	0.139	0.036	2.143	0.551	3.358	0.864
	Liberia	0.198	0.051	0.069	0.558	0.747	1.256	1.680
G2	Sudan	0.202	0.607	0.152	3.433	0.860	5.929	1.484
	Mali	0.239	1.634	0.124	8.859	0.673	17.596	1.337
	Mauritania	0.257	3.070	0.473	5.831	0.900	14.858	2.294
	Madagascar	0.263	0.317	0.271	3.076	2.630	5.779	4.941
	Equatorial Guinea	0.294	0.001	0.481	0.003	1.800	0.006	3.228
	Cameroon	0.297	1.602	0.129	5.559	0.446	8.886	0.713
	Côte d'Ivoire	0.329	2.989	0.236	11.656	0.919	19.750	1.557
	Benin	0.360	3.954	0.110	16.612	0.463	26.705	0.745
	Congo, Dem. Rep.	0.364	0.047	0.040	0.522	0.445	0.901	0.768
	Zimbabwe	0.373	8.397	0.056	48.586	0.327	98.968	0.666
G3	Tanzania	0.412	0.503	0.142	2.863	0.806	5.198	1.463
	Guinea	0.448	0.329	0.096	1.133	0.333	2.342	0.688
	Burkina Faso	0.448	1.810	0.054	9.520	0.288	16.982	0.513
	Niger	0.503	0.525	0.400	0.344	0.262	0.744	0.566
	Senegal	0.515	2.919	0.137	8.900	0.419	18.485	0.869
	Ghana	0.516	0.773	0.150	2.814	0.544	4.498	0.870
	Somalia	0.606	0.434	2.506	0.496	2.861	1.411	8.147
	Togo	0.611	3.930	0.167	7.019	0.299	12.008	0.511
	South Africa	0.659	48.221	1.365	55.109	1.559	127.255	3.601
	Sierra Leone	0.675	0.451	0.200	2.608	1.161	6.226	2.770

Quantile	Country	Population density	Fertilizer/ CPA	Capital/ CPA	Fertilizer/ Ar	Capital/ Ar	Fertilizer/ Ar_{adj}	Capital/ Ar_{adj}
G4	Guinea-Bissau	0.759	0.773	0.859	1.500	1.789	4.011	4.786
	Gambia, The	0.980	1.958	0.121	5.706	0.342	12.888	0.773
	Swaziland	1.005	8.661	0.556	27.634	1.775	69.736	4.480
	Nigeria	1.065	2.719	0.433	5.250	0.837	9.656	1.539
	Ethiopia	1.128	4.477	0.111	14.724	0.367	29.081	0.724
	Uganda	1.181	0.186	0.195	0.396	0.420	0.682	0.723
	Kenya	1.190	8.731	0.255	24.529	0.710	49.595	1.436
	Malawi	1.463	9.053	0.209	19.400	0.446	35.836	0.823
	Burundi	4.189	2.445	0.532	2.268	0.494	5.275	1.148
	Rwanda	5.571	0.502	0.593	0.337	0.392	0.843	0.980

Source: Author's calculations based on agricultural intensity index analysis.

Notes: Ar = arable land in hectares; Ar_{adj} = Ar adjusted for quality: equivalent hectares of land well suited for rainfed crops; fertilizer = amount in kg; capital = investment used for crop production in 2004–2006 US$ constant prices; CPA = potential land suitable for crop production in hectares; population density = total rural population divided by total potential agricultural area in hectares adjusted for quality (TPA_{adj}).

maximum observed values of fertilizer per hectare of CPA are 8–9 kg. Only 4 of the 10 countries in G4 appear among the group of countries using the highest levels of fertilizer (Malawi, Kenya, Swaziland, and Ethiopia). Other densely populated countries in G4, such as Nigeria, The Gambia, and Burundi, use less fertilizer per hectare of CPA than countries with much lower population pressure, such as Senegal, Côte d'Ivoire, Mauritania, Togo, and Benin.

Similar variability for similar levels of population density is observed in the use of capital. For example, Rwanda, the country with the highest population density in SSA, uses less fertilizer than Sudan, while Uganda, another highly populated country, shows one of the lowest levels of fertilizer use in SSA.

Average values for the use of inputs per hectare show no clear patterns across groups of population density (Table 5.4). Fertilizer and capital use per hectare of CPA is much lower in G1, but there is no clear pattern among the other three groups. For example, fertilizer and capital use per hectare is highest in G3, and no large differences are observed between G4 and G2. The normally used measures of fertilizer per hectare of arable land show almost no differences in the use of fertilizer and capital among G2, G3, and G4, and relatively small differences between these groups and G1, if we compare them with the differences observed when CPA is used. Correlation coefficients in the last rows of Table 5.4 confirm the large variability in the use of fertilizer for similar levels of population density. Only the measures that use CPA show

TABLE 5.4 Population density and inputs per hectare of different measures of agricultural area by quantile of population density (average values for 1995–2000)

Indicators	Population density	Fertilizer/ CPA	Capital/ CPA	Fertilizer/ Ar	Capital/ Ar	Fertilizer/ Ar_{adj}	Capital/ Ar_{adj}
Quantile							
G1	0.12	0.292	0.052	4.5	0.626	9.093	1.265
G2	0.30	2.262	0.207	10.4	0.946	19.938	1.773
G3	0.54	5.990	0.522	9.1	0.853	19.515	2.000
G4	1.85	3.950	0.386	10.2	0.757	21.760	1.741
Correlation between population density and averaged values, 1995–2000	1.000	0.435	0.507	0.522	–0.048	0.628	0.306
P-value		0.565	0.493	0.478	0.952	0.372	0.694
Correlation between population density and pooled country-year values	1.000	0.186	0.235	0.023	–0.119	0.035	–0.063
P-value		0.000	0.000	0.549	0.000	0.360	0.154

Source: Author's calculations based on agricultural intensity index analysis.

Notes: Ar = arable land in hectares; Ar_{adj} = Ar adjusted for quality: equivalent hectares of land well suited for rainfed crops; fertilizer = amount in kg; capital = investment used for crop production in 2004–2006 US$ constant prices; CPA = potential land suitable for crop production in hectares; population density = total rural population divided by total potential agricultural area in hectares adjusted for quality (TPA_{adj}); G1 = countries with population density less than 0.2/ha; G2 = countries with population density greater than 0.2/ha and up to 0.4/ha; G3 = countries with population density greater than 0.4/ha and up to 0.75/ha; G4 = countries with population density greater than 0.75/ha.

the expected sign, although correlation is low (0.18 and 0.23 for fertilizer and capital, respectively). The measures using arable land show very low and insignificant coefficients in the case of fertilizer, or significant but negative coefficients in the case of capital.

Paths to increase intensification

We now look at the paths followed by different countries to increase intensification. The first part of Table 5.5 presents the decomposition in levels of total output per hectare of potential agricultural land for countries grouped by quantile of population density, while the second part shows total growth for each component for the period 1995–2011.

The differences in output per hectare of TPA observed in Table 5.1 can be explained first by looking at the crop and livestock components. The proportion of land suitable for crop production is similar across density groups, which means that on average there should not be significant differences between groups in the contribution of crop production to total output per hectare of potential land. We focus on crop production, as it is the driver of growth and intensification in all groups.

What explains the differences in the observed levels of crop output per hectare of CPA between groups? Only a small part of these differences is explained by the levels of crop output per hectare of harvested land. For instance, Yc/Ah for G4 is $631, while productivity of harvested land for G1 is $440, or 70 percent of G4's value. Conversely, crop output per hectare of potential arable land is only $24 for G1, or 7 percent of G4's value ($358). The differences between groups are explained by the proportion of potential arable land that is harvested (Ah/CPA), and by cropping intensity measured by the ratio of harvested to arable land. There is a vast potential to expand crop production in G1 and G2 countries, where only 7 and 20 percent of land suited for crop production is used, respectively. These values increase to about 43 and 54 percent in G3 and G4, respectively. Differences in cropping intensity (Ah/Ar) between groups are smaller, and they appear to be significant only between G4 and the rest (0.96 compared with 0.69 in G1 and 0.79 and 083 in G2 and G3, respectively).

Intensity in livestock production is mostly driven by the number of animals per hectare of TPA, as differences in output per head of animal stock are small. Output per animal is $81 in G4, $88 in G1, and $97 in G3. On the other hand, the number of animals per hectare of TPA is 0.06 in G1 and increases with population density, reaching 0.51 in G4.

The growth rates of the different components of agricultural intensity are presented in the bottom half of Table 5.5. Countries in G1 and G2 increased production in recent years by incorporating new arable land into crop production and by increasing cropping intensity, the ratio of harvested to arable land. With less land available and Ah/Ar close to 1, G3 and G4 are better suited to increase production using land-saving technologies that result in output growth per hectare of harvested land.

Figure 5.1 uses growth rates of the ratio of arable land used relative to potential arable land, cropping intensity, and output per hectare of harvested land from Table 5.5 to show the contribution of each of these variables to growth of output per hectare of potential crop land. The importance for G3 and G4 of increasing output per hectare (50 percent) and for G1 and G2 of cultivating more land and increasing cropping intensity is clear from the figure (70 percent of total growth).

Based on Figure 5.1, we can determine the intensification path that countries in SSA seem to follow. At very low levels of population density, the main contribution to intensification comes from increasing cropping intensity,

TABLE 5.5 Decomposition of total output per hectare of potential agricultural land (2008–2011) and growth rates of its different components by quantile of population density, 1995–2011

	Total	Crop contribution [a]						Livestock contribution		
Indicators	YT/TPA	Yc/TPA	CPA/TPA	Yc/CPA	Ar/CPA	Ah/Ar	Yc/Ah	YL/TPA	YL/SK	SK/TPA
Yield (US$/ha)										
G1	26	21	0.73	24	0.07	0.71	440	5	88	0.06
G2	80	66	0.70	81	0.20	0.79	475	14	83	0.16
G3	104	81	0.64	123	0.43	0.83	445	23	97	0.26
G4	286	245	0.71	358	0.52	0.96	631	42	81	0.51
Average	124	103	0.70	146	0.31	0.82	498	21	87	0.25
Growth rate (%)										
G1	54	61	—	60	12.67	25.41	11	31	−5	35.70
G2	34	30	—	30	8.35	5.54	4	61	23	21.83
G3	56	55	—	48	15.03	5.81	21	63	23	41.66
G4	68	68	—	74	29.84	0.86	25	70	7	55.11
Average	58	57	—	60	19.47	7.82	16	64	11	43.85

Source: Author's calculations based on agricultural intensity index analysis.

Notes: [a] The product of (1+growth rate) of Ah/CPA, Ah/Ar, and Yc/Ah equals (1+Yc/CPA growth rate). — = data not available; YT = total output in 2004–2006 US$; Yc = crop output in 2004–2006 US$; YL = livestock output in 2004–2006 US$; TPA = total potential agricultural area in hectares; CPA= potential land suitable for crop production in hectares; Ah = harvested area in hectares; Ar = arable land in hectares, including annual and permanent crops and land fallowed for less than five years; SK = animal stock in cow equivalents; G1 = countries with population density less than 0.2/ha; G2 = countries with population density greater than 0.2/ha and up to 0.4/ha; G3 = countries with population density greater than 0.4/ha and up to 0.75/ha; G4 = countries with population density greater than 0.75/ha.

which it is still low compared with other groups. Incorporation of new land into production and increasing yields show similar contributions (around 20 percent of total growth).

Several factors could explain differences in the rate at which new land is brought into production between G1 and G2. For example, very low densities and remoteness could be playing a role in countries in forest-based production systems and in some of the large semiarid countries in G1. Using available land next to roads and population centers more intensively could be the strategy when infrastructure is poor and returns to public investments are low. With higher population pressure, the contribution of new land to production increases, as is the case in G2, reducing the contribution of increased cropping intensity but keeping the contribution of yields at a similar level as in G1. When population density reaches 0.5 people per hectare as in G3, the contribution of yields suddenly jumps from 20 percent to 50 percent, and the importance of cropping intensity reduces substantially. At the highest levels of

FIGURE 5.1 Contribution of new arable land, cropping intensity, and output per hectare of harvested area to growth of crop output per hectare of potential cropland by quantile of population density, 1995–2011

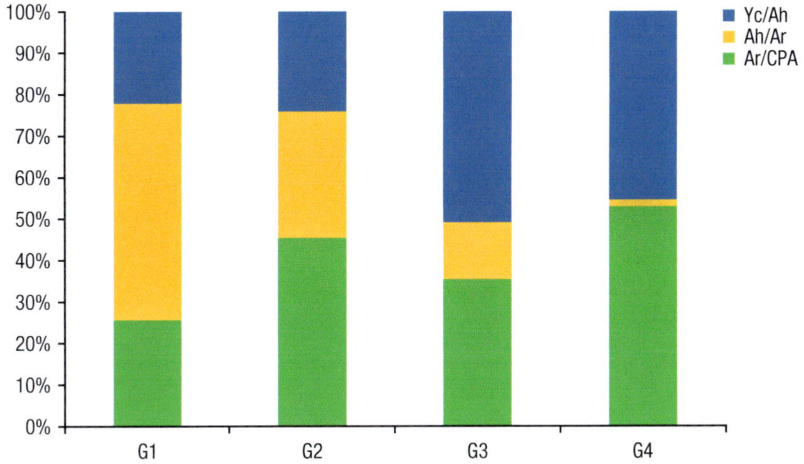

Source: Author's calculations and illustration based on agricultural intensity index analysis.

Notes: Ah = harvested area in hectares; Ar = arable land in hectares, including annual and permanent crops and land fallowed for less than five years; CPA= potential land suitable for crop production in hectares; Yc = crop output in 2004–2006 US$; G1 = countries with population density less than 0.2/ha; G2 = countries with population density greater than 0.2/ha and up to 0.4/ha; G3 = countries with population density greater than 0.4/ha and up to 0.75/ha; G4 = countries with population density greater than 0.75/ha.

population density, the contribution of cropping intensity becomes insignificant, but the contribution of yields remains at about 50 percent. Incorporation of new arable land still plays a significant role, increasing production even at high levels of population density.

These patterns of intensification should be reflected in the relative prices of land and labor at different levels of population density. To check this, we use shadow prices from data envelopment analysis distance functions used for total factor productivity estimation in previous chapters (Figure 5.2). It is interesting to notice that the relative price of labor is highest for G2, possibly accelerating the incorporation of new land into production as a way to increase labor productivity.

Figure 5.3 complements Figure 5.1, showing the evolution of the contribution of different factors to output per hectare of CPA in crop production. The contrast between G1 and the other three groups is clear. Cropping intensity and yields is where the action is in G1. In all other groups, the curve for

FIGURE 5.2 Shadow price of labor relative to land at different levels of population density, 1995–2011

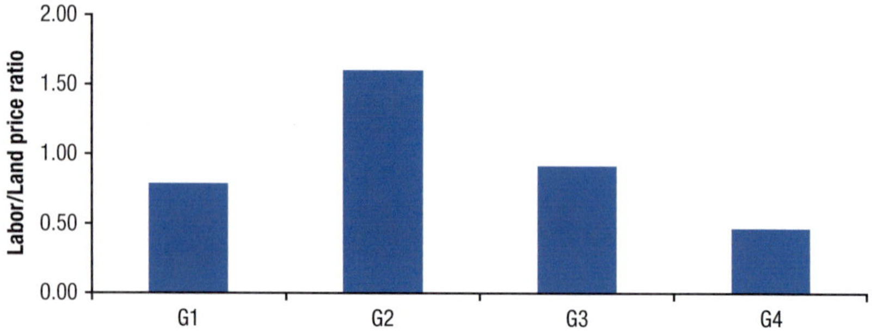

Source: Author's calculations and illustration based on agricultural intensity index analysis.

Notes: Shadow prices are obtained from data envelopment analysis estimates of distance functions used to calculate total factor productivity. G1 = countries with population density less than 0.2/ha; G2 = countries with population density greater than 0.2/km and up to 0.4/ha; G3 = countries with population density greater than 0.4/ha and up to 0.75/ha; G4 = countries with population density greater than 0.75/ha.

cropping intensity shows little growth and is replaced by the incorporation of new arable land into production (Ar/CPA), including G4.

According to these results, between 1995 and 2011, countries in G1 and G2 increased output following a "land-abundant" path that includes (1) incorporating more land into crop production and (2) increasing cropping intensity by reducing fallow periods and/or double cropping. Beyond densities of 0.5 people per hectare of potential agricultural land, the contribution of yields substantially increases. What is still puzzling is the persistence of the contribution of new land to production, even at the highest levels of population density. We provide more information on these issues by looking at the intensification paths followed by individual countries.

Table 5.6 shows the structure of intensity of all countries, while Figure 5.4 decomposes the contribution of increased arable land, cropping intensity, and yields to growth in cropping intensity, like in Figure 5.1, but in this case at the country level. The importance of incorporating new land into production even at high levels of population density is clear in G4. Six countries show a contribution of about 40 percent or more to crop production coming from incorporating arable land, including Rwanda, the country with the highest population density. The exceptions are Burundi, with less than 10 percent contribution of new land, and Swaziland, with no arable land incorporated

FIGURE 5.3 **Patterns of the contribution of different components to growth of total crop output per hectare of potential crop land by quantile of population density, 1995–2011**

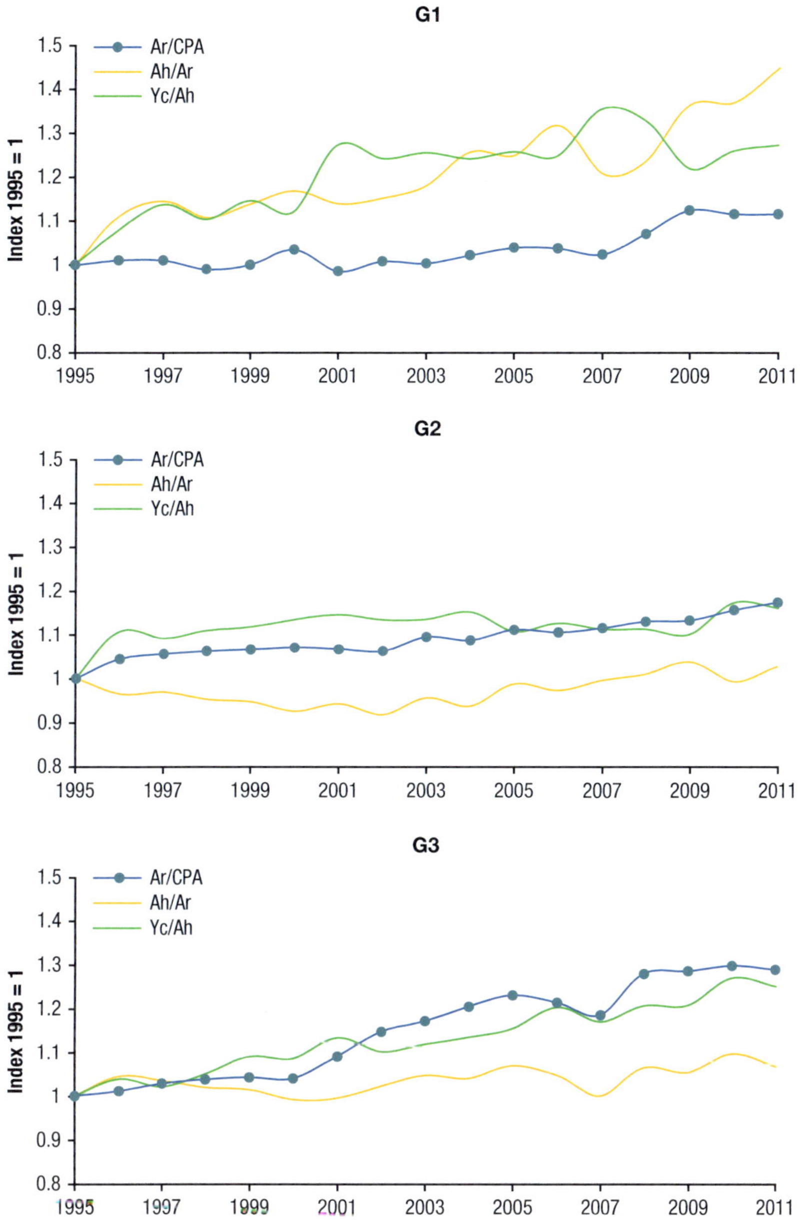

(continued)

FIGURE 5.3 (continued)

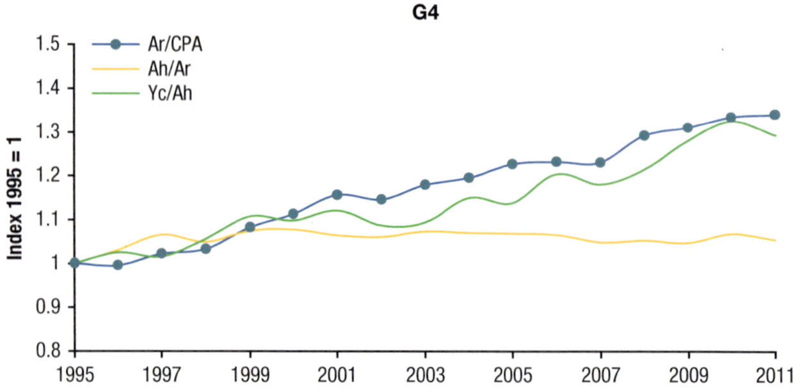

Source: Author's calculations and illustration based on agricultural intensity index analysis.

Notes: Ah = harvested area in hectares; Ar = arable land in hectares, including annual and permanent crops and land fallowed for less than five years; CPA= potential land suitable for crop production in hectares; Yc = crop output in 2004–2006 US$; G1 = countries with population density less than 0.2/ha; G2 = countries with population density greater than 0.2/ha and up to 0.4/ha; G3 = countries with population density greater than 0.4/ha and up to 0.75/ha; G4 = countries with population density greater than 0.75/ha.

into production. Kenya is an intermediate case, with a contribution of about 20 percent of new land to growth.

A similar pattern is observed in G3 in terms of the importance of the incorporation of new land into crop production. Note that the contribution of new land is particularly high in West African countries, such as Burkina Faso, Ghana, Togo, and Senegal. There is more variability than in G4 in the contribution of yields, and there is still a significant contribution of cropping intensity, especially in the countries with the lowest population density levels within the group.

Expected patterns are observed in G1 and G2 countries. Large low-density countries with forest-based production systems, such as the Central African Republic, Republic of the Congo, and Gabon, and arid low-density Botswana and Namibia increased cropping intensity (Ah/Ar) instead of bringing new land into production. The contribution of new arable land increases in countries with the lowest population density in G1 (Chad, Mozambique, and Liberia). It extends to countries in G2, with such exceptions as Mauritania, a country with limited possibilities to expand cropped area; an outlier in agricultural production, such as Equatorial Guinea; and the cases of Democratic Republic of the Congo and Zimbabwe, a result that is probably related to the political situation in those countries.

TABLE 5.6 Decomposition of total output per hectare of potential agricultural land (2008–2011), and contribution of its different components to growth by country, 1995–2011

			Crop contribution [a]						Livestock contribution		
Quantile	Country	YT/TPA	CPA/TPA	Yc/TPA	Yc/CPA	Ar/CPA	Ah/Ar	Yc/Ah	YL/TPA	YL/SK	SK/TPA
G1	Central Afr. Rep.	16	0.98	9	9	0.04	0.49	488	7	69	0.10
	Namibia	4	0.20	1	3	0.06	0.46	100	4	82	0.05
	Botswana	4	0.17	0	1	0.03	0.71	60	4	62	0.06
	Gabon	67	0.94	62	66	0.15	0.51	854	5	92	0.05
	Zambia	23	0.93	17	18	0.06	0.58	498	6	94	0.07
	Angola	38	0.86	32	37	0.06	1.00	615	6	85	0.07
	Congo, Rep.	26	0.99	23	23	0.05	0.61	749	3	75	0.04
	Chad	14	0.43	9	20	0.12	0.79	213	6	47	0.12
	Mozambique	30	0.91	25	27	0.08	1.04	313	5	124	0.04
	Liberia	35	0.91	31	33	0.07	0.90	514	4	148	0.03
G2	Sudan	56	0.55	16	29	0.18	0.66	241	39	119	0.32
	Mali	40	0.36	24	66	0.21	0.87	359	16	95	0.17
	Mauritania	12	0.04	2	38	0.28	0.82	169	10	72	0.14
	Madagascar	59	0.75	45	61	0.10	0.88	697	14	72	0.20
	Equatorial Guinea	31	0.79	29	37	0.20	0.46	405	1	80	0.01
	Cameroon	151	0.87	130	150	0.29	0.77	671	21	86	0.24
	Côte d'Ivoire	182	0.99	175	175	0.25	1.02	676	7	79	0.09
	Benin	195	0.99	182	184	0.27	0.92	735	14	54	0.25
	Congo, Dem. Rep.	31	0.93	30	32	0.08	0.84	485	1	72	0.02
	Zimbabwe	44	0.76	27	35	0.16	0.70	312	17	99	0.17
G3	Tanzania	91	0.93	69	73	0.20	0.96	388	22	69	0.32
	Guinea	64	0.59	52	88	0.25	0.96	367	11	48	0.23
	Burkina Faso	97	0.96	68	71	0.25	1.04	271	29	57	0.51
	Niger	39	0.20	19	94	1.16	1.06	76	20	80	0.24
	Senegal	82	0.74	62	81	0.31	0.71	365	21	65	0.32
	Ghana	299	0.97	285	294	0.37	0.89	898	14	84	0.16
	Somalia	27	0.04	2	66	0.50	0.72	181	24	90	0.27
	Togo	157	0.96	134	140	0.54	0.63	409	23	91	0.25
	South Africa	109	0.21	51	239	0.53	0.41	1,100	58	318	0.18
	Sierra Leone	77	0.80	68	86	0.23	0.96	395	8	68	0.12

(continued)

TABLE 5.6 (continued)

		Crop contribution [a]							Livestock contribution		
Quantile	Country	YT/TPA	CPA/TPA	Yc/TPA	Yc/CPA	Ar/CPA	Ah/Ar	Yc/Ah	YL/TPA	YL/SK	SK/TPA
G4	Guinea-Bissau	130	0.79	104	130	0.32	0.91	446	26	59	0.44
	Gambia, The	134	0.92	113	120	0.49	0.97	254	21	41	0.52
	Swaziland	169	0.63	127	203	0.19	0.80	1,331	42	95	0.44
	Nigeria	344	0.89	306	341	0.52	1.04	634	39	91	0.42
	Ethiopia	43	0.39	30	77	0.34	0.95	240	12	23	0.53
	Uganda	304	0.93	245	262	0.52	0.87	579	59	105	0.57
	Kenya	120	0.38	62	164	0.31	0.84	628	57	130	0.44
	Malawi	358	0.88	323	366	0.54	1.02	662	35	113	0.31
	Burundi	353	0.77	326	427	0.73	0.93	629	27	56	0.47
	Rwanda	906	0.54	810	1488	1.29	1.27	902	96	100	0.96

Source: Author's calculations based on agricultural intensity index analysis.

Notes: [a] The product of (1+growth rate) of Ah/CPA, Ah/Ar, and Yc/Ah equals (1+Yc/CPA growth rate). YT = total output in 2004–2006 US$; Yc = crop output in 2004–2006 US$; YL = livestock output in 2004–2006 US$; TPA = total potential agricultural area in hectares; CPA= potential land suitable for crop production in hectares; Ah = harvested area in hectares; Ar = arable land in hectares, including annual and permanent crops and land fallowed for less than five years; SK = animal stock in cow equivalents.

Intensification and fertilizer use

Analysis in previous sections shows that the framework developed by Boserup (1965), Ruthenberg (1980), and others with a focus on population density is a powerful tool to explain agricultural growth and intensification in Africa. Although the overall results obtained can be explained by the conceptual framework used in this study, some issues with policy implications are still a puzzle. The most important of these is fertilizer use and its low correlation with population density in the African context.

Table 5.7 shows that there is no correlation between fertilizer use per hectare of arable land and population density. On the other hand, it also reports a significant but low correlation between population density and fertilizer per hectare of potential cropland. A possible interpretation of this relation is that population pressure on natural resources increases fertilizer use, but not necessarily the amount used per hectare of arable land. In other words, and as it happens with other inputs, incorporating more land into production could increase overall fertilizer use, but at the same rates of application per hectare of arable land.

Table 5.7 depicts the correlation between two measures of fertilizer use per hectare and the different factors contributing to intensification in crop

FIGURE 5.4 Contribution of new arable land, cropping intensity, and output per hectare of harvested area to growth of crop output per hectare of potential cropland by country and quantile of population density, 1995–2011

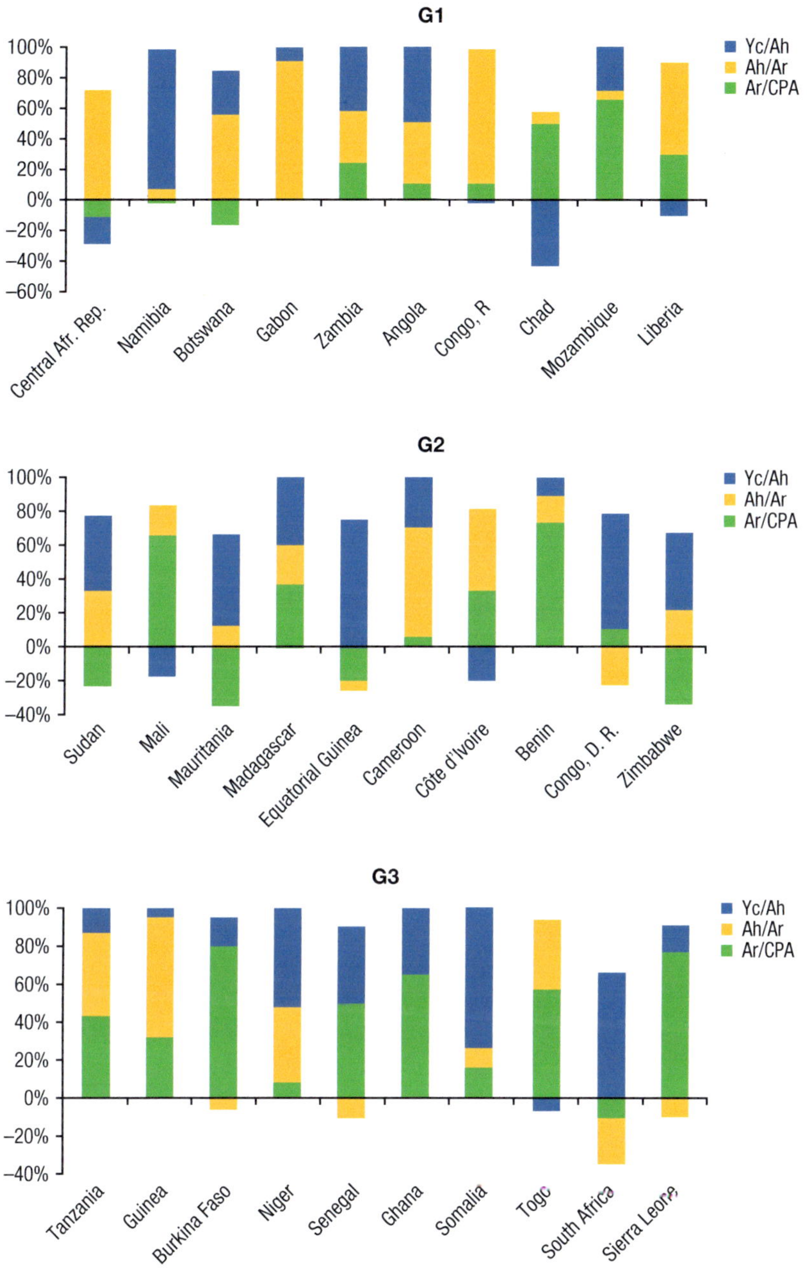

(continued)

FIGURE 5.4 (continued)

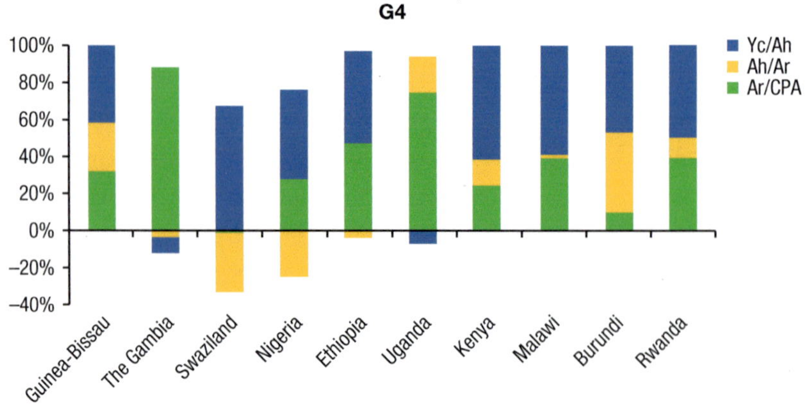

Source: Author's calculations and based on agricultural intensity index analysis.

Notes: Ah = harvested area in hectares; Ar = arable land in hectares, including annual and permanent crops and land fallowed for less than five years; CPA= potential land suitable for crop production in hectares; Yc = crop output in 2004–2006 US$; G1 = countries with population density less than 0.2/ha; G2 = countries with population density greater than 0.2/ha and up to 0.4/ha; G3 = countries with population density greater than 0.4/ha and up to 0.75/ha; G4 = countries with population density greater than 0.75/ha.

production. A comparison of the overall correlation of the two fertilizer measures with the different variables shows that fertilizer per hectare of CPA is correlated with population density, with crop output per hectare of CPA, with arable land use per hectare of CPA, and with output per hectare of harvested area. There is a similar pattern of correlation in the case of fertilizer per hectare of arable land, except that no correlation exists with the proportion of arable land used. This could mean that higher use of available arable land is not related to higher fertilizer use per hectare.

Focusing on fertilizer/CPA values in Table 5.7, we observe that correlations within groups show some contrasting patterns. There is no correlation between population density and fertilizer per hectare within the two extreme groups (G1 and G4). Within G1, high levels of fertilizer use are related to the proportion of arable land use relative to CPA, but not to yields. In the case of G4, the opposite is observed: the proportion of arable land used is not correlated with fertilizer use, but higher yields are expected in countries with high levels of fertilizer use.

There are also some interesting contrasts between G2 and G3. As in G1, fertilizer use in both groups is correlated to the proportion of potential arable land being used, but yields relate to fertilizer use only in G3. The role of cropping intensity is also different in G2 and G3, as it is positively correlated

TABLE 5.7 Correlation coefficients of different components of intensification and fertilizer use

	Fertilizer/CPA					Fertilizer/Ar				
Indicators	G1	G2	G3	G4	All	G1	G2	G3	G4	All
Population density	−0.05	0.25	0.45	0.01	0.18	−0.18	0.27	0.40	−0.33	0.02
p-value	0.49	0.00	0.00	0.92	0.00	0.02	0.00	0.00	0.00	0.55
Yc/CPA	$0.22	$0.31	$0.69	$0.08	$0.30	$−0.24	$0.11	$0.67	$−0.27	$0.12
p-value	0.00	0.00	0.00	0.29	0.00	0.00	0.14	0.00	0.00	0.00
Ar/CPA	0.30	0.26	0.19	−0.01	0.29	−0.19	0.02	0.12	−0.46	0.05
p-value	0.00	0.00	0.01	0.94	0.00	0.01	0.83	0.11	0.00	0.22
Ah/Ar	−0.15	0.31	−0.71	0.13	−0.04	−0.15	0.23	−0.66	−0.26	−0.09
p-value	0.04	0.00	0.00	0.09	0.24	0.05	0.00	0.00	0.00	0.02
Yc/Ah	$0.11	$0.08	$0.71	$0.18	$0.37	$−0.22	$0.01	$0.72	$0.45	$0.37
p-value	0.15	0.27	0.00	0.02	0.00	0.00	0.91	0.00	0.00	0.00

Source: Author's calculations based on agricultural intensity index analysis.

Notes: Ah = harvested area in hectares; Ar = arable land in hectares, including annual and permanent crops and land fallowed for less than five years; CPA = potential land suitable for crop production in hectares; fertilizer = amount in kg; Yc = crop output in 2004–2006 US$; population density = total rural population divided by total potential agricultural area in hectares adjusted for quality (TPA_{adj}); G1 = countries with population density less than 0.2/ha; G2 = countries with population density greater than 0.2/ha and up to 0.4/ha; G3 = countries with population density greater than 0.4/ha and up to 0.75/ha; G4 = countries with population density greater than 0.75/ha.

with fertilizer use in G2, and high and negatively correlated with fertilizer use in G3.

The analysis so far has shown that we expect high yields in densely populated countries in SSA (those in G3 and G4), and that yields should be correlated with relatively high levels of fertilizer use per hectare. So what explains the observed variability between population density, output per hectare, and fertilizer use among these countries?

Table 5.8 shows population densities, output per hectare of harvested land, and fertilizer per hectare of arable land sorted by output per hectare. On average, Rwanda, Burundi, Uganda, and Nigeria use less than 4 kg/ha of fertilizer with a population density of three people/ha, compared with 22 kg/ha in Malawi, Kenya, and Ethiopia. Why do countries with the highest levels of population density, such as Rwanda, Burundi, Uganda, and Nigeria, use low levels of fertilizer compared with other countries in the same range of

TABLE 5.8 Population densities, output per hectare of harvested land, and fertilizer per hectare of arable land, 1995–2011

Quantile	Country	Population density	Output per hectare (Yc/Ar)	Fertilizer (kg per hectare)
G1	Central African Republic	0.047	$483	0.3
	Namibia	0.068	$91	1.1
	Botswana	0.078	$68	17.5
	Gabon	0.104	$807	5.4
	Zambia	0.115	$451	14.7
	Angola	0.117	$443	4.1
	Congo, Rep.	0.127	$754	4.7
	Chad	0.139	$248	4.0
	Mozambique	0.183	$307	2.8
	Liberia	0.198	$569	0.5
G2	Sudan	0.202	$237	3.6
	Mali	0.239	$376	10.9
	Mauritania	0.257	$148	6.5
	Madagascar	0.263	$631	2.9
	Equatorial Guinea	0.294	$335	0.0
	Cameroon	0.297	$610	5.7
	Côte d'Ivoire	0.329	$735	11.1
	Benin	0.360	$730	7.7
	Congo, Dem. Rep.	0.364	$497	0.6
	Zimbabwe	0.373	$394	26.0
G3	Tanzania	0.412	$365	3.1
	Guinea	0.448	$372	1.0
	Burkina Faso	0.448	$274	9.4
	Niger	0.503	$67	0.4
	Senegal	0.515	$325	7.6
	Ghana	0.516	$772	4.5
	Somalia	0.606	$162	0.5
	Togo	0.611	$420	6.0
	South Africa	0.659	$919	55.4
	Sierra Leone	0.675	$329	1.1

Quantile	Country	Population density	Output per hectare (Yc/Ar)	Fertilizer (kg per hectare)
G4	Guinea-Bissau	0.759	$379	3.6
	Gambia, The	0.980	$274	5.9
	Swaziland	1.005	$1,237	33.7
	Nigeria	1.065	$547	7.7
	Ethiopia	1.128	$195	14.3
	Uganda	1.181	$588	1.2
	Kenya	1.190	$543	25.4
	Malawi	1.463	$508	25.3
	Burundi	4.189	$605	1.7
	Rwanda	5.571	$671	4.6

Source: Author's calculations based on agricultural intensity index analysis.

Notes: Ar = arable land in hectares, including annual and permanent crops and land fallowed for less than five years; Yc = crop output in 2004–2006 US$; population density = total rural population divided by total potential agricultural area in hectares adjusted for quality (TPA_{adj}).

population density, such as Malawi, Kenya, and Ethiopia? Why do some countries in G1 and G2 that use relatively high levels of fertilizer produce more output per hectare of potential agricultural land (Zambia, Botswana, Mali, Côte d'Ivoire, and Zimbabwe)? Why do Burkina Faso and Senegal use twice as much fertilizer as other countries in G3?

As developing a model explaining fertilizer use in SSA is beyond the scope of this study, we compare the mean values of several variables that are expected to be related to fertilizer use (and in general to intensification). These are variables representing the importance of the country's target markets (domestic and international), infrastructure, and quality of natural resources. A "tropicality" index (TI) that intends to capture agroecological conditions for production in different countries is calculated as the output of root crops and fruits divided by cereal output. A high TI is an indicator of the relative advantage of the country to produce root crops, fruits, and other tropical tree crops typical of tree crop, forest-based, and cereal-root crop mixed farming systems. As root and tree crops in Africa respond less to fertilizer and are expected to benefit less than cereals from research and development (R&D) spillovers, we expect a high TI to be associated with low levels of fertilizer use.

We look first at the differences in fertilizer use among countries in G4, the group of densely populated countries. Table 5.9 shows the variables expected to affect fertilizer use for the four countries using the lowest levels of fertilizer in the group: Burundi, Rwanda, Uganda, and Nigeria. The values of the different variables for these four countries are compared with the average values

TABLE 5.9 Variables expected to affect fertilizer use, showing countries with high population density and low fertilizer use (all countries in G4)

Indicators	Burundi	Rwanda	Uganda	Nigeria	Average	Rest of G4	
Fertilizer/ha	1.5	11.9	2.1	12.0	6.9	19.6	
Population density per km	4.2	5.6	1.2	1.1	3.0	1.1	
Tropicality index[a]	12.1	14.4	5.8	3.6	9.0	1.5	
Income per capita	$392	$940	$1,064	$1,775	$1,043	$1,308	
Market size[b]	266	2,500	283	2,084	1,283	786	
% of urban population	10.5	18.5	13.1	48.7	22.7	27.6	
Exports/output	$0.1	$0.1	$0.1	$0.0	$0.1	$0.4	
Road density per km	6.2	7.3	5.1	2.2	5.2	2.6	
Travel time (hrs)[c]	4.7	4.7	4.5	3.5	4.4	6.2	
R&D intensity[d]	$2.6	$0.8	$1.8	$2.5	$1.9	$1.9	
Potential arable land[e]	0.8	0.5	0.9	0.9	0.8	0.7	
						Two-sample *t*-test[f]	
Ratios average/rest of G4	Burundi	Rwanda	Uganda	Nigeria	Average	*t*-test	p-value
Fertilizer/ha	0.08	0.61	0.11	0.61	0.35	−5.1382	0.000
Population density per km	3.85	5.12	1.09	0.98	2.76	6.2399	0.000
Tropicality index	7.95	9.45	3.81	2.36	5.89	9.993	0.000
Income per capita	$0.30	$0.72	$0.81	$1.36	$0.80	−1.2168	0.229
Market size	0.34	3.18	0.36	2.65	1.63	2.2223	0.030
% of urban population	0.38	0.67	0.47	1.76	0.82	−1.3998	0.166
Exports/output	$0.18	$0.15	$0.26	$0.05	$0.16	−4.8893	0.000
Road density per km	2.36	2.75	1.94	0.85	1.97	6.7229	0.000
Travel time (hrs)	0.77	0.75	0.73	0.57	0.70	−3.8666	0.000
R&D intensity	$1.39	$0.45	$0.96	$1.33	$1.03	−0.0142	0.989
Potential arable land	1.15	0.82	1.40	1.34	1.18	3.2045	0.002

Source: Author's calculations based on agricultural intensity index analysis.

Notes: [a] Measured as the ratio of outputs of different crops: TI = (cassava+other roots+fruits)/(maize+millet+rice+sorghum). [b] Urban population × GDP per capita/cropland equivalent. [c] Travel time to towns of 50,000 people. [d] Public expenditure in agricultural R&D in 2004–2006 US$ per hectare of quality-adjusted cropland. [e] The ratio of potential land suitable for crop production and total potential agricultural area, CPA/TPA. [f] Student *t*-test of the differences between means. G4 = countries with population density greater than 0.75/ha.

of the rest of the group. A two-sample *t*-test is included to check for statistically significant differences between the means of the two groups. Burundi, Rwanda, Uganda, and Nigeria, with better infrastructure (positive and significant difference in road density and negative and significant difference in travel time), better quality of natural resources (larger proportion of high-quality land for crop production in total agricultural area), a larger domestic market, and higher population density than the rest of the group, were expected to use more fertilizer than what they are actually using. No significant differences between groups were found in income per capita, urbanization, and R&D investment.

On the other hand, countries using low fertilizer levels export less than other countries (not surprising, as three of the four countries in the group are landlocked). According to the TI index, these countries are also producers of root crops, fruits, and tree crops, rather than cereals. Differences in agroecology could be part of the explanation of the differences in fertilizer use between groups.

The case of high fertilizer use by Burkina Faso and Senegal in G3 is shown in Table 5.10. These countries employ on average 8 kg/ha of fertilizer nutrients, compared with only 2 kg/ha in other countries in the group.

Income per capita, exports, R&D investment, and travel time to towns of 50,000 or more inhabitants are significantly different from those in other countries in G3, and help to explain differences in fertilizer use among groups. On the other hand, Burkina Faso and Senegal have poorer infrastructure and lower population density than other countries in the group—all factors that are expected to have a negative effect on the use of fertilizer. No significant differences were found in urbanization or in the size of the domestic market. As in the previous case, the TI is significantly lower for these two countries, which is probably related to the fact that the savanna agroecology, more favorable to producing cereals and cash crops, will demand higher levels of fertilizer for production.

The last case is the one of relatively high levels of fertilizer use among low-density countries. Table 5.11 shows that Botswana, Côte d'Ivoire, Mali, Zambia, and Zimbabwe use on average 15 kg/ha of fertilizer, compared with only 4 kg/ha among other countries in G1 and G2. Even without including Botswana, which could be seen as an outlier in this group of countries, average fertilizer use is 12 kg/ha, which is significantly higher than in the rest of the group. Both groups show similar population densities, income per capita, market size, urbanization, and infrastructure. Conversely, high-fertilizer users have less potential for agricultural production, as the potential area suitable for crop production is smaller than in the group of low-fertilizer users (the difference is not highly significant). Factors that appear to favor fertilizer use are a low TI and a higher share of exports in total production.

The recurrence of significant differences in TI between high and low fertilizer users suggests that production systems and agroecology play an important role in the low fertilizer use in Africa, all other things being unchanged.

TABLE 5.10 Variables expected to affect fertilizer use, showing countries with intermediate levels of population density and high fertilizer use (all countries in G3)

Indicators	Burkina Faso	Senegal	Average	Rest of G3	
Fertilizer/ha	10.78	5.51	8.14	2.31	
Population density per km	0.45	0.52	0.48	0.54	
Tropicality index[a]	0.05	0.56	0.30	2.22	
Income per capita	$900	$1444	$1,172	$901	
Market size[b]	167	769	468	549	
% of urban population	24	42	33	35	
Exports/output	$0.12	$0.16	$0.14	$0.10	
Road density per km	1.62	1.56	1.59	1.89	
Travel time (hrs)[c]	4.07	3.64	3.86	6.67	
R&D intensity[d]	$0.37	$1.10	$0.73	$0.55	
Potential arable land[e]	0.56	0.32	0.44	0.34	
				Two-sample *t*-test[f]	
Ratios Average/Rest of G3	Burkina Faso	Senegal	Average	*t*-test	p-value
Fertilizer/ha	4.67	2.39	3.53	11.78	0.0000
Population density per km	0.83	0.96	0.89	–3.66	0.0003
Tropicality index	0.02	0.25	0.14	–3.58	0.0005
Income per capita	$1.00	$1.60	$1.30	3.60	0.0004
Market size	0.30	1.40	0.85	–0.93	0.3524
% of urban population	0.70	1.20	0.95	–1.04	0.3020
Exports/output	$1.20	$1.58	$1.39	2.99	0.0033
Road density per km	0.86	0.83	0.84	–1.59	0.1148
Travel time (hrs)	0.61	0.55	0.58	–4.44	0.0000
R&D intensity	$0.68	$2.00	$1.34	4.68	0.0000
Potential arable land	1.64	0.93	1.28	2.06	0.0410

Source: Author's calculations based on agricultural intensity index analysis.

Notes: [a] Measured as the ratio of outputs of different crops: TI = (cassava+other roots+fruits)/(maize+millet+rice+sorghum). [b] Urban population × GDP per capita/cropland equivalent. [c] Travel time to towns of 50,000 people. [d] Public expenditure in agricultural R&D in 2004–2006 US$ per hectare of quality-adjusted cropland. [e] The ratio of potential land suitable for crop production and total potential agricultural area, CPA/TPA. [f] Student *t*-test of the differences between means. G3 = countries with population density greater than 0.4/ha and up to 0.75/ha.

TABLE 5.11 Variables expected to affect fertilizer use, showing countries with low population density and high fertilizer use (all countries in G1 and G2)

Indicators	Botswana	Côte d'Ivoire	Mali	Zambia	Zimbabwe	Average	Rest of G1–G2
Fertilizer/ha	18.98	9.60	8.04	13.00	8.02	11.53	3.68
Population density per km	0.08	0.33	0.24	0.12	0.37	0.23	0.19
Tropicality index[a]	0.02	7.44	0.20	0.52	0.62	1.76	9.24
Income per capita	$10,022	$1,284	$951	$1,377	$310	$2,789	$2,459
Market size[b]	2,292	459	196	127	72	629	869
% of urban population	60.00	49.43	34.69	35.48	37.52	43.43	45.27
Exports/output	$0.92	$0.58	$0.07	$0.00	$0.34	$0.38	$0.15
Road density per km	0.58	2.57	0.36	2.42	3.39	1.86	2.08
Travel time (hrs)[c]	14.23	4.32	13.97	10.68	4.99	9.64	9.85
R&D intensity[d]	$2.16	$0.91	$0.48	0$.09	$1.34	$1.00	$0.66
Potential arable land[e]	0.03	0.68	0.17	0.53	0.33	0.35	0.44

Ratios average/rest of G1–G2	Botswana	Côte d'Ivoire	Mali	Zambia	Zimbabwe	Average	Two-sample *t*-test[f]	
							t-test	p-value
Fertilizer/ha	5.162	2.610	2.186	3.535	2.182	3.135	9.835	0.0000
Population density per km	0.401	1.690	1.225	0.592	1.918	1.165	1.737	0.0847
Tropicality index	0.002	0.805	0.021	0.056	0.067	0.190	–2.989	0.0033
Income per capita	$4.076	$0.522	$0.387	$0.560	$0.126	$1.134	$0.533	$0.5951
Market size	2.636	0.528	0.225	0.146	0.083	0.723	–0.903	0.3684
% of urban population	1.325	1.092	0.766	0.784	0.829	0.959	–0.686	0.4937
Exports/output	$6.098	$3.865	$0.478	$0.000	$2.234	$2.535	$5.912	$0.0000
Road density per km	0.280	1.236	0.176	1.164	1.629	0.897	–0.645	0.5199
Travel time (hrs)	1.445	0.439	1.418	1.085	0.507	0.979	–0.287	0.7747
R&D intensity	$3.254	$1.373	$0.731	$0.143	$2.030	$1.506	$1.238	$0.2197
Potential arable land	0.075	1.547	0.394	1.202	0.759	0.795	–1.988	0.0489

Source: Author's calculations based on agricultural intensity index analysis.

Notes: [a] Measured as the ratio of outputs of different crops: TI = (cassava+other roots+fruits)/(maize+millet+rice+sorghum). [b] Urban population × GDP per capita/cropland equivalent. [c] Travel time to towns of 50,000 people. [d] Public expenditure in agricultural R&D in 2004–2006 US$ per hectare of quality-adjusted cropland. [e] The ratio of potential land suitable for crop production and total potential agricultural area, CPA/TPA. [f] Student *t*-test of the differences between means. G1 = countries with population density less than 0.2/ha; G2 = countries with population density greater than 0.2/ha and up to 0.4/ha.

TABLE 5.12 Comparison of average values of fertilizer use per hectare between the maize mixed farming system and other farming systems (2005–2011)

Farming system	Coefficient	Std. Err.	*t*-statistics	P>*t*
Root crop & cereal-root crop mixed	−16.9	4.94	−3.42	0.002
Tree crop & rice-tree crop	−16.9	6.51	−2.60	0.014
Forest based	−16.8	7.17	−2.34	0.026
Highland perennial & highland temperate mixed	−12.7	7.17	−1.77	0.086
Pastoral & agropastoral millet/sorghum	−14.3	5.12	−2.79	0.009
Constant term	21.7	3.93	5.52	0.000
Number of observations	39			
F(5, 33)	2.75			
P-value	0.03			
R-squared	0.29			
Adjusted R-squared	0.19			
Root mean square error	10.39			

Source: Author's calculations based on agricultural intensity index analysis.

Note: Results are differences with respect to mean fertilizer per hectare in the maize mixed farming system and those listed in the first column.

Going back to the typology and the production systems discussed in previous chapters, we expect lower levels of fertilizer use in the root crop and tree crop systems relative to cereal-based systems. Table 5.12 shows a comparison of the means of fertilizer per hectare in countries grouped by production system, which confirms this as a plausible hypothesis. The biggest difference in the use of fertilizer between maize mixed systems and others occurs with root crop and tree crop systems.

Where in SSA can we expect good agroecological conditions for a cereal Green Revolution? Table 5.13 shows the distribution of total land in SSA suited to crop production by country and production system. We assume that the maize mixed and highland temperate mixed systems have a comparative advantage for cereal production. Of the total land suited for crop production in SSA (under cultivation or not), 18 percent has an advantage for cereal production. About 70 percent of this land is located in 5 countries (Tanzania, Ethiopia, Zambia, Zimbabwe, and Mozambique), while 95 percent is in 10 countries (Kenya, Uganda, Sudan, Democratic Republic of the Congo, and Malawi).

We conclude that there is not a unique path of agricultural intensification in the region. The agroecological conditions for the expansion of a package of high-yielding cereal varieties and fertilizer are limited, and even when these

TABLE 5.13 Total land suitable for crop production under maize mixed and highland temperate mixed systems, compared with other systems by country

	Maize mixed and highland temperate mixed systems		Other systems			
Country	Crop area	Population density	Crop area	Population density	Total crop area	% MM-HT
Tanzania	42,800	0.34	24,200	0.51	67,000	63.9
Ethiopia	28,000	0.95	18,500	0.38	46,500	60.2
Zambia	28,000	0.15	28,500	0.09	56,500	49.6
Zimbabwe	21,100	0.32	5,557	0.12	26,657	79.2
Mozambique	15,600	0.15	48,600	0.24	64,200	24.3
Kenya	13,100	1.12	6,692	0.21	19,792	66.2
Uganda	10,900	1.09	6,267	2.03	17,167	63.5
Sudan	10,100	0.07	95,200	0.13	105,300	9.6
Congo, Dem. Rep.	8,080	0.24	87,200	0.26	95,280	8.5
Malawi	5,666	1.35	1,156	2.30	6,822	83.1
South Africa	4,908	0.72	19,200	0.15	24,108	20.4
Angola	1,477	0.45	70,300	0.06	71,777	2.1
Central Afr. Rep.	983	0.00	48,100	0.05	49,083	2.0
Swaziland	975	0.66	23	0.45	998	97.7
Namibia	605	0.10	13,900	0.02	14,505	4.2
Cameroon	247	0.36	25,600	0.25	25,847	1.0
Nigeria	213	0.43	74,000	0.96	74,213	0.3
Lesotho	142	0.66	23	1.04	165	86.4
Botswana	16	0.03	8,105	0.02	8,122	0.2
Other	—	—	277,700	0.57	277,700	0.0
Total of Sample	**192,913**	**0.48**	**858,823**	**0.53**	**1,051,736**	**18.3**

Source: Author's calculations based on agricultural intensity index analysis.
Notes: — = data not available; MM = maize mixed; TH = temperate highland.

conditions are met, differences in relative prices and in economic and institutional constraints will require different technological packages adapted to the needs of the different countries. At low levels of population density, agricultural output and labor productivity result from increased cropping intensity and incorporation of new land into production, with relatively low contribution of increased land productivity. With high population density, the contribution of land productivity increases. However, rather than being related to fertilizer technologies, it is the result of production systems based on crops (tree and root crops) that use land more intensively and are less responsive

FIGURE 5.5 Number of SSA countries with more than 50 percent of their national agriculture in a particular farming system, 2005–2011

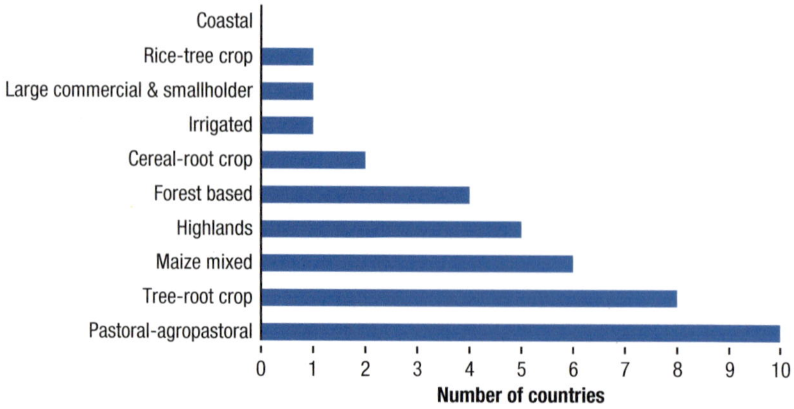

Source: Author's calculations based on agricultural intensity index analysis.
Note: SSA = Africa south of the Sahara.

than cereals to fertilizer, as is the case in the tree-root crop, cereal-root crop, forest based, and coastal systems, and in the banana+roots subsystem of the highland system.

The remarkable heterogeneity in farming systems observed within countries is also a major constraint for the development and diffusion of new technologies (including fertilizer-cereal technologies), assuming that these technologies need to be tailored to the relevant agroecological characteristics and production systems of the different countries. Figure 5.5 shows this heterogeneity by depicting the number of countries with more than 50 percent of their national agriculture in different farming systems, as defined in Chapter 4. Only 6 countries show more than 50 percent of national agriculture in the mixed-maize system, the most favorable system for the fertilizer-cereal technology. Among the major production systems, 10 countries have more than 50 percent of their national agriculture in the pastoral-agropastoral system, while 8 countries have more than 50 percent in the tree-root crop system.

Heterogeneity within these groups is still high, with different subsystems and large geographic distances between countries. For example, the pastoral-agropastoral system includes three subsystems. One of these subsystems includes West African countries only (Chad, Mali, Niger, and Senegal). The other two subsystems group southern African countries (Botswana, Namibia, and Angola), one West African country (Mauritania), and one East African country (Somalia). Substantial heterogeneity in the production environment constrains the possibility of R&D spillovers within and between countries, increasing research costs and thus making development and diffusion of new technologies more difficult.

Discussion and Conclusions

Although no definitive conclusions can be reached with these simple cross-country comparisons, our results suggest some hypotheses that could be tested with more detailed information at the country level in future studies.

First, we have shown the intensification paths followed by SSA countries in recent years. Countries with population densities below 0.4 people per hectare of total agricultural land both increased output following a clear "land-abundant" path that includes more land incorporated into crop production, and increased cropping intensity by reducing fallow periods and/or double cropping. Beyond densities of 0.5 people per hectare of potential agricultural land, the contribution of yields substantially increases. However, the incorporation of new land into production still contributes to output growth, even at the highest levels of population density. This means that at least 50 percent of SSA countries clearly have abundant land, showing evidence of high labor prices relative to those of land, while the other half is more land constrained. However, in most cases bringing new land into production is still a significant component of output growth.

Second, we find that the relation between fertilizer use per hectare and population density is positive as expected, but low, implying that there is high variability in the use of fertilizer at different levels of population density. A first explanation for this finding is the observed fact that within a wide range of population densities, fertilizer is not a major factor contributing to output growth (in most SSA countries). At low population densities, population pressure on natural resources increases fertilizer use, but not necessarily fertilizer intensity per hectare of arable land. As it happens with other inputs at low population densities, incorporating more land into production could increase overall fertilizer use, but at the same rates of application per hectare of arable land. In other words, fertilizer is an instrument for land expansion and not for yield increases. There is no correlation between fertilizer use and output per hectare at these density levels. It is only at high-density levels that fertilizer use is correlated with increased yields.

Another factor explaining the low correlation between fertilizer per hectare and population density is related to the different agroecologies and the production systems in SSA. Lower levels of fertilizer use are expected in root crop– and tree crop–based systems, compared with levels in maize mixed and highland farming systems, with comparative advantages for cereal production. Of the total land suited for crop production in SSA (under cultivation or not), only 18 percent is better suited for cereals than for root and tree crops, with 70 percent of this land located in five countries: Tanzania, Ethiopia, Zambia, Zimbabwe, and Mozambique.

Adding to this, the remarkable heterogeneity in farming systems is also a major constraint for the development and diffusion of new technologies (including fertilizer-cereal technologies), assuming that these technologies need to be tailored to the relevant agroecological characteristics and production systems of the different countries. In this context, we cannot expect a unique path of agricultural intensification in the region. The agroecological conditions for the expansion of a package of high-yielding cereal varieties and fertilizer are limited, and even when these conditions are met, differences in relative prices and in economic and institutional constraints will require different technological packages adapted to the needs of the different countries.

If true, the policy implications of these results are significant. First, expecting the Asian-style Green Revolution in SSA is, at best, misguided. The agroecological possibilities for it are limited, and low population densities in regions with advantages for cereal production do not make the technology attractive, unless they are complemented by capital investments that increase labor productivity. The best possibilities of success for the fertilizer technology at present using a fertilizer-focused technology package are in Ethiopia, Kenya, Uganda, and Malawi. These countries have more than 60 percent of potential agricultural land in favorable agroecologies and high population densities in those areas. For other cereal-producing countries, the best strategy seems to be the promotion of labor-saving technologies that accelerate the incorporation of new land into production and create incentives for increased fertilizer use in the future as the countries approach their land frontier.

Finally, for the 60 percent of land under root crop, tree crop, highland, and forest-based farming systems, SSA will need to develop its own Green Revolution—one that increases output of root and tree crops in the most productive agroecologies. This strategy will require more investment in agricultural R&D, as international spillovers for these crops and ecologies are expected to be lower than those for cereals. It will also require opening new markets, especially for staple crops, such as cassava, which are nontradables and constrained to small domestic markets.

References

Acemoglu, D. 1998. "Why Do New Technologies Complement Skills? Directed Technical Change and Wage Inequality." *Quarterly Journal of Economics* 113: 1055–1090.

———. 2002. "Directed Technical Change." *Review of Economic Studies* 69: 781–810.

———. 2007. "Equilibrium Bias of Technology." *Econometrica* 75: 1371–1410.

———. 2009. *When Does Labor Scarcity Encourage Innovation?* Working Paper 14809. Cambridge, MA: National Bureau of Economic Research.

Angelsen, A. 1999. "Agricultural Expansion and Deforestation: Modeling the Impact of Population, Market Forces and Property Rights." *Journal of Development Economics* 58: 185–218.

AU (African Union). 2006. *Abuja Declaration on Fertilizer for an African Green Revolution.* The Africa Fertilizer Summit, African Union Special Summit of the Heads of State and Government, Abuja, Nigeria, June 13. Addis Ababa, Ethiopia.

Binswanger, H. P. 1986. "Evaluating Research System Performance and Targeting Research in Land-Abundant Areas of Sub-Saharan Africa." *World Development* 14 (4): 469–475.

Binswanger, H. P., and P. Pingali. 1988. "Technological Priorities for Farming in Sub-Saharan Africa." *World Bank Research Observer* 3 (1): 81–98.

Binswanger, H. P., and V. W. Ruttan. 1978. *Induced Innovation: Technology, Institutions and Development.* Baltimore: Johns Hopkins University Press.

Boserup, E. 1965. *The Conditions of Agricultural Growth: The Economics of Agricultural Change under Population Pressure.* London: Allen and Unwing.

Brookfield, H. C. 1972. "Intensification and Disintensification in Pacific Agriculture: A Theoretical Approach." *Pacific Viewpoint* 13: 30–48.

———. 2001. "Intensification and Alternative Approaches to Agricultural Change." *Asia Pacific Viewpoint* 42: 181–192.

Carr, D. L. 2004. "Proximate Population Factors and Deforestation in Tropical Agricultural Frontiers." *Population and Environment* 25: 585–612.

Carswell, G. 1997. *Agricultural Intensification and Rural Sustainable Livelihoods: A "Think Piece."* Brighton, UK: Institute of Development Studies.

Chilonda, P., and J. Otte. 2006. "Indicators to Monitor Trends in Livestock Production at National, Regional and International Levels." *Livestock Research for Rural Development* 18 (8): Article #117. Accessed April 3, 2015. http://www.lrrd.org/lrrd18/8/chil18117.htm.

Darity W. A., Jr. 1980. "The Boserup Theory of Agricultural Growth: A Model for Anthropological Economics." *Journal of Development Economics* 7 (2): 137–157.

Demont, M., P. Jouve, J. Stessens, and E. Tollens. 2007. "Boserup versus Malthus Revisited: Evolution of Farming Systems in Northern Côte d'Ivoire." *Agricultural Systems* 93: 215–228.

Dorsey, B. 1999. "Agricultural Intensification, Diversification, and Commercial Production among Smallholder Coffee Growers in Central Kenya." *Economic Geography* 75: 178–195.

FAO (Food and Agriculture Organization of the United Nations). 2013. "FAO Country Profiles: Ghana." http://www.fao.org/countryprofiles/index/en/?iso3=GHA.

Goldman, A. 1993. "Agricultural Innovation in Three Areas of Kenya: Neo-Boserupian Theories and Regional Characterization." *Economic Geography* 69 (1): 44–71.

Goldman, A., and J. Smith. 1995. "Agricultural Transformations in India and Northern Nigeria: Exploring the Nature of Green Revolutions." *World Development* 23 (2): 243–263.

Gollin, D., R. Jedwab, and D. Vollrath. 2013. "Urbanization with and without Industrialization." Unpublished paper. http://dept.econ.yorku.ca/seminars/gjv_040213.pdf.

Gray, L. C., and M. Kevane. 2001. "Evolving Tenure Rights and Agricultural Intensification in Southwestern Burkina Faso." *World Development* 29: 573–587.

Haggblade, S., and P. Hazell, eds. 2010. *Successes in African Agriculture: Lessons for the Future.* Baltimore: Johns Hopkins University Press.

Hansen, B. 1969. "Employment and Wages in Rural Egypt." *American Economic Review* 59 (3): 298–313.

Hayami, Y., and V. W. Ruttan. 1970. "Factor Prices and Technical Change in Agricultural Development: The United States and Japan, 1880–1960." *Journal of Political Economy* 78: 1115–1141.

———. 1985. *Agricultural Development: An International Perspective*, 2nd ed. Baltimore: Johns Hopkins University Press.

Heston, A., R. Summers, and B. Aten. 2012. "Penn World Table Version 7.1." Philadelphia: University of Pennsylvania Center for International Comparisons of Production, Income and Prices.

Hillocks, R. J. 2002. "Cassava in Africa." In *Cassava: Biology, Production and Utilization,* edited by R. J. Hillocks, J. M. Thresh, and A. C. Bellolli, 41–54. Wallingford, CT: CABI.

Jayne, T., D. Mather, N. Mason, and J. Ricker-Gilbert. 2013. "How Do Fertilizer Subsidy Programs Affect Total Fertilizer Use in Sub-Saharan Africa? Crowding Out, Diversion, and Benefit/Cost Assessments." *Agricultural Economics* 44 (6): 687–703.

Jayne, T. S., and S. Rashid. 2013. "Input Subsidy Programs in Sub-Saharan Africa: A Synthesis of Recent Evidence." *Agricultural Economics* 44: 547–562.

Lambin, E. F., M. D. A. Rounsevell, and H. J. Geist. 2000. "Are Agricultural Land-Use Models Able to Predict Land-Use Intensity?" *Agricultural Ecosystem and Environment* 82: 321–331.

Lewis, W. A. 1954. "Economic Development with Unlimited Supplies of Labour." *The Manchester School* 22 (2): 139–191.

Malmberg, B., and B. Tegenu. 2006. *Population Pressure and Dynamics of Household Livelihoods in an Ethiopian Village—An Elaboration of the Boserup-Chayanovian Framework*. Stockholm: Författaren samt Kulturgeografiska Institutionen.

Maredia, M. K., D. Byerlee, and P. Pee. 2000. "Impacts of Food Crop Improvement Research: Evidence from Sub-Saharan Africa." *Food Policy* 25: 531–559.

Morris, M. L., V. A. Kelly, R. J. Kopicki, and D. Byerlee. 2007. *Fertilizer Use in African Agriculture: Lessons Learned and Good Practice Guidelines*. Washington, DC: World Bank.

Morrison, K. D. 1996. "Typological Schemes and Agricultural Change: Beyond Boserup in Precolonial South India." *Current Anthropology* 37: 583–608.

Netting, R. M., M. P. Stone, and G. D. Stone. 1989. "Kofyar Cash-Cropping: Choice and Change in Indigenous Agricultural Development." *Human Ecology* 17 (3): 299–319.

Otsuka, K., and F. Place. 2014. *Changes in Land Tenure and Agricultural Intensification in Sub-Saharan Africa*. WIDER Working Paper 2014/051. Tokyo: United Nations University.

Pascual, U., and E. B. Barbier. 2006. "Deprived Land-Use Intensification in Shifting Cultivation: The Population Pressure Hypothesis Revisited." *Agricultural Economics* 34: 155–165.

Pingali, P. L. 2012. "Green Revolution: Impacts, Limits and the Path Ahead." *Proceedings of the National Academy of Sciences of the United States of America* 109: 12302–12308.

Place, F., and K. Otsuka. 2000. "Population Pressure, Land Tenure and Tree Resource Management in Uganda." *Land Economics* 76: 233–251.

Reenberg, A. 2001. "Agricultural Land Use Pattern Dynamics in the Sudan-Sahel—Towards an Event-Driven Framework." *Land Use Policy* 18: 309–319.

Ruthenberg, H. 1980. *Farming Systems in the Tropics*. Oxford, UK: Clarendon Press.

Schultz, T. W. 1964. *Transforming Traditional Agriculture*. New Haven, CT: Yale University Press.

Stone, G. D. 2001. "Theory of the Square Chicken: Advances in Agricultural Intensification Theory." *Asia Pacific Viewpoint* 42: 163–180.

Tachibana, T. T., M. Nguyen, and K. Otskuka. 2010. "Agricultural Intensification versus Extensification: A Case Study of Deforestation in the Northern-Hill Region of Vietnam." *Journal of Environmental Economics and Management* 41: 44–69.

Tiffen, M., M. Mortimore, and F. Gichuki. 1994. *More People, Less Erosion: Environmental Recovery in Kenya*. Chichester, UK: Wiley.

Turner, B. L., and A. M. Ali. 1996. "Induced Intensification: Agricultural Change in Bangladesh with Implications for Malthus and Boserup." *Proceedings of the National Academy of Sciences of the United States of America* 93: 14984–14991.

Turner B. L., and S. B. Brush, eds. 1987. *Comparative Farming Systems*. New York: Guilford Press.

Turner, B. L., and M. Fischer-Kowalskic. 2010. "Ester Boserup: An Interdisciplinary Visionary Relevant for Sustainability." *Proceedings of the National Academy of Sciences of the United States of America* 107 (51): 21963–21965.

Woodhouse, P. 2009. "Technology, Environment and the Productivity Problem in African Agriculture: Comment on the World Development Report 2008." *Journal of Agrarian Change* 9 (2): 263–276.

World Bank. 2013. "World Development Indicators." http://data.worldbank.org/data-catalog/world-development-indicators.

Chapter 6

FACTORS INFLUENCING THE EFFECTIVENESS OF PRODUCTIVITY-ENHANCING INTERVENTIONS: AN ASSESSMENT OF SELECTED PROGRAMS

Joseph Karugia, Stella Massawe, Paul Guthiga, Maurice Ogada, Manson Nwafor, Pius Chilonda, and Emmanuel Musaba

Introduction

The preceding chapter has shown that agricultural intensification paths differ markedly in different farming systems and are influenced significantly by population density; yield-enhancing technologies, such as fertilizer, may have been used more as instruments for expanding cultivated area, rather than for increasing yield; and further in-country analysis is needed to sharpen the policy implications of these findings. This chapter assesses selected productivity-enhancing interventions across Africa, with the aim of distilling lessons on key factors contributing to the effectiveness of such interventions. *Intervention* as used here stands for programs, projects, strategies, or other agricultural development initiatives, and *effectiveness* refers to the extent to which the objectives of the intervention were achieved, taking into account their relative importance (OECD 2010).

Several agricultural interventions have been implemented across the continent over recent decades, with the aim of addressing constraints to agricultural productivity. Some interventions performed relatively well and were able to achieve their objectives to a large extent, whereas others barely achieved their intended objectives. Differences in performance could be attributed to variations in adherence to important factors that need to be considered during the design and implementation of the interventions, as well as to the sustainability of their effects. There is a lot to learn, not only from interventions that were effective or successful in achieving their objectives, but also from those that were not effective or failed to achieve their objectives. The lessons summarized in this chapter are expected to inform the design and implementation of future agricultural productivity-enhancing interventions in Africa. They are also expected to be useful to highlight areas where ongoing interventions may require adjustment to enhance their effectiveness. Unlike the preceding chapters, in which quantitative methods are used, a qualitative approach is used here to synthesize information from different literature sources about measures of the effectiveness of the interventions.

In the next section, we describe the conceptual framework that we used in guiding our assessment of the effectiveness of interventions. This is followed by a detailed discussion of the methodology used, after which the findings are presented and discussed. The final section presents conclusions and recommendations.

Conceptual Framework for Understanding Factors for Assessing the Effectiveness of Interventions

This subsection is divided into three parts: (1) a discussion of the definition of success of productivity-enhancing interventions; (2) our conceptual framework, which uses multiple criteria to assess the performance of productivity-enhancing interventions; and (3) the components of the conceptual framework.

Defining the success of productivity-enhancing interventions

Defining the success of agricultural productivity-enhancing interventions is a fundamental conceptual issue to be addressed in the process of developing a conceptual framework for assessing the performance of such interventions. Various indicators are used to define this success, including measures of whether the intervention has led to increased production and yields as a result of alleviation of productivity constraints. Interventions that increase agricultural labor productivity also fall into this category.

A different and broader perspective considered involves measures of whether the productivity-enhancing programs or projects contributed to (1) introducing enterprises (for example, new high-value enterprises); (2) improving standards of living of the beneficiaries through increased income, employment opportunities, food availability, and dietary diversification; (3) promoting value addition and market linkages; and (4) reducing postharvest losses.

Although these definitions of success are clear, the evaluation of whether or how extensively an intervention has actually led to an increase in the relevant indicator of success is outside the scope of this chapter. Thus, while the interventions selected were also those whose impacts were partly evaluated, they have different levels of rigor in their counterfactual designs and identification strategies to demonstrate a causal link between the intervention and measure of success. This chapter discusses this later when presenting the methodology and selected interventions.

Conceptual framework

The use of such terms as "success stories" is becoming increasingly common in development discourse. However, one needs to be cautious when classifying an intervention as "successful" or "failed," given the analytical rigor that is required to demonstrate causation or lack of it in order to reliably categorize interventions into either of these two classes. Because any particular intervention often has multiple objectives, the analysis can quickly become cumbersome if one assumes that (1) "successful" interventions are those that performed well in all dimensions necessary for achieving and sustaining productivity gains, or (2) "failed" interventions are those that performed poorly in all dimensions. Hence, the conceptual framework (Figure 6.1) used here for assessing the "success" or "failure" of productivity-enhancing interventions focuses on key nodes along the broader project implementation pathway, rather than achievement of the final outcomes themselves.

Development of the conceptual framework was informed by a wide range of literature on development theories and rural development (Uphoff 1986; Rondinelli 1986; Baccarini and Collins 2003; Fonchingong and Fonjong 2003; Boussard et al. 2005; Gawler 2005; NFSD 2005; Poulton and Dorward 2008; Sahee Foundation 2008; Spielman and Pandya-Lorch 2009; TANGO International 2009; Haggblade and Hazell 2010). Consultations with several national agricultural and rural development practitioners in different countries were also conducted to generate greater confidence in the results. This is discussed further in the methodology section of this chapter.

Five thematic areas are considered in the conceptual framework for examining the performance of a productivity-enhancing intervention: (1) definition of the productivity problem or constraint; (2) targeting; (3) design, focusing on such areas as intervention strategy, implementation mechanism, and related factors; (4) sustainability; and (5) overarching supportive factors or crosscutting or conditioning factors. The arrows in Figure 6.1 indicate influence across thematic areas. A one-way directional arrow indicates one-way influence (that is, a factor influences another only), while a two-way directional arrow indicates that a factor influences another and vice versa. Although not distinguished in the framework presented in Figure 6.1, it is important to note that whereas some crosscutting factors, such as

FIGURE 6.1 Factors influencing the success or failure of agricultural productivity-enhancing interventions

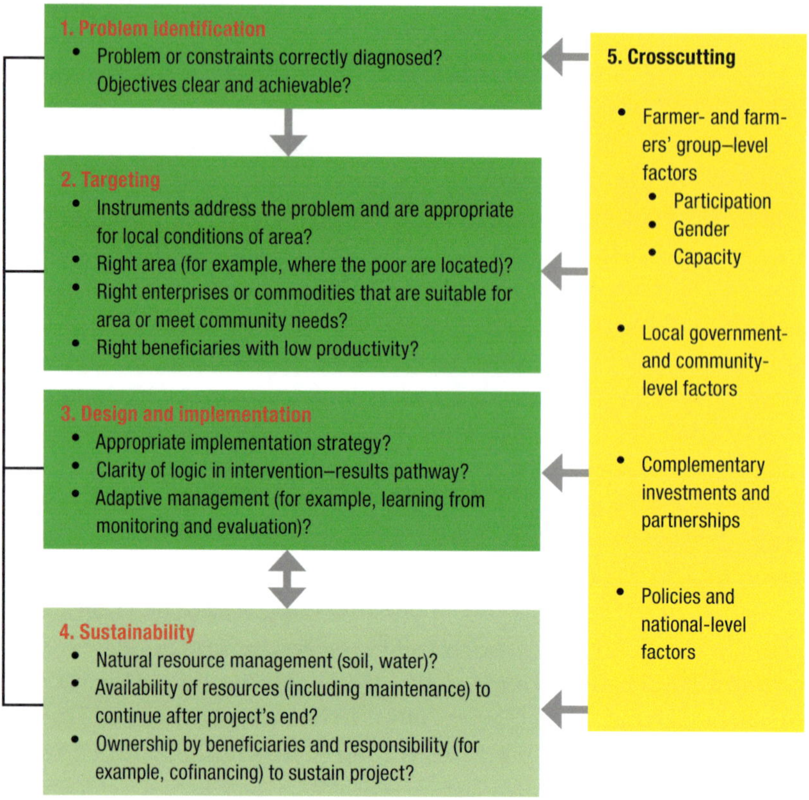

Source: Authors' conceptualization based on literature review and consultations.

participation of beneficiaries, may directly influence all the other four thematic areas, other crosscutting factors, such as funding and complementary interventions, may directly influence a subset of the other four thematic areas only.

We now discuss each of the thematic areas in the conceptual framework to flesh out how they interact with each other and influence the overall "success" or "failure" of productivity-enhancing interventions within the broader project implementation pathway.

PROBLEM IDENTIFICATION

The problem identification stage is fundamental to the entire implementation process, whether it is for a productivity-enhancing intervention or for any other development project (Touwen 2001). Some key questions for consideration here are whether the problem was well understood and defined before designing the intervention, and whether the objectives that were set are relevant and achievable. Because the intended beneficiaries are expected to be better informed about their situation and local conditions, their participation at this stage in diagnosing the problems and constraints is particularly critical. Therefore, some of the criteria to consider in the evaluation are how the beneficiaries were engaged at this stage and the sources of information used. For example, was a survey conducted to collect baseline information on relevant factors that affect farmers' production decisionmaking and productivity?

As we have seen from the preceding chapters, there is substantial heterogeneity in the paths of intensification, technology adoption, and productivity, which results from differences in the production environments faced by different farmers in different areas. Because of such variations across different groups of beneficiaries, even in the same locality, another consideration in the evaluation is how the objectives of the project account for the needs and constraints of different groups of beneficiaries (Touwen 2001; NFSD 2005; Sahee Foundation 2008), which significantly influences the design of the intervention and targeting.

TARGETING

Appropriate targeting is a crucial factor for the success of productivity-enhancing interventions. In the preceding chapter's review of input subsidy programs (ISPs) in Africa, for example, we saw that poor targeting of ISPs has not led to an overall increase in fertilizers (which is a major objective of ISPs), because the subsidized fertilizer has crowded out commercial fertilizer (Jayne et al. 2013). Therefore, how the intended beneficiaries are reached (whether through the commodities or enterprises they are involved in, where they are geographically located, or other mechanisms) is an important issue to deal with in a manner that accounts for differences in socioeconomic, agroecological, and other relevant factors of the target population (Nubukpo and Galiba 1999; Boussard et al. 2005).

As the preceding chapters suggest, technologies that can work in high-potential areas, for example, are different from those suited to low-potential areas. Production of high-input, perishable products, such as milk and horticultural products, for example, are more amenable to areas with high market access, while production of low-input, nonperishable commodities may be more suitable in remote areas (Pender, Place, and Ehui 2006).

DESIGN AND IMPLEMENTATION

The design of the intervention includes definition of the overall goal, objectives, beneficiaries, and implementation strategy (NFSD 2005; Rondinelli 1986; Sahee Foundation 2008; TANGO International 2009). Several issues need to be considered here, including (1) the technical, managerial, and financial capacities to implement and sustain the intervention; (2) the roles and responsibilities of different actors involved in implementing the project; and (3) the existence or lack of incentives to undertake the intervention.

Important questions here include:

- What is the basis on which the project was developed?
- Was a feasibility study or situation analysis conducted, and were the results used in designing the project?
- Is the project feasible?
- Does the implementing agency have the capacity to run the project as planned?
- Are the beneficiaries able to absorb the benefits from project activities?
- Are the overall project timelines clearly defined and realistic?
- Are the cost estimates accurate and realistic?
- How well is the project suited to achieve the desired outcome?
- Were the appropriate design and strategy used?
- Is the theory of change well constructed?
- Is the intervention based on reasonable assumptions?
- What are the coordination mechanisms during and after the project's life?
- What challenges are likely to affect the project, and what are the appropriate strategies to address them?
- Are any activities planned to ensure future sustainability (Baccarini and Collins 2003; NFSD 2005; Sahee Foundation 2008; Tango International 2009)?

Overall management is an important aspect too, as problems of poor management are often cited as being responsible for program failure in Africa (White 1986).

SUSTAINABILITY

An intervention can be said to be sustainable if there is indication that it will have lasting benefits for an extended period after the main part of the implementation has been completed (Gawler 2005; TANGO International 2009). Sustainability has social, economic, and environmental dimensions. A project is socially sustainable if it is supported by the existing social structures and institutions, and if it guarantees the health and safety of individuals, households, and communities. It is economically sustainable if it is economically viable over time, adaptable with minimal cost, affordable, and supported by local and external economic realities. Environmental sustainability, however, implies that the project avoids overexploitation of renewable resources, protects and enhances biodiversity, optimizes soil and water conservation, reduces pollution, manages wastes effectively, or guarantees energy and/or water and energy efficiency (TANGO International 2009). Therefore, a key question for the evaluation is, what factors are in place to sustain the benefits of the intervention?

Various aspects of the social and economic dimensions include balance or complementarity in the use of locally available resources and materials versus external capital and inputs, which is important for local ownership and participation (Fonchingong and Fonjong 2003; Sahee Foundation 2008). Regarding financial sustainability, projects that are expected to generate financial income (or internally generated revenue) are expected to be self-sustainable. Some of the issues to look at will be the type of approaches that have been put in place to ensure the financial sustainability of the intervention (Sahee Foundation 2008; TANGO International 2009).

On environmental sustainability, a key question is, what complementary interventions have been put in place to address any negative or damaging environmental side effects, which may or may not reduce the benefits of the agriculture productivity-enhancing intervention itself, in either the immediate or the long run? A typical scenario in the case of an irrigation intervention, for example, is soil salinity or increased incidence of waterborne diseases and pests for both humans and animals.

CROSSCUTTING FACTORS

The following crosscutting factors, which are expected to influence performance in the themes discussed above, are presented under four main categories: farmer- and farmers' group–level factors, local government- and community-level factors, complementary investments and partnerships, and policies and national-level factors.

Farmer- and farmers' group–level factors

We discuss several factors, including participation of beneficiaries, particularly gender and capacity. Participation of beneficiaries has been found to be crucial for the different stages of project implementation (Uphoff 1986; Baccarini and Collins 2003; Gawler 2005; Noble et al. 2005; Sahee Foundation 2008; TANGO International 2009). Including the needs and preferences of beneficiaries during the design and implementation of development interventions enhances local acceptability and the long-term sustainability of the interventions (Uphoff 1986; Gawler 2005; Noble 2005). Therefore, the quality of participation, which is more than merely informing the beneficiaries of what is happening or going to happen, is important (Pretty 1995).

Several questions to be considered include:

- Who are the right beneficiaries to be involved?
- What is their capacity to engage effectively in the intervention?
- If their capacity is weak, what improvements are feasible within the scope of the intervention?
- What form will the participation process take?
- When is the right time to involve the beneficiaries?

The issue of gender is important, because evidence shows that there are differences in household welfare outcomes because of gender differences in access to factors of production, inputs, technologies, and other productive resources that affect output, productivity, and related development outcomes (Tadelle and Ogle 2001; Mapiye et al. 2008; Kristjanson et al. 2010; Peterman et al. 2010). A common observation manifested from such differences is choice of agricultural enterprise between men and women. Gender differences in productivity constraints are observed across Africa (SOFA Team 2011; SOFA Team and Doss 2011), suggesting that female farmers face more constraints, resulting in lower yields than those of their male counterparts (Seeley et al. 2004; Kristjanson et al. 2010; Peterman 2010; SOFA Team 2011; SOFA Team and Doss 2011; Croppenstedt, Goldstein, and Rosas 2013). Thus, a key question for the evaluation is how the different constraints faced by men and women are internalized in the productivity-enhancing intervention to maximize its benefits and their distribution. The gender issue can be extended to include other groups of beneficiaries, children and youths, the aged, etc.

How well farmers or beneficiaries are able to participate depends on their capacity to understand the various aspects of the project and engage effectively (Sahee Foundation 2008). Thus, developing the skills and competencies of the beneficiaries to take greater control of the project may be an important aspect to consider, even if capacity building is not an explicitly stated objective of the intervention. The source of capacity building (central government, development agent, or local authority) may not be of consequence, but tapping into indigenous knowledge could prove useful. Building the capacity of the beneficiaries to be able to effectively manage the project may contribute to a strong exit strategy (Baccarini and Collins 2003; NFSD 2005; Noble et al. 2005; Andersen et al. 2006; Hyvari 2006; Muller and Turner 2007; Khang and Moe 2008; Sahee Foundation 2008).

Many development projects are implemented through farmers' groups (organic or induced) as achieving collective action to provide support or diffuse technology. This may be argued as a way of achieving economies of scale or reducing transaction costs (Poole and de Frece 2010), or certain activities may require some level of collective action, such as integrated watershed development, canal irrigation, and conservation of common property resources (Uphoff 1986; Noble et al. 2005). However, the question of what benefits farmers' groups and collective action bring to such interventions is ultimately complex and empirical. For example, a farmers' group may suffer from elite capture and social exclusion problems (Feder et al. 2010), which may lead to choices that may be inconsistent with considerations of equity. Based on Feder et al. (2010) and Mansuri and Rao (2013), key questions for the evaluation here include:

- Do group leaders act in ways that support or undermine the larger interests of the farmers they claim to represent?
- Do they maximize rents, or do they lead with the collective welfare of the farmers in mind?
- Are there specialized groups, such as those exclusively for female farmers or disadvantaged people?
- Are such farmers represented in the leadership council or other decision-making bodies for the group?

Local government- and community-level factors

Several of the issues discussed under the farmers' groups also apply here. However, because the composition of the community is much larger than farmers' groups, the problems of elite capture and social exclusion are likely to be more prevalent at this level. Thus, having strong and committed local government officials and community leaders may be critical for the success of productivity-enhancing interventions (Penning de Vries 2005). In particular, such leadership may be necessary for mobilizing people and resources and ensuring the best use of those resources (Spielman and Pandya-Lorch 2009). These committed and dedicated leaders are sometimes referred to as "champions," implying how they can make a difference in society in several ways, including pushing an issue to the forefront of the public's consciousness, demonstrating what can be done in the face of seemingly insurmountable challenges, or mobilizing the political and financial capital to overcome inertia (Spielman and Pandya-Lorch 2009).

Issues of the participation and capacity building of farmers discussed earlier are also relevant at this higher community and local government level, in terms of empowerment of the community to own the project. As Mansuri and Rao (2013) find, however, inducing local community participation is a difficult, often unpredictable, and potentially contentious undertaking and, because several factors come into play, a successful project in one context may fail miserably in another. Therefore, some of the issues to consider are the specific modalities put in place for inducing participation and the willingness and ability of the community to adapt to changes in expected funding and state support and expected outcomes.

Complementary investments and partnerships

The main issue here is considering how other ongoing or planned investments or projects may mitigate (compete against) or complement (enhance) the intervention. For example, a project that seeks to promote high-value horticultural production in a remote area with poor access to markets, limited post-harvest-handling facilities, and lack of agroprocessing plants may not likely be successful. Therefore, a key question to consider in the evaluation here is, what investments or partnerships may enhance or undermine achievement of the objectives of the intervention?

Increasingly, many interventions tend to adopt an integrated approach, whether for addressing one or multiple constraints. Therefore, planners and implementers have had to leverage strategic partnerships and use multistakeholder approaches, ideally involving actors who have comparative advantage in different aspects of the intervention (Diagne et al. 2010). Public–private partnerships are one such approach being used in agriculture and rural development (Sahee Foundation 2008; Druilhe and Barreiro-Hurlé 2012). Because of the complexities and difficulties in coordinating across agencies, there are many examples of such integrated interventions that have failed to achieve their objectives. Therefore, other key questions to consider, especially in the case of integrated, multistakeholder development intervention are, what are the modalities for coordinating the different partners, and how do the different parts that are integrated contribute to achieving each objective?

Policies and national-level factors

The success or failure of an intervention also depends on policies and other national-level factors. Although such factors as national land tenure policy or infrastructure development may exert the same force everywhere in the country, they may have different effects on different interventions in different locations, depending on how each intervention or locality relates to policy.

For example, land tenure policy may have little bearing on an intervention that seeks to promote technologies for maize (annual crop) production on farmers' own fields compared with one that seeks to promote technologies for tree crop production whether on farmers' own fields or previously uncultivated lands. Another example of the differential effects of a national policy is with the ISPs, which are typically characterized as being chronically late in the delivery of program fertilizer (Jayne and Rashid 2013). Because of different planting seasons associated with different agroecologies in a country, it is likely that the delivery will be late for some and timely for others. This example also highlights the importance of timing, which is especially critical in seasonal agriculture (Dorward et al. 2006). Therefore, some key questions to consider in the evaluation of the influence of national-level factors include: How will different policies, programs, regulations, or reforms affect achievement of the objectives of the intervention, and how can community leaders be empowered to provide relevant local public goods where national policies or programs fall short of the community needs?

Methodology

We employed both quantitative and qualitative methods of collecting and analyzing data. We used *quantitative* methods to collate and analyze data on the *outcomes* of productivity-enhancing interventions, and *qualitative* methods to synthesize information on the *effectiveness* of the interventions from different literature sources. Because most of the available information on the performance of the interventions was in narrative form, a higher proportion of the analysis in this chapter was achieved through qualitative approaches. By combining quantitative and qualitative analysis, we are able to examine some potential impacts of the interventions that were analyzed in the literature based on group interviews of beneficiaries of the interventions (Khandker, Koolwal, and Samad 2010). The qualitative methods also are more suitable in situations that involve a small number of case studies (Patton 1986), and can provide a deep understanding of the impact pathways (Copestake, Johnson, and Wright 2004).

As stated in the introduction to this chapter, the performance criteria of the "success" or "failure" of productivity-enhancing interventions used here focus on key nodes along the broader project implementation pathway, rather than on achievement of the final outcomes themselves. Nevertheless, the issue of attribution is important, and we note that qualitative methods generally on their own may not provide robust information about attributing the observed performance to the intervention (Catley et al. 2008; Petticrew and Roberts 2006). Defining and measuring the appropriate counterfactual are at the core of a rigorous impact evaluation, for which qualitative methods on their own are generally less effective. Although qualitative analyses are suitable for in-depth analysis of small numbers of case studies, small sample sizes may place limits on the extent to which the findings can be generalized to the larger population.

For the combined quantitative and qualitative analysis used here, we employed a multicriteria evaluation (MCE) technique (Voogd 1982) based on the project implementation pathway nodes presented in the conceptual framework. MCE techniques have been used widely for different policy purposes in different contexts, such as (1) identifying the best manager for a project (Zavadskas et al. 2008); (2) identifying effective options or solutions in natural resource management (Munda, Nijkamp, and Rietveld 1994; Abdul et al. 2011); (3) identifying the best location for an investment (Lin et al. 1997); (4) identifying strategic options at the planning stage of a project (Linton, Walshand, and Morabito 2002; Raju and Kumar 2005; De Brucker, Macharis, and Verbeke 2011); and (5) monitoring and evaluating progress in implementing development projects (Karamouz et al. 2002). The MCE technique can be applied as an *ex ante* evaluation tool, particularly for making strategic choices, or an ex post evaluation tool, particularly for evaluating multiple outputs and outcomes.

In this chapter, the MCE technique is applied as an ex post evaluation tool on the multiple criteria presented in the conceptual framework, assuming equal weights for each criterion. Although different weights may be used (Saaty 1987; Wang, Jing, and Zhang 2009), we chose equal weights for simplicity, because the data (interventions considered) were from different studies conducted at different times on different countries, which complicates the choice of an unequal weighting system.

Data and sources of information

The main sources of data on productivity-enhancing interventions were from a wide range of peer-reviewed published literature and gray literature in the form of project implementation progress reports, technical reports, project evaluation reports, documentation of case studies, briefs on success stories, and external evaluation reports. About 110 potential projects representing interventions that aimed to address a broad range of agricultural productivity issues (for example, inputs, extension, irrigation and water management, crop and animal health) were identified, and then relevant literature on them was assembled. After reviewing the literature, we dropped the majority of the projects (95 in total), because of lack of adequate information on the different criteria presented in the conceptual framework. Therefore, we retained 25 projects for the analysis. (Table 6A.1 in the appendix to this chapter contains detailed descriptions of the projects, including their location.)

Even among the 25 projects, the analytical rigor on the effectiveness of the intervention varied. For example we found that stronger analyses and more credible evidence on the effectiveness of interventions were generated by independent evaluations than those based on self reporting by project implementers. For 17 of the projects out of the total 25 analyzed, we were able to access external evaluation reports, in addition to independent analytical and scientific papers, peer-reviewed journal articles, and other forms of external technical reviews. Only in 3 projects were the dominant sources of information from internal evaluations and self-reporting by project implementers or funders.

In our review of all the 110 potential projects, we found a general tendency to document "success" stories more often than unsuccessful or "failed" cases. For the few cases that were deemed failures, there was not much information to evaluate them according to the criteria presented in the conceptual framework. To avoid having only "successful" projects in the sample without variation in the indicators to be analyzed, we considered different levels of success.

To complement the literature review that was used in developing the conceptual framework in Figure 6.1, a short questionnaire was sent to several in-country experts on national agricultural and rural development. (Appendix Table 6A.2 presents details on the survey and instrument.) These experts were asked to identify three productivity-enhancing interventions that were considered to have worked well in their countries and three that did not do well. They were then asked to provide reasons for their performance ratings of the identified interventions. Later, the experts were convened at a workshop in Nairobi, Kenya, to brainstorm further on the topic of using a checklist of questions regarding performance of past agricultural productivity projects. (Appendix Table 6A.6 presents details on the checklist and the experts consulted.) The research team also visited one of the productivity-enhancing project sites in Yatta district in Kenya—the Operation *Mwolyo*[1] Out (OMO) project (Table 6A.1), to gain a practical sense for evaluating the performance of the various interventions.

Measurement of project effectiveness

Guided by the analytical framework depicted in Figure 6.1, we evaluated the performance of the 25 productivity-enhancing interventions on 13 criteria or indicators—representing the four main themes and the four crosscutting factors. Table 6.1 outlines the relationship between the empirical indicators and the conceptual factors, and appendix Table 6A.3 presents details on the pathways through which these indicators influence productivity.

Based on our assessment of the extent to which the issues associated with each indicator were dealt with, we defined two levels of performance for each indicator: (1) well done to very well done, and (2) moderate to low performance. (Appendix Table 6A.4 provides details on the information considered for scoring each of the 13 indicators.) We used a Likert scale to convert the qualitative information to numerical values. Assuming that the strength or intensity of opinions, perceptions, or attitudes can be measured and is linear—that is to say, on a continuum from, for example, strongly positive to strongly negative—the Likert scale was used to convert the ordinal measures to numeric values, as summarized in Table 6.2.

To assess performance in meeting the overall productivity objective or target, we compared what was achieved against stated targets. In cases where there was no stated target for the project, we used the national productivity target, as stated in the government's agricultural strategy. Overall productivity performance was rated as follows:

1 *Mwolyo* means relief food in the local Kamba language spoken in Yatta district, Kenya.

TABLE 6.1 Conceptual factors and empirical indicators used in performance assessment

Conceptual factors	Empirical indicators
Main themes	
1. Problem identification	1. Problem definition
2. Targeting	2. Choice of instruments (commodity, solution)
	3. Suitability of instruments
3. Design and implementation	4. Design and timing of implementation
4. Sustainability	5. Environmental sustainability
	6. Financial sustainability
Crosscutting themes	
5. Farmer- and farmers' group–level factors	7. Community participation
	8. Gender consideration
	9. Capacity building
	10. Organized groups
6. Local government- and community-level factors	11. Leadership and dedication
7. Complementary investments and partnerships	12. Investments and partnerships
8. Policies and national-level factors	13. Policies and political stability

Source: Authors' construction based on literature review and consultations.

TABLE 6.2 Likert scales and associated scores

Likert level	Description	Score	Symbol
1	Good to very good	2	++
2	Moderate to low	1	••

Source: Authors' construction based on literature review and qualitative performance assessment.

- If less than 50 percent of the target was met, performance was rated as "very poor."
- If between 50 and 74.9 percent of the target was met, performance was rated as "poor."
- If between 75 and 99.9 percent of the target was met, performance was rated as "moderate."
- If between 100 and 149.9 percent of the target was met, performance was rated as "good."
- If the target was exceeded by more than 150 percent, performance was rated as "very good."

TABLE 6.3 Interventions selected for assessment by countries and farming systems

Name of project/intervention	Countries	Farming systems
Agricultural Sector Development Programme—irrigation component (ASDP-irrigation)	Tanzania	Maize mixed, tree crop
Agriculture Productivity Enhancement Programme (APEP)	Uganda	Highland perennial, maize mixed
Animal Health Services Rehabilitation Programme (AHSRP)	Kenya	Maize mixed, highland perennial
Cassava Enterprise Development Project (CEDP)	Nigeria	Tree crop, coastal, root crop
Conservation Agriculture Project 1 (CAP1)	Zambia	Maize mixed, cereal-root crop mixed
Crop Crisis Control Project (C3P)	Burundi, Democratic Republic of the Congo, Kenya, Rwanda, Uganda, Tanzania	Maize mixed, root crop, highland perennial, forest based
East Africa Dairy Development Project (EADD)	Kenya, Uganda, Rwanda	Maize mixed, highland perennial
FARM Africa Dairy Goat Improvement Project (FADGIP)	Burundi, Ethiopia, Kenya, Rwanda, Tanzania, Uganda	Maize mixed, highland mixed
Farm Input Subsidy Program (FISBP)	Malawi	Maize mixed
Farmer Input Support Program (FISPP)	Zambia	Maize mixed, cereal-root crop mixed
Fodder Trees and Shrubs Project (FTSP)	Kenya, Rwanda, Tanzania, Uganda	Cereal-root crop mixed, maize mixed
Fuve Panganai Irrigation Scheme (FPIS)	Zimbabwe	Maize mixed
Kaleya Irrigation Project (KIP)	Zambia	Maize mixed

Results and Discussions

Description of case study interventions

Table 6.3 provides a summary of the case study interventions and the countries and farming systems within which they were implemented based on classification in Chapters 3 and 4. (Appendix Table 6A.1 provides more details.) The maize mixed farming system was predominant, because it accounts for a large area in East and southern Africa, where the majority of the case studies are located. In fact, only 5 of the 25 projects exclusively fell outside the maize mixed system, including the Cassava Enterprise Development Project (CEDP) in Nigeria, the System of Rice Intensification (SRI) project in Rwanda, the Specialty Coffee Program (SCP) in Rwanda, the National Agricultural

Name of project/intervention	Countries	Farming systems
Kenya Dairy Development Programme (KDDP)	Kenya	Maize mixed
National Agricultural Advisory Services (NAADS)	Uganda	Highland perennial, maize mixed
National Agricultural Extension Intervention Program (NAEIP)	Ethiopia	Highland perennial, highland temperate mixed, pastoral-agropastoral
New Rice for Africa (NERICA) upland rice	Uganda	Highland perennial
Operation *Mwolyo* Out (OMO)	Kenya	Maize mixed
Participatory Irrigation Development Programme (PIDP)	Tanzania	Maize mixed, root crop
Push–Pull Technology (PPT)	Kenya, Tanzania, Uganda, Ethiopia	Root crop, pastoral-agropastoral, maize mixed, highland temperate mixed
Regional Land Management Unit (RELMA)	Eritrea, Ethiopia, Kenya, Tanzania, Uganda, Zambia	Maize mixed, root crop, highland temperate mixed
Sasakawa Global 2000 Agricultural Program (SG2000-AP)	Ghana, Sudan, Tanzania, Benin, Togo, Mozambique, Eritrea, Guinea, Burkina Faso, Malawi, Mali, Nigeria, Ethiopia, Uganda, Zambia	Maize mixed, highland temperate mixed, pastoral
Specialty Coffee Program (SCP)	Rwanda	Highland perennial
System of Rice Intensification (SRI)	Rwanda	Highland perennial
Wei Wei Integrated Development Project (WWIDP)	Kenya	Maize mixed

Source: Authors' construction based on literature review and farming system classification in Chapter 4.

Extension Intervention Program (NAEIP) in Ethiopia, and the New Rice for Africa (NERICA) upland rice program in Uganda. We attempted to use the farming systems to make inferences about project targeting, but given the large geographical scope of each farming system, doing so proved to be difficult. Furthermore, because most of the projects were addressing similar constraints across different farming systems, lessons on project effectiveness could apply to a broad range of farming systems.

Performance of the interventions

PERFORMANCE AGAINST 13 INDICATORS OF EFFECTIVENESS IN IMPLEMENTATION

Table 6.4 summarizes the qualitative scores for each of the 13 indicators (appendix Table 6A.5 provides more details), and Table 6.5 provides the

TABLE 6.4 **Performance of the interventions in meeting criteria for effectiveness in implementation**

Project/ intervention	Problem definition	Choice of instruments	Suitability of instruments	Design and timing of implementation	Environmental sustainability	Financial sustainability	Community participation	Gender consideration	Capacity building	Organized groups	Leadership and dedication	Complementary investments and partnerships	Policies & national-level factors	Total score	Percentage score
AHSRP	••	++	••	••	••	••	••	••	++	••	••	••	••	15	58
APEP	++	++	++	++	••	••	++	++	++	++	++	++	••	23	89
ASDP-irrigation	++	++	++	••	••	••	••	••	••	••	++	••	••	17	65
C3P	++	++	++	++	••	••	++	••	++	++	++	++	++	23	89
CAP1	++	++	++	++	••	••	++	••	++	++	++	••	••	21	81
CEDP	++	++	++	++	++	••	++	++	++	••	••	++	••	22	85
EADD	++	++	••	••	••	••	++	++	++	++	••	++	••	20	77
FADGIP	++	++	++	++	++	••	++	++	++	++	++	++	••	24	92
FISBP	++	++	••	••	••	••	++	++	••	••	++	••	••	18	69
FISPP	++	++	••	••	••	••	••	••	••	••	++	++	++	18	69
FPIS	++	++	++	••	••	••	••	++	••	••	++	••	••	18	69
FTSP	++	++	++	++	++	++	++	++	++	++	++	++	••	25	96
KDDP	++	++	++	++	••	••	++	++	++	••	++	++	••	22	85
KIP	++	++	++	++	••	••	++	++	++	++	++	++	++	24	92
NAADS	++	++	••	••	••	••	++	••	••	++	••	••	++	18	69
NAIEP	++	++	••	••	••	••	••	••	••	••	++	++	••	17	65
NERICA	++	++	++	++	••	••	++	++	++	++	++	++	••	23	89
OMO	++	++	++	++	++	++	++	++	++	++	++	••	••	24	92
PIDP	••	++	••	••	••	••	++	++	++	++	++	++	••	20	77
PPT	++	++	++	++	++	++	++	++	++	++	++	++	••	25	96
RELMA	••	++	++	••	++	••	••	++	++	++	++	++	++	22	85
SCP	++	++	++	++	••	••	++	++	++	••	++	++	••	22	85
SG 2000-AP	++	++	++	++	++	••	++	++	++	++	++	++	••	24	92
SRI	++	++	++	++	++	••	++	••	••	++	++	••	••	21	81
WWIDP	++	++	++	++	++	++	++	++	++	++	++	++	••	25	96
Total	**1.9**	**2.0**	**1.7**	**1.6**	**1.4**	**1.2**	**1.8**	**1.7**	**1.7**	**1.6**	**1.8**	**1.7**	**1.2**	**21**	**82**

Source: Authors' construction based on literature review and qualitative performance assessment.

Notes: See Table 6.3 for the full names of the projects. ++ = 2 points and •• = 1 point. Percentage score is the "Total score" as a percent of the highest possible total score of 26 (13 indicators multiplied by 2) (Table 6.3).

TABLE 6.5 Overall performance in implementing the interventions

Quartile	Performance	Number of interventions
1st	75.0 percent and above	18
2nd	50.0–74.9 percent	7
3rd	25.0–49.9 percent	None
4th	Less than 25 percent	None

Source: Authors' compilation based on literature review and qualitative performance assessment.

overall performance based on aggregate scores by quartiles. Together, the results show that the majority of the projects performed well, with 18 of the total 25 interventions having an overall score of 75 percent and above. The remaining 7 interventions were in the 50–75 percent quartile (Table 6.5).

The results in Table 6.4 show that choice of commodity/instrument was the criterion on which all of the interventions performed perfectly; followed by problem definition, with an average score of 1.9; and community participation and leadership and dedication, with scores of 1.8 each. Two of the criteria had the lowest average score of 1.2—financial sustainability and policies and national-level factors. Environmental sustainability was also a challenge for many of the projects, with only nine projects receiving a score of 2.

Looking at the total scores obtained, the Push–Pull Technology project of the International Centre of Insect Physiology and Ecology (ICIPE), the Fodder Trees and Shrub Project implemented by the World Agroforestry Centre, the Farm Africa Dairy Goat Improvement Project in Meru, Kenya, and the Wei Wei Integrated Development Project scored the highest, with 96 percent. In contrast, the Animal Health Services Rehabilitation Programme scored the lowest, with 58 percent.

PERFORMANCE IN MEETING OVERALL PRODUCTIVITY TARGET

As the results summarized in Table 6.6 indicate, eight of the projects had a good or very good performance rating in terms of meeting or surpassing the stated productivity targets, nine had a moderate performance rating, and eight had a poor or very poor rating. (Appendix Table 6A.6 provides details on the productivity achievements for each project.) The key question now is, how does performance in implementation vis-à-vis the 13 criteria examined earlier influence performance in meeting the overall productivity target? This is the topic of the upcoming subsection.

TABLE 6.6 Distribution of projects in meeting the overall productivity target

Criteria	Performance rating	Project/intervention	Number of projects	Percentage of projects
Less than 50% of target	Very poor	AHSRP, KDDP, FPIS, NAEIP	4	16
50.0–74.9% of target	Poor	CAP1, FISPP, PIDP, SCP	4	16
75.0–99.9% of target	Moderate	APEP, ASDP-irrigation, EADD, FADGIP, FISBP, FTSP NAADS, NERICA, RELMA	9	36
100.0–149.9% of target	Good	C3P, CEDP, KIP, SRI	4	16
150% or more of target	Very good	SG 2000-AP, OMO, PPT, WWIDP	4	16
Total			25	100

Source: Authors' compilation based on literature review and performance assessment.

Note: See Table 6.3 for the full names of the projects and appendix Table 6A.6 for details on the productivity achievements for each project.

FACTORS INFLUENCING PERFORMANCE IN MEETING THE OVERALL PRODUCTIVITY TARGET

To better understand how the 13 criteria contribute to performance in meeting the overall productivity target, we organized the projects into three groups, calculated the average score for each indicator within each group, and analyzed the differences across the groups. Group 1 (G1) is made up of the eight projects that performed poorly or very poorly, group 2 (G2) consists of the nine projects that performed moderately, and group 3 (G3) contains the eight projects that met or exceeded the overall productivity target. The results presented in Table 6.7 show that interventions that achieved their intended overall productivity objectives generally performed better in each criterion compared, except in one criterion—choice of instruments—where the performance was the same across the three groups. In fact, the largest difference in average performance across the groups was in environmental sustainability, suggesting this is a serious problem for most agricultural projects that do not perform well.

Comparing the average scores across the three groups in Table 6.7, six of the criteria stand out in terms of having large differences of at least 0.4 points in the average scores among the groups: suitability of instruments, design and timing of implementation, environmental sustainability, financial sustainability, community participation, and organized groups. Therefore, how well implementation of a productivity-enhancing intervention performs in these six criteria seems to exert the largest influence on the overall performance of the intervention in meeting its productivity target. The small differences in the average scores across the two groups in the other criteria (particularly

TABLE 6.7 Performance in indicators of implementation by performance in overall productivity

Group scores and comparisons	Problem definition	Choice of instruments	Suitability of instruments	Design and timing of implementation	Environmental sustainability	Financial sustainability	Community participation	Gender consideration	Capacity building	Organized groups	Leadership and dedication	Investments and partnerships	Policies and political stability
Average score													
G1: Poor or very poor (8)	1.8	2.0	1.5	1.4	1.0	1.0	1.5	1.5	1.6	1.3	1.9	1.6	1.1
G2: Moderate (9)	1.9	2.0	1.7	1.4	1.3	1.1	1.8	1.8	1.7	1.8	1.8	1.7	1.2
G3: Good or very good (8)	2.0	2.0	2.0	2.0	1.8	1.4	2.0	1.8	1.9	1.9	1.9	1.8	1.3
Difference in average scores between:													
G1 and G2	0.1	0.0	0.2	0.1	0.3	0.1	0.3	0.3	0.0	**0.5**	−0.1	0.0	0.1
G2 and G3	0.1	0.0	0.3	**0.6**	**0.4**	0.3	0.2	0.0	0.2	0.1	0.1	0.1	0.0
G1 and G3	0.3	0.0	**0.5**	**0.6**	**0.8**	**0.4**	**0.5**	0.3	0.3	**0.6**	0.0	0.1	0.1

Source: Authors' computations based on literature review and qualitative performance assessment.

Notes: Figures in parentheses are the number of projects in the group (Table 6.6 provides a list of projects by rating). Numbers in bold face show the criterion for which there exists a difference of at least 0.4 points in the average scores among pairs of the three groups.

choice of instrument, leadership and dedication, complementary investments and partnerships, and policies and political stability) suggest that they may not be as influential in the success of projects.

Generally, most productivity-enhancing interventions perform well in defining the problem and selecting relevant instruments to use in addressing the productivity constraints, which explains why there were little or no differences in the average scores across the groups with regard to these criteria. Overall, the poor performance in the environmental and financial sustainability criteria in G1 and G2 (which is 17 out of the 25 projects studied) raises the question as to whether the productivity gains attained by the projects can be sustained in the long run.

Lessons on environmental sustainability can be drawn from the eight projects that scored the highest points on this analysis. Taking the OMO project, for example, its implementation relies on locally available resources and technologies that are based on indigenous knowledge and are amenable to the local community. The project uses low-cost and readily available technologies, such as foot pumps, *zai* pits,[2] manure, and water pans or small dams.

We now discuss in a bit more detail the six factors that seem to exert the largest influence on overall productivity performance: suitability of instruments, design and timing of implementation, environmental sustainability, financial sustainability, community participation, and organized groups. The OMO project represents a very good intervention, not only in terms of performance of the individual factors, but also in terms of how the interactions and interdependence among the factors are critical for overall success. The theory of change for this intervention seems to have been well grounded on addressing the primary constraint of access to water for agricultural production, which before the project resulted in unpredictable yields and frequent crop failure. With effective water-harvesting technologies from the project, farmers have been able to cultivate their plots more than once a year and increase their yields. In addition to increasing their land productivity, farmers are now growing high-value crops that enable them to generate greater income (Box 6.1). Other factors that seem to have helped significantly include motivating the community to participate actively and working in groups or cooperatives, in addition to obtaining necessary financial support for value-addition activities, including a development agency's donation of bakery equipment, an oven, and a solar dryer for the community (Murgo 2015).

For several of the projects that did not perform well, we found that underestimation of the cost of the intervention; delays in implementation processes, including late disbursement of funds, lengthy tendering processes, and delays in completing contractual agreements; low capacity of implementing officials; and weak monitoring and evaluation (M&E) systems were the major factors contributing to low performance (appendix Table 6A.5 provides details). The overall poor performance of the Animal Health Services Rehabilitation Programme in Kenya is an example of the negative influence that lack of commitment from different levels of government can have on overall project implementation (Box 6.2). In this example, the theory of change was based on the implementing government agency's commitment to change, its management structure, and its mode of operations. However, the agency continued to conduct business as usual and, with the central government failing to provide its counterpart funding as planned, several key items needed for operating the project (for example, vehicles, staff, and equipment) could not be procured.

2 A *zai* pit is basically a planting hole that is created with manure, grass, and topsoil in a manner that collects rainwater to sustain the plant for a long period of time.

BOX 6.1 Operation *Mwolyo* Out intervention and selected performance indicators

Operation *Mwolyo* Out (OMO) promoted water-harvesting and related technologies to help farmers intensify their production on a 0.4-hectare (ha) piece of land: 0.2 ha for food security (maize) and 0.2 ha for wealth creation (high-value crops). Before the project, mostly only maize was grown, and harvest was uncertain, with an average maize yield of 112.5–225.0 kg/ha. With the OMO project, farm households are now growing both food and high-value cash crops, and their food security and incomes have substantially improved, as shown in the table below.

Average returns to a modest project member

Plot size (in hectares)	Crop	Harvest per season	Seasons per year	Value (KES)
0.20	Maize	12 (90-kg bags)	2	72,000
0.05	Onions	KES 120,000	3	360,000
0.05	Watermelon	KES 50,000	2	100,000
0.05	French beans	KES 45,000	3	135,000
0.05	Sweet potatoes	KES 50,000	2	100,000
0.40	All crops			767,000

Source: Authors' compilation based on literature review and field interviews (December 5, 2011).
Notes: KES = Kenyan shilling; kg = kilogram.

Consequently, most of the targets were not met. For example, vaccinations against endemic diseases were carried out in only 50 percent of the target communities, only 15 percent of the projected cattle stock was dipped, and clinical cases recorded as treated were a mere 10 percent of project estimates.

Some key observations affecting the findings

In our search for productivity-enhancing interventions, we came across more and better documentation of success stories than failures. This trend has several plausible explanations.

First, for obvious reasons, there may be bias toward reporting successful projects among funders and implementers of projects. This creates a situation where the opportunity to learn from failures is limited—likely resulting in repetition of past mistakes, and in turn leading to failures that could have been avoided or reduced. Increasingly, the importance of learning from failures as well as from successes is coming to the fore in the literature on

BOX 6.2 Kenya Animal Health Services Rehabilitation Programme

Project description. The Kenya Animal Health Services Rehabilitation Programme aimed to improve the delivery of animal health services to smallholder livestock owners nationwide. This goal was to be achieved by strengthening the management structure and operations of the Department of Veterinary Services (DVS). Four international agencies—the International Fund for Agricultural Development, International Development Association, Organization of Petroleum Exporting Countries, and United Nations Development Programme—contributed funds to the project.

Overall performance rating. A detailed review of the project concluded that it had performed poorly.

Reasons cited for failure. The main reasons cited for the poor performance were delays in project start-up, shortages of Government of Kenya counterpart funds, delays in procurement of goods and services, and inadequate staff and capacity of the Project Management Support Unit (PMSU). Regarding delays in procurement, for example, only 51 percent of the total project funds was used, even after a two-year extension was granted.

Theory of change and resulting lack of change. The major underlying cause for the poor performance seems to be the lack of change in DVS's management structure and operations—that is, DVS conducted business as usual. The main interventions targeted were to (1) introduce modernized management practices in DVS to provide improved animal health services to livestock owners, (2) establish an effective monitoring and evaluation system, (3) provide appropriate training to staff, and (4) make DVS's overall operations cost-effective. Specific staff appointments were to be made within DVS to make this work, and PMSU was to receive additional staff and equipment to provide the necessary oversight of these management innovations. The design of these innovations was to be based on studies carried out at the beginning of the project. Almost all of the studies were delayed, and when eventually completed, were often not accepted by DVS. As such, recommended staff appointments were not made, PMSU staffing or capacity was barely changed, and no useful project monitoring took place.

Underachievements. Vaccinations against endemic diseases were carried out in only 50 percent of the targeted population, 15 percent of the projected cattle stock was dipped, and clinical cases recorded as treated were a mere 10 percent of project estimates. There was no increase in the diagnostic and surveillance work of the veterinary laboratories. Only the tsetse control trials showed a marked positive response to project interventions.

External factors. External factors that were likely not taken into account and may have influenced project performance include the poor state of Kenya's rural roads (which contributes to higher transport cost) and the ability of farmers to pay for the services (which affects the program's cost recovery). The ability of farmers to pay for the services may have been affected by the timing of payment by the Kenya Cooperative Creameries for farmers' milk supplies.

Source: IFAD (1993).

development studies, M&E, and accountability. As such, creating a culture where it is acceptable to admit failure, learn, and innovate in order to continuously progress is critical (Lewis 2011). Therefore, in addition to understanding and replicating what worked well, it will be important for development analysts, funders, and implementers to also focus on and learn from what did not work well, or what could have been done differently to enhance program performance.

The second point is that, although there are more and better-documented cases of successful productivity-enhancing interventions than failures, a key question that arises is to what extent these successful interventions can be replicated in different locations, in order to sustainably raise the levels of agricultural productivity in different parts of Africa. Perhaps there are just too few projects with credible impact evaluations across the continent that constitute a critical mass of good things to reliably scale up and out.

For example, we started out with 110 potential projects and then settled on 25 that we assessed to have reliable impact data, although the analytical rigor of the impact evaluation methods varied. The majority of the 25 projects are located in East and southern Africa, and in the maize mixed farming system, with only 5 of the 25 projects falling exclusively outside it. The maize mixed farming system accounts for only 15.7 percent of the total population in Africa south of the Sahara and 11.5 percent of the total crop area (Table 4.2 in Chapter 4). This suggests that more projects with rigorous impact evaluations in different parts of the continent are needed to generate the necessary critical mass of knowledge on scalable practices, technologies, and interventions.

Conclusions and Implications

Several agricultural productivity-enhancing interventions have been implemented in different parts of the continent. This chapter presented lessons on key factors contributing to the success (or lack of it) in implementing such interventions. Of 110 potential projects identified to be used in this study, we selected 25 that had better documentation on their implementation, but different levels of impact evaluation rigor. The interventions addressed a wide range of constraints, such as extension, institutional capacity, input subsidies, environmental degradation, and water resources. We used mostly qualitative methods to (1) assess performance based on 13 factors capturing different stages along the broad project implementation pathway, and (2) evaluate the influence of the factors in achieving the overall productivity target.

We find that projects that scored high in most of the factors performed better in achieving the overall productivity target, compared with projects that had low scores for the majority of the factors. The likelihood of achieving the overall productivity target seems to be influenced most by six of the factors—suitability of instruments, design and timing of implementation, environmental sustainability, financial sustainability, community participation, and organized groups. A common feature across many projects that did not score well was performance in environmental sustainability, suggesting this may be a serious problem for most agricultural projects implemented in the continent. We find that the overall environmental and financial sustainability of the interventions is dependent on the long-term commitment of actors (farmers, communities, local and higher levels of government, donors, and other development stakeholders).

To extend the findings to different biophysical and socioeconomic environments, we tried to analyze the data for different farming systems, as defined in Chapter 4. However, most of the 25 interventions analyzed were implemented in the maize mixed farming system. Thus, the findings are likely to be limited to this farming system only, and to the extent that the rigor of the impact evaluations of the projects was sufficient. While there are more and better-documented success stories than failures, because most the potential projects (77 percent) were dropped from our analysis (because of poor documentation or lack of analytical rigor), the issue of the applicability of the findings to other contexts is critical. There is need to learn from failures as well as from successes. A solid recommendation emanating from the findings and observations in this chapter is that more investment in project M&E systems is required to enable more rigorous impact analysis to be undertaken for more projects ("successful" or "failed") in different parts of the continent, so as to generate the necessary critical mass of knowledge on scalable practices, technologies, and interventions.

Appendix for Chapter 6

TABLE 6A.1 Productivity-enhancing interventions, their locations and objectives, and sources of information

Name*	Objectives (main constraint addressed or instrument used to increase productivity)	Remarks (origin and replication)
Animal Health Services Rehabilitation Programme (AHSRP) Kenya Sources: World Bank (1990); IFAD (1993)	To strengthen the capacity of the Department of Veterinary Services (DVS) for effective delivery of animal health services.	Replication has occurred in animal health interventions in Kenya, Tanzania, Ethiopia, South Sudan, and Uganda, with adjustments to correct for the weaknesses observed in AHSRP.
Agriculture Productivity Enhancement Programme (APEP) Uganda Source: USAID (2007)	To enhance agricultural productivity by promoting use of improved agricultural inputs and addressing marketing challenges by linking smallholder farmers to markets.	Replication has occurred in other agriculture projects in Uganda (for example, the Kaweri Coffee Farmers Alliance Support Project and the Livelihoods and Enterprises for Agricultural Development Project). Replication is also taking place in Ethiopia.
Agricultural Sector Development Programme—Irrigation component (ASDP-irrigation) Tanzania (dry areas) Sources: MAFC (2010); URP (2011)	To rehabilitate existing irrigation systems and establish new ones.	Replication is taking place in other parts of the country in the rehabilitation of existing irrigation schemes there.
Cassava Enterprise Development Project (CEDP) Nigeria (south and southeast) Source: PCU (2007)	To increase the productivity of cassava by reducing the impact of cassava mosaic disease and enhancing marketing and post-harvest handling of cassava.	Replication has occurred in other cassava-producing countries, such as Uganda and Tanzania.
Conservation Agriculture Project 1 (CAP1) Zambia (maize and cotton belts) Source: Haggblade and Tembo (2003)	To address low soil fertility and water constraints.	Replication has occurred in Kenya, Tanzania, Malawi, Uganda, and Zimbabwe.
Crop Crisis Control Project (C3P) Burundi, Democratic Republic of the Congo, Kenya, Rwanda, Uganda, and Tanzania (Association for Strengthening Agricultural Research in eastern and central Africa region) Source: Kimenye and Bombom (2009)	To control the spread of cassava mosaic and banana wilt diseases.	Uptake and adoption have been reported in Burundi, Democratic Republic of the Congo, Madagascar, Rwanda, and Tanzania.
East Africa Dairy Development Project (EADD) Kenya, Uganda, and Rwanda Sources: Gaitano (2011); Mutinda (2011); TANGO International (2011); Baltenweck and Mutinda (2013)	To increase milk yield, reduce milk perishability, and address milk-marketing problems.	Phase 2 of the project is being implemented. Tanzania has been included in the second phase.

(continued)

TABLE 6A.1 (continued)

Name*	Objectives (main constraint addressed or instrument used to increase productivity)	Remarks (origin and replication)
FARM Africa Dairy Goat Improvement Project (FADGIP) Burundi, Ethiopia, Kenya, Rwanda, Tanzania, and Uganda (semiarid lands and high-potential areas with crop livestock mixed systems) Sources: Ayele and Peacock (2003); Peacock (2005); Farm Africa (2007); Ojango et al. (2010)	To improve milk yield and the growth rate of the East African indigenous goat through introduction of improved breeds.	Replication has occurred in other mixed systems and semiarid lands in Kenya.
Farm Input Subsidy Program (FISBP) Malawi (all zones) Sources; Poulton and Dorward (2008); Denning et al. (2009); Dorward, Chirwa, and Jayne (2010)	To increase access to and use of yield-enhancing agricultural inputs.	Has been replicated with adjustments in countries such as Zambia, Tanzania, and Kenya.
Farmer Input Support Program (FISPP) Zambia (all zones) Sources: CSPR, Zambia (2011); Jayne et al. (2011); Burke (2012)	To increase access to and use of yield-enhancing agricultural inputs.	Has been replicated from Malawi.
Fodder Trees and Shrubs Project (FTSP) Kenya, Rwanda, Tanzania, and Uganda Sources: Franzel and Wambugu (2007); Place et al. (2009)	To improve the quality and quantity of feed resources by developing and disseminating high-protein fodder species for dairy animals.	Replication is taking place through dairy development projects. EADD phases 1 and 2 are examples of projects that are replicating this technology.
Fuve Panganai Irrigation Scheme (FPIS) Zimbabwe Sources: Mazungu (1999); Mangwezi (2011); Chazovachii (2012)	To address water constraints caused by recurrent droughts.	Replication is taking place through other irrigation projects in Zimbabwe.
Kaleya Irrigation Project (KIP) Zambia (Kafue River Basin in the south) Sources: AfDB (2010); EU (2010); Bangwe and van Koppen (2012); Illovo Sugar (2014)	To increase smallholder sugarcane production under irrigation.	Has been replicated in the Magobbo and Manyonyo projects in Zambia.
Kenya Dairy Development Programme (KDDP) Kenya Sources: Land O' Lakes (2008); USAID (2008); Ouma et al. (2007)	To improve the dairy value chain (including milk yields, dairy product demand, and industry efficiencies).	Has been replicated in another project, the Kenya Dairy Sector Competitiveness Program (2008–2013), funded by the same donor and implemented by Land O' Lakes.

Name*	Objectives (main constraint addressed or instrument used to increase productivity)	Remarks (origin and replication)
National Agricultural Advisory Services (NAADS) Uganda (all zones) Sources: ITAD (2008); Benin et al. (2007); Benin et al. (2011); EPRC (2011); Okoboi et al. (2011)	To improve the delivery of extension services by developing a demand-driven, farmer-led agricultural service delivery system targeting the poor subsistence farmers, so as to increase access to and use of yield-enhancing technologies.	Replication is occurring through the activities of nongovernmental organizations, with adjustments to correct for identified weakness.
National Agricultural Extension Intervention Program (NAEIP) Ethiopia Sources: Gebreselassie (2006); Byerlee et al. (2007)	To improve extension services and increase access to and use of yield-enhancing agricultural inputs.	Developed on the documented success of the Sasakawa Global 2000 Agricultural Program.
New Rice for Africa (NERICA) upland rice Guinea, Côte d'Ivoire, Sierra Leone, and Uganda Sources: WARDA (2001); Kijima, Sserunkuuma, and Otsuka (2006); Kijima (2008); Diagne et al. (2010)	To increase adoption of a hybrid variety of rice adapted to local conditions.	Developed by the West Africa Rice Development Association (WARDA) and adopted in various countries.
Operation *Mwolyo* Out (OMO) Kenya (Mwala district and semiarid parts of Kenya) Sources: Masika (2011); Field interview (December 5, 2011)	To address water constraints through adoption of water-harvesting technologies (water pans or small dams, *zai* pits).	Replication is happening in other villages in the Machakos district.
Participatory Irrigation Development Programme (PIDP) Tanzania (dry subhumid zone in maize mixed farming system) Source: IFAD (2007)	To address water constraints through irrigation.	Interventions under other rice projects in the country have been developed based on lessons from this project.
Push–Pull Technology (PPT) Ethiopia, Kenya, Tanzania, and Uganda (East Africa in areas where cereals are produced) Sources: ICIPE (2003); Khan et al. (2006); Fischler (2010)	To address biotic (insect pests, parasitic weed *Striga*) and abiotic (land degradation and poor soil fertility) constraints.	Replication has occurred in several districts within East Africa, where conditions and challenges are similar to those of the original project sites.
Regional Land Management Unit (RELMA) Eritrea, Ethiopia, Kenya, Tanzania, Uganda, and Zambia (arid and semiarid areas in East and southern Africa) Sources: RELMA (2005); Gathiru and Ong (2006); Erikson (2008)	To address water scarcity through a range of improved land management practices and water-harvesting technologies.	Replication with adjustments is occurring in Tanzania, Kenya, Botswana, Malawi, Mozambique, Rwanda, Uganda, and Zimbabwe.

(continued)

TABLE 6A.1 (continued)

Name*	Objectives (main constraint addressed or instrument used to increase productivity)	Remarks (origin and replication)
Sasakawa Global 2000 Agricultural Program (SG2000-AP) Ghana, Sudan, Tanzania, Benin, Togo, Mozambique, Eritrea, Guinea, Burkina Faso, Malawi, Mali, Nigeria, Ethiopia, and Uganda Sources: Nubukpo and Galiba (1999); Dowswell (2011)	To increase adoption of yield-enhancing inputs and improved farming practices.	Initially started in the subhumid zones of West Africa (Ghana and Benin) and expanded to the semiarid areas. Replication took place in the other project countries. The NAIEP in Ethiopia (see above) was developed on the documented success of the SG2000-AP.
Specialty Coffee Program (SCP) Rwanda Sources: Chemonics International (2006); Boudreaux (2010) Abramovich and Zook (2015)	To improve the quality of coffee produced and the value chain.	Initially, new coffee-washing stations were built in a few districts (for example, Maraba, Karaba, and Gashonga districts). Maraba coffee growers received fair trade certification and began to grow shade-grown coffee, which has become a model for the Rwandan coffee industry and has been replicated countrywide. Replication is also taking place in Ethiopia.
System of Rice Intensification (SRI) Rwanda (marshland areas of Kibaza in Bugesera district and Rwabutazi in Kihere district) Source: IFAD (2009)	To increase rice yields by promoting adoption of the SRI, which involves changing the management of plants, soil, water, and nutrients.	Originated in Madagascar. Replication has occurred in other parts of Rwanda, in Sierra Leone, and in the Gambia.
Wei Wei Integrated Development Project (WWIDP) Kenya (arid areas in west Pokot district) Source: Mugova and Mavunga (2000)	To address food insecurity problems through improved soil and water management.	Has been replicated in the Arror irrigation scheme in the Marakwet district of Kenya.

Source: Authors' compilation.

* Name of intervention, countries, agroecology or production environment, sources.

TABLE 6A.2 Instrument used to collect information from agricultural and rural development practitioners

To facilitate effective discussions during the workshop, you are kindly requested to address the questions below.

1. Please list three to five cases/examples of interventions that successfully increased agricultural productivity in your country.
2. Please list three to five cases/examples of interventions that were unsuccessful (failed) in increasing agricultural productivity in your country.
3. In your opinion, why did the projects you have identified in question 1 above succeed? (Please fill in the table below.)

Name of the program/project	Factors for success
1.	
2.	
3.	
4.	
5.	

4. In your opinion, why did the projects you have identified in question 2 above fail? (Please fill in the table below.)

Name of the program/project	Reasons for failure (lack of success)
1.	
2.	
3.	
4.	
5.	

TABLE 6A.3 Agricultural productivity impact pathways: How the 13 factors identified in the conceptual framework affect productivity

Factor	Pathway of influence
1. Problem definition	Identifying the problem correctly raises the probability of using the appropriate productivity-enhancing instruments or technologies. Also, with proper diagnosis, stakeholders may be better informed, may help set the relevant objectives and targets, and may demand the appropriate instruments for achieving the objectives and targets.
2. Choice and suitability of instruments	Because farmers in different locations face different biophysical and socioeconomic constraints, and different approaches have different impacts in different locations (Chapter 5), when instruments (technologies, commodities, enterprises, institutional capacity building, etc.) that suit the production environment and needs of the local communities or beneficiaries are used, then the chances of the project having a positive influence on productivity are higher.
3. Target population	Related to the choice and suitability of instruments, this factor has to do more with reaching a specific target population. For example, a universal input subsidy (that is, subsidizing the price of the input sold in the market) is not likely to increase overall use of the input, because farmers who would have purchased the fertilizer without the subsidy may likely substitute what they would have purchased commercially with what they obtain via the subsidy (that is, crowding out—Jayne et al. 2013). It may be more effective to provide the input directly (for example, via coupons) to farmers who would not have purchased the input without the subsidy.
4. Design and timing of implementation	Design refers to having a plan of what will be done when and with what resources, based on a sound logical framework and including monitoring and evaluation to help make informed mid-course changes in implementation. Thus, having a good plan and implementing activities in a timely fashion raise the probability of staying on course to achieve intended impacts. Delays in procurement, for example, may lead to unused or underused resources and may cause suboptimal achievements.
5. Environmental sustainability	This is a long-term concept, whose effect may manifest after completion of the project. For example, an irrigation project may not be environmentally sustainable if it causes salinization of the soil that negatively impacts yield by more than the positive productivity or income effect of the irrigation. Thus, having measures in place that sustain productivity gains during and after completion of the project is critical.
6. Financial sustainability	Although having adequate financial resources to implement the intervention is critical (which means having a solid budget and funding commitments), like environmental sustainability, what happens after the project is important. For example, most of the productivity-enhancing interventions include providing farmers with free or subsidized inputs to demonstrate their profitability, so that farmers can purchase and use them on their own when the project is complete. If farmers are unable to purchase the inputs on their own because of lower-than-expected profitability, then the project is not financially sustainable. Thus, ensuring financial availability during the project and financial viability after project completion is critical, which ties in with the choice and suitability of instruments.
7. Community participation	Because farmers and communities are expected to be better informed about their own constraints, potentials, and production environments, involving them in all stages of the project will likely enhance the accuracy of defining the problem, the choice of solutions, their ownership of the project, and their commitment to implementing it accordingly. This also influences the sustainability of the project.

Factor	Pathway of influence
8. Gender consideration	Because there are gender differences in agricultural production and productivity, primarily because of differences in access to productive assets and markets, taking such gender differences into account in the project design and implementation enhances the chances of addressing the constraints that each homogeneous group faces, thereby increasing the likelihood of attaining the overall productivity objective. This is closely related to the target population factor, which is more general.
9. Capacity building	The capacity (especially technical and managerial skills) of farmers and communities influences several of the factors discussed above. Whereas farmers may be better informed about their production environment, etc., they may not have the technical skills to analyze complex interrelated factors and to manage them accordingly. Therefore, building their capacity in a manner that complements their indigenous knowledge will likely not only increase the speed of innovation and adoption, but also strengthen their commitment to manage the project sustainably.
10. Organized groups	Collective action by farmers (through farmers' groups, associations, cooperatives, etc.) has several benefits, including reduction in transaction costs, better access to credit and markets, and better negotiation outcomes, which in turn reduce production costs and increase the output and value of production. Because groups that form organically are more likely to work well and last longer (primarily because of trust issues), they are more likely to have sustainable productivity effects than those that form only to take advantage of the project handouts.
11. Leadership and dedication	Having in place a strong and dedicated leadership in the community is important in several aspects, particularly in implementing the project according to plan, making informed decisions to stay on course to achieve the intended impacts, and ensuring fair participation of members in the project and equitable distribution of the benefits. Because of issues of elite capture (Feder et al. 2010) and the history of mismanagement by cooperatives (Kherallah et al. 2002), having a leadership council with representation of different and marginalized groups is more likely to have greater buy-in to the project and sustainable productivity effects than those that represent a few or a homogeneous group. This also influences the relationship with and support from higher levels of government, which are discussed later.
12. Complementary investments and partnerships	Because many factors affect the performance of agriculture in complex ways, and because projects tend to focus on or address a few of those factors only, the outcomes of the project will depend on how the factors outside the control of the project influence or interact with the project components. The effects of those influences or interactions may be positive or negative. Therefore, projects that internalize these effects and put in place measures that minimize the negative effects and enhance the positive effects are more likely to be successful at raising agricultural productivity than those that do not. For example, increasing the access of farmers to yield-enhancing inputs (such as expensive inorganic fertilizers) will likely result in higher output and yield. However, if farmers do not get favorable markets for their produce and lose out, then they will likely not continue to use the fertilizers. The same logic holds for forging partnerships.
13. Policies and political stability	Policies and national-level factors, including infrastructure development, macroeconomic management, and political stability, affect farmers' decisions and their outcomes. The major policies and factors include land tenure and natural resource management; input (fertilizer, seed, mechanization, etc.) policies; and market development (price support, buffer stock, etc.). Here too, projects that internalize these factors and put in place measures that enhance the positive effects and minimize the negative effects are more likely to be successful at raising agricultural productivity than those that do not.

Source: Authors' compilation from literature review and consultations.

TABLE 6A.4 Description of methodology used in rating performance against the criteria of successful project implementation

Indicator 1: Problem definition
Measures:
- Clear and concise description of the problem or productivity constraint and objectives of the intervention.
- Indication that the project responds to local and national priorities, as identified by national planning documents, policy and strategy documents, etc.
- Problem definition is informed by technical analysis (baseline study, feasibility study, cost-benefit analysis, etc.).
- Participatory needs assessment is undertaken with the intended beneficiaries.

Performance rating: High to very high if there are at least three of the above; very low to moderate otherwise.

Indicator 2: Choice of the commodity/instruments
Measures:
- Evidence that the project is focused on priority commodities for food security and income generation in the area, based on what people are already doing.
- Evidence that the proposed productivity solution (commodity/enterprise) is suitable for the area (agroecological conditions are appropriate).
- Indication that the commodity is identified as a priority in the national strategy documents or in empirical studies.

Performance rating: High to very high if there are at least two of the above; very low to moderate otherwise.

Indicator 3: Suitability of instruments
Measures:
- Evidence that the right beneficiaries are targeted and reached (the vulnerable, the poor, women, etc.).
- Indication that the project considered the local-level socioeconomic factors (for example, local demand, culture, religion, beliefs, preferences in the project design).
- Evidence that the intervention can address the productivity constraint identified under the problem definition phase.

Performance rating: High to very high if there are at least two of the above; very low to moderate otherwise.

Indicator 4: Design and timing of implementation
Measures:
- Clear articulation of the suitability of the project intervention in addressing the identified productivity problem.
- Clear roles and responsibilities of the different project implementers and partners.
- Timely implementation.
- Implementation agency has the necessary technical, managerial, and financial capacities to implement the intervention.
- Built-in mechanisms address challenges likely to affect project performance.
- Evidence that monitoring and evaluation were used to inform project implementation.

Performance rating: High to very high if there are at least three of the above; very low to moderate otherwise.

Indicator 5: Environmental sustainability
Measures:
- Evidence that the project undertook an environmental impact assessment.
- Evidence that intervention is environmentally friendly or incorporates environmental protection measures (for example, avoids overexploitation of natural resources, reduces pollution of water and air).
- Indication that the intervention resulted in significant improvement in natural resources.

Performance rating: High to very high if there are at least two of the above; very low to moderate otherwise.

Indicator 6: Financial sustainability
Measures:
- Evidence that financial resources are adequate to implement the project as planned.
- Indication of transparent and accountable use of financial resources.
- Indication of adequate managerial capacity after the end of the project.
- Presence of a well-defined exit strategy (that is, adequate measures/activities are in place to sustain the activities after the project ends).
- Indication that the target communities are still accessing the benefits of the project after its lifetime.

Performance rating: High to very high if there are at least three of the above; very low to moderate otherwise.

Indicator 7: Community participation
Measures:
- Evidence of participation/involvement of the local communities (beneficiaries) at the planning, design, and implementation phases.
- Presence of strong support and commitment to project objectives by local government officials and community leaders.
- Willingness and ability of the community to adapt to changes as a result of the intervention.

Performance rating: High to very high if there are at least two of the above; very low to moderate otherwise.

Indicator 8: Gender consideration
Measures:
- Evidence of gender consideration in project design and implementation through clear articulation of involvement of men, women, and youths (as seen in the project documents).
- Evidence that gender issues were mainstreamed in activities.
- Evidence of benefits that were accessed by women, men, and youths (for example, access to services, capacity-building support).

Performance rating: High to very high if there are at least two of the above; very low to moderate otherwise.

Indicator 9: Complementary investments and partnerships
Measures:
- Evidence that the project sought/established effective partnerships with others in implementing the project (based on the number of partners and funders).
- Indication that a multidisciplinary approach was adopted in implementing the project.
- Evidence of the availability of complementary interventions (for example, an irrigation project shows evidence of the availability of seeds, fertilizers, and markets by the project or project partners).
- Clear modalities for coordinating different partners.
- Clear roles and responsibilities of the different partners.

Performance rating: High to very high if there are at least three of the above; very low to moderate otherwise.

Indicator 10: Capacity building
Measures:
- The project has an explicit objective to build the capacity of beneficiaries (through technical or institutional support).
- Evidence that the project trained beneficiaries and other relevant partners.
- Evidence that knowledge transfer products, such as videos, training manuals, and brochures, were disseminated.

Performance rating: High to very high if there are at least two of the above; very low to moderate otherwise.

Indicator 11: Organized groups
Measures:
- Evidence that the project tapped into and built upon social networks and groups.
- Presence of an objective to build or strengthen farmers' groups.
- Evidence that the beneficiary groups had the capacity to engage effectively in the project.

Performance rating: High to very high if there are at least two of the above; very low to moderate otherwise.

Indicator 12: Leadership and dedication
Measures:
- Evidence that the project sought and obtained the support of the local leaders in the implementation process.
- Evidence that the government provided a conducive environment for the intervention.
- Evidence that community leaders were empowered to provide relevant local public goods to support the intervention.
- Evidence of government support for the initiative through financial or in-kind support.
- Evidence that the local communities contributed to the project (for example, through in-kind support).

Performance rating: High to very high if there are at least two of the above; very low to moderate otherwise.

(continued)

TABLE 6A.4 (continued)

Indicator 13: Policies and national-level factors
Measures:
- Evidence that national policies and regulations supported the project or provided an opportunity for the project's objectives to have an impact.
- Evidence that the existing climate supported the project's implementation.
- Existence of conducive macroeconomic conditions.
- Existence of a conducive political environment.

Performance rating: High to very high if there are at least two of the above; very low to moderate otherwise.

Source: Authors' compilation from literature review and consultations.

TABLE 6A.5 Summary of review of performance in implementation of selected agricultural productivity-enhancing interventions in Africa south of the Sahara

Here, each of the 25 productivity-enhancing interventions evaluated in the study are presented in relation to the 13 project implementation performance indicators. The methodology used in rating performance relative to each indicator is presented in Table 6A.4.

A: Animal Health Services Rehabilitation Project (AHSRP), Kenya

Problem definition: Aimed to address institutional challenges affecting the delivery of animal health services, which was in a poor state and a constraint to production and productivity. Innovations were to be carried out based on the studies that were commissioned at the beginning of the project. However, most of the studies were delayed, and when eventually completed, they were often not accepted by the responsible government agency because of their poor quality.
Rating: Low to moderate

Choice of the commodity/instruments: Instruments used included disease-control campaigns, provision of clinical services and field extension services, rehabilitation and re-outfitting of veterinary laboratories, and enhancement of surveillance activities and support to pilot trials of tsetse-fly and tick-borne disease control. These instruments were appropriate for improving animal health services delivery.
Rating: High to very high

Suitability of instruments: Targeting was not properly done and the project did not reach the poor and vulnerable populations. Most of the beneficiaries were the more influential members of the communities. Implementation of the project also faced challenges resulting from wrong assumptions. For example, it was assumed that animals could be vaccinated throughout the year; however, it later was discovered that herds could be vaccinated only at specific times of the year because of the nomadic lifestyle of the livestock owners.
Rating: Low to moderate

Design and timing of implementation: The project faced various implementation problems, including poor management capacity, delays in project start-up and procurement, and failure to adequately staff and equip the Project Management Support Unit. Project disbursement was very slow. A two-year extension was granted, but only 51 percent of total project funds was used.
Rating: Low to moderate

Environmental sustainability: There was no articulation of how any environmental issues were to be handled.
Rating: Low to moderate

Financial sustainability: The costs of the project were underestimated: the initial International Fund for Agricultural Development (IFAD) appraisal projected a total cost of $19.3 million, whereas the final International Development Association (IDA)/World Bank appraisal finished at $70.5 million. The government's agreed contribution of $41.52 million did not materialize because of budgetary constraints.
Rating: Low to moderate

Community participation: Evaluation findings indicate that the beneficiaries were not consulted adequately at the design and implementation stages, and this negatively affected the project. For example, failure to consult smallholders and pastoralists on the period of vaccination resulted in a low number of cattle vaccinated.
Rating: Low to moderate

Gender consideration: There is no explicit information that the project paid attention to gender aspects.
Rating: Low to moderate

Complementary investments and partnerships: Four international agencies (IFAD, IDA, the Organization of Petroleum Exporting Countries, and United Nations Development Programme [UNDP]) committed to contribute funds to the project. There were no clear modalities of coordinating different partners.
Rating: Low to moderate

Capacity building: The project aimed to enhance the capacity of the animal health workers. A total of 170 staff members of the Department of Veterinary Services (DVS) were trained, and four fellowships were provided for veterinary officers to study for Master of Science degrees in veterinary epidemiology and economics.
Rating: High to very high

Organized groups: The design of the project mainly focused on supporting DVS. Minimal efforts were targeted at supporting livestock keeper groups.
Rating: Low to moderate

Leadership and dedication: Efforts by many District Veterinary Officers and field staff to implement project activities did not materialize because of inadequate operational support (for example, vehicles and resources to cover operating expenses) from the central office.
Rating: Low to moderate

Policies and national-level factors: The introduction of cost recovery, coupled with the privatization of communal dips and clinical veterinary services, seems to have negatively influenced use of these services by livestock owners, which contributed to more and extensive disease outbreaks. Other external factors, such as droughts and market forces, also contributed to a decline in the production of meat and milk during the project period.
Rating: Low to moderate

B: Agricultural Productivity Enhancement Program (APEP), Uganda

Problem definition: The program aimed at expanding rural economic opportunities and creating economies of scale to catalyze transformation of agriculture from low-input and low-output subsistence farming to higher-yielding and commercially competitive agriculture. This was informed mainly by lessons from the Investment in Developing Export Agriculture project that had been implemented in Uganda for 10 years from 1995 to 2004. It focused on addressing key agriculture challenges in Uganda—that is, low productivity, high-postharvest losses, and poor marketing. The project was consistent with the Government of Uganda's Poverty Eradication Action Plan, Plan for Modernisation of Agriculture, and Medium-Term Competitiveness Strategy.
Rating: High to very high

Choice of the commodity/instruments: The program focused on priority commodities for food security and income generation in Uganda, such as cotton, maize, coffee, sesame, upland rice, sunflowers, barley, flowers, vanilla, and bananas. These commodities were identified as priority commodities in strategy documents for Uganda.
Rating: High to very high

Suitability of instruments: The program targeted producer organizations for each of the commodities. A wide range of instruments was used to meet project objectives, including production-to-market transactions, improvements in input distribution and technology transfer, strengthening of producer organizations, and development of competitive agricultural and rural enterprises. These strategies are known to contribute to enhancing productivity and stimulating agricultural trade.
Rating: High to very high

(continued)

TABLE 6A.5 (continued)

Design and timing of implementation: The program was well designed and adequately staffed by qualified staff to deliver on its work plan. It had adequate resources for project implementation. Monitoring and evaluation (M&E) was used as a tool for learning throughout the project. In addition to the funding from the United States Agency for International Development (USAID), the project was allocated $665.5 million by the Government of Uganda, the Danish International Development Agency (DANIDA), IFAD, the European Union, the West Africa Rice Development Association (WARDA), and IDA.
Rating: High to very high

Environmental sustainability: Procedures for environmental sustainability were adhered to for coffee-related activities. There is no clear evidence on strategy for environmental sustainability in other crops.
Rating: Low to moderate

Financial sustainability: Financial sustainability was not ensured after the life of the project. Although there were efforts to ensure the sustainability of the project's services after the life of the project through partnering with local organizations, there are still sustainability challenges. The activities of the producer groups have not been sustainable after the end of the project because of financial constraints. Agriculture commercialization is still a challenge among the smallholder farmers.
Rating: Low to moderate

Community participation: Local communities participated during planning and implementation of the project. For example, the program effectively used a producer organization's model in technology transfer. It identified a lead farmer who could direct the farmer field schools (FFS), which acted as a substitute for formal agricultural training. The commitments of the local communities to adopt the new technologies promoted by the program were clear. They implemented the improved farming techniques promoted by the program and organized themselves into groups for the purpose of bulking their produce to facilitate marketing.
Rating: High to very high

Gender consideration: Participation of women was encouraged in all aspects of building the farm enterprise, including being lead farmers, site coordinators, and farm committee executives. Women producers also benefited from training. For example, the program organized training workshops on gender mainstreaming in agriculture, and females constituted 58 percent of 3,000 farmers who were exposed to improved banana production and maintenance practices.
Rating: High to very high

Complementary investments and partnerships: A wide range of partnerships was forged and exploited to further the program's objectives. In addition to the government and other partners, APEP attracted the support of other donors and leveraged some of their resources to finance program activities. The program had more than 15 partners from both the public and the private sectors, including the Government of Uganda, the Japan International Cooperation Agency (JICA), the Royal Netherlands Embassy, DANIDA, IFAD, the National Cooperative, the Business Association, and the National Agricultural Advisory Services (NAADS). The roles of the partners were clearly defined. The program addressed various constraints along the crop value chain. The activities were implemented by different actors in a coordinated manner.
Rating: High to very high

Capacity building: The program provided technical and financial assistance to farmers' groups (including producers as trade groups) and associations. It trained farmers in various areas, such as crop production techniques, soil and water management, formation of groups, marketing, gender awareness, postharvest and storage practices, business planning, financial management, and improved market information using such approaches as demonstration plots, lead farmer training, and extension visits. Extension manuals and videos were produced and disseminated. APEP's approach created a "critical mass of capable local producers (smallholder farmers) and support industries, such as input suppliers, and [linked] them to local, regional, and international markets."
Rating: High to very high

Organized groups: The program strengthened producer organizations. Farmers were encouraged to form producer organizations to enhance their abilities to access input credit, undertake bulk marketing, and improve net farmgate prices.
Rating: High to very high

Leadership and dedication: The local communities (individual farmers and community leaders) supported the program and were dedicated to undertake program activities. Group leaders were trained and championed program activities in collaboration with the program team.
Rating: High to very high

Policies and national-level factors: The implementation of the program benefited from favorable political stability in the country. However, the program suffered a setback because of extreme climatic conditions. It was affected by droughts and occasional heavy rains and floods that damaged infrastructure in several places in the country. There was an outbreak of a fungal disease, which affected cotton production. The program was also affected by inadequate implementation of agricultural policies and regulations and lack of adequate storage facilities for crops. Although the prices of food staple commodities were high during the program period, the prices for high-value commodities promoted by the program, such as vanilla and flowers, were depressed.
Rating: Low to moderate

C: Agricultural Sector Development Programme—irrigation component (ASDP-irrigation), Tanzania

Problem definition: The program aimed at increasing water availability for agricultural production, mostly following the Agricultural Sector Development Strategy (ASDS) participatory process, which prioritized enhancement of crop production through irrigation development and improvement.
Rating: High to very high

Choice of the commodity/instruments: Instruments included rehabilitation and management of low-cost smallholder irrigation schemes for producing food security crops (for example, rice and maize), which could also serve as cash crops and also be considered as strategic commodities. There are concerns that the program paid too much attention to rice and left out horticulture, which could have added value to increasing income and food security.
Rating: High to very high

Suitability of instruments: By design, ASDP is well aligned to National Strategy for Growth and Reduction of Poverty targets. Therefore, it is meant to target poor districts. Assessment of placement of the ASDP irrigation scheme shows that ASDP investments were well targeted to districts with severe poverty. On average, the per capita area of ASDP irrigation schemes was 92 square meters (m^2) in districts with very severe poverty, compared with only 52 m^2 in districts with low poverty. Additionally, 40 percent of the ASDP irrigation scheme area was located in districts with very severe or severe poverty, even though such districts accounted for only 35 percent of the total population.
Rating: High to very high

Design and timing of implementation: There were various challenges with coordination of program activities and implementation, including delays in procurement, financial management, and low capacity of the implementing agencies at both the Agricultural Sector Lead Ministries and the local level. These challenges caused delays in implementation of activities. ASDP was mainstreamed in the existing government system of financing public expenditures, but detailed assessment to identify bottlenecks in the system was not undertaken at appraisal. Staffing in the Prime Minister's Office–Regional Administration and Local Government was sufficiently strengthened.
Rating: Low to moderate

Environmental sustainability: Although an environmental impact assessment (EIA) was conducted, measures to mitigate the negative effects of irrigation have been inadequate. An EIA conducted on the program indicates that salinity is building up in some schemes, leading to decline in yields. This problem is yet to be addressed effectively.
Rating: Low to moderate

Financial sustainability: About 75 percent of the resources was allocated at the Local Government Authority level and 25 percent at the national level. Most of the resources were invested in infrastructure development and rehabilitation, with little allocation for strengthening the weak operation and maintenance (O&M) mechanisms in the irrigation schemes. For example, the third ASDP implementation report observed no or a very small O&M budget in most of the schemes visited. Most of the success of the ASDP interventions seems to have largely resulted from input subsidies, whose sustainability is questionable.
Rating: Low to moderate

(continued)

TABLE 6A.5 (continued)

Community participation: Irrigation schemes are owned, managed, and operated by the targeted beneficiaries, but there are no clear measures for financing the irrigation costs. Some community groups have outlined plans for regular maintenance of the infrastructure through membership fees. Limited market opportunities limit their ability to contribute to the fees effectively. Furthermore, private-sector participation in the delivery of agricultural services at the local level was generally low.
Rating: Low to moderate

Gender consideration: Although males, females, and youths are involved in the irrigation activities, gender mainstreaming in ASDP interventions has been inadequate.
Rating: Low to moderate

Complementary investments and partnerships: The National Input Voucher System (NAIVS), an input subsidy program, has contributed to the success of the irrigation project. However, implementation challenges and targeting problems of the NAIVS have limited the project's ability to realize its full potential. There is indication that the quantity of fertilizer available to farmers is less than the amount required, and targeting tends to exclude the poor.
Rating: Low to moderate

Capacity building: Although some capacity-building activities have been conducted, program planning and implementation capacity at district and subdistrict levels are still weak, and training provided to farmers, extension officers, and the private sector is limited. An impact evaluation study showed that even though about 35 percent of ASDP funds allocated was used for farmer training and extension services, the funds were limited to production activities only, leaving out training on marketing and postproduction activities.
Rating: Low to moderate

Organized groups: Although farmers organized into groups, not all of them have taken full advantage of learning from their colleagues in the group, as most farmers do not appreciate the benefits of proper use of inputs. (They use either too little fertilizer or none at all.) As a result, these farmers are hardly benefiting from the irrigation interventions. Similarly, several marketing constraints prohibit farmers from benefiting from collective action.
Rating: Low to moderate

Leadership and dedication: This program was implemented at a time when there was a high level of government commitment and attention to agriculture in support of ASDS and other key initiatives in Tanzania, including Kilimo Kwanza, Big Results Now, and the Southern Agricultural Growth Corridor of Tanzania. These initiatives have support from various stakeholders, including the local communities. Studies have indicated that the performance of traditional, improved traditional, modern, and rainwater harvesting-based schemes in Tanzania can be improved at the field level through building the capacity of farmers and empowering and enabling them to secure full ownership of the schemes. The government is working to empower local communities.
Rating: High to very high

Policies and national-level factors: There are good policies to support irrigation, but their implementation is inadequate due to lack of government resources at the national level. The government has developed a national irrigation policy to provide direction to the implementation of irrigation interventions and to ensure the optimal availability of land and water resources for agricultural production and productivity, so as to contribute effectively to food security and poverty reduction, as stipulated in the National Strategy for Growth and Reduction of Poverty (or MKUKUTA) (URP 2011). According to the National Irrigation Policy, 2009, the national target of increasing the area under irrigation has not been met because of inadequate financial resources. Also, various policies constrain agricultural marketing, especially for maize and rice, affecting the profitability of farmers. Frequent export bans and unpredictable importation of rice have been cited to distort the marketing of these key commodities. Finally, water availability during the dry season has been a perennial problem in the irrigation schemes.
Rating: Low to moderate

Source: Authors' evaluation based on URP (2011) and Nkonya et al. (2013).

D: Cassava Enterprise Development Programme (CEDP), Nigeria

Problem definition: This program was nationally important, given that Nigeria is the largest producer of cassava in the world. The program aimed at addressing the problem of low cassava productivity by reducing the impact of cassava mosaic disease in selected communities in the southern states of Nigeria. Problems of limited marketing opportunities and inadequate postharvest-handling options were also targeted. A baseline study, which was carried out in a participatory manner, was useful in identifying the prevailing productivity constraints, which include diseases, poor agronomic practices, and low use of agrochemicals.
Rating: High to very high

Choice of commodity/instrument: Various instruments were used: introduction of disease-resistant and high-yielding cassava varieties, training on improved agronomic practices to increase production, disease-control interventions to reduce the impact of cassava mosaic disease, development and expansion of postharvest processing, and marketing and agroenterprise development.
Rating: High to very high

Suitability of instruments: The program targeted the poor, who are mostly located in impoverished states situated in the Niger Delta region. Interventions were targeted to resource-poor producers (mostly women), micro- and small-scale processors, fabricators, traders, agribusiness entrepreneurs, and consumers.
Rating: High to very high

Design and timing of implementation: The International Institute of Tropical Agriculture (IITA) had a qualified coordinator to coordinate implementation of program activities. The program had appropriate staff for program implementation. A program management committee consisting of representatives of the three partners (USAID, SPDC, and IITA) oversaw the program activities. The advisory arm of CEDP was a seven-member stakeholder committee/cassava enterprise association.
Rating: High to very high

Environmental sustainability: The program aimed at addressing unsustainable land management practices. Farmers were trained on soil fertility management and sustainable soil management.
Rating: High to very high

Financial sustainability: CEDP trained local nongovernmental organizations (NGOs), and most of the funding for this program was from development partners. Farmers who were subsidized heavily during the program's lifetime found difficulties in making a profit in the absence of program support. Most of the processing factories that were set up with support from the program are not operational (they have broken down). Some groups have financial problems because of the unprofitable nature of the product they were producing and its disconnect from the end markets.
Rating: Low to moderate

Community participation: Local communities were involved adequately in program implementation and decisionmaking. This was evident from the composition of the program steering committee, which was made up of representatives from the Federal Ministry of Agriculture and Rural Development, state governments, the National Agricultural Research and Extension System, producers' associations, NGOs, and the private sector. Participatory tools and techniques were used to explore relevant issues on cassava enterprise. The technologies promoted by the program were well received by many farmers, who adopted improved techniques for cassava production and postharvest handling.
Rating: High to very high

Gender consideration: CEDP targeted women and youths, who play important roles in cassava processing and marketing. The program had a gender specialist who made sure that gender issues were mainstreamed in CEDP activities. The project supported enterprises that created new jobs for women and youths.
Rating: High to very high

Complementary investments and partnerships: Public–private partnership was highly applied in this program, which was funded by USAID and the Shell Petroleum Development Company, and was implemented by IITA. Many partners were involved in the program, as mentioned above, and they were well coordinated and had clear roles.
Rating: High to very high

(continued)

TABLE 6A.5 (continued)

Capacity building: The program focused on strengthening the human and institutional capacities of different groups of beneficiaries to produce, process, and market cassava efficiently, as well as on increasing private-sector investment in cassava production, processing, storage, and marketing. Extension materials and other knowledge products were distributed to the project beneficiaries.
Rating: High to very high

Organized groups: CEDP supported collective action by providing technical and financial support to producers', processors', and traders' groups. However, not all groups did well; some suffered from internal group disagreements, management problems, and poor accountability.
Rating: Low to moderate

Leadership and dedication: The level of government support for the cassava subsector was high. It was spearheaded by the Presidential Initiative on Cassava, which was launched in Nigeria in 2003 and brought cassava and its potential to the limelight. However, there were challenges in obtaining contributions from the local communities, and there was resistance from some youth groups.
Rating: Low to moderate

Policies and national-level factors: The cassava industry has been affected by inconsistent government market policy. Prices for cassava are very unstable. There is weak market information: farmers do not know where their products are sold or for how much, or which are the best links for disposing of produce. Enterprises faced difficulties in selling cassava to flour millers because of nonenforcement of the policy to use 10 percent of cassava flour in bread. The country was affected by political problems and poor infrastructure, especially in the Niger Delta region.
Rating: Low to moderate

E: Conservation Agriculture Project 1 (CAP1), Zambia

Problem definition: The project aimed to address the constraints of low productivity caused by water problems and poor soil quality by using conservation agriculture (CA) technologies and practices. Promotion of CA is a priority activity in Zambia, and is stipulated within the 2004–2015 Zambian National Agricultural Policy. The project was informed by various studies that had indicated that CA technology does work in Zambia and could contribute to productivity enhancement.
Rating: High to very high

Choice of the commodity/instruments: The project focused on maize and cotton, key crops for food security and income generation. Different tillage methods, such as basins and ripping were promoted; tuber, grain, and legume plant materials were distributed; and tree planting (*Faidherbida albida* and *Jatropha curcas*) was promoted. The poorer segments of the population benefited most from CAP1.
Rating: High to very high

Suitability of instruments: The above instruments, which are known to improve soil properties and increase productivity, are appropriate for the geographical location, specifically agroecological zones suitable for maize production.
Rating: High to very high

Design and timing of implementation: The project was well designed and had adequate funding from the Norwegian Embassy in Zambia. The project was implemented by the Conservation Farming Unit (CFU) of the Zambian National Farmers Union. It had adequate staff. It built up an extension system based on regional coordinators who were CFU staff, farm coordinators, contact farmers, and associate farmers. NORAGRIC (a department of International Environment and Development Studies at the Norwegian University of Life Sciences) was given the role of monitoring the project's implementation. The project had a good extension strategy and benefited from the extensive knowledge of CA within CFU. M&E was used to guide the project's implementation.
Rating: High to very high

Environmental sustainability: The project aimed to contribute to environmental sustainability through adopting conservation farming and reforestation, increasing carbon content, and having each farmer plant 200 *Faidherbida albida* trees after five years. However, these goals were not fully achieved. Adoption of CA still faces challenges resulting from labor constraints, inadequate access to improved inputs, the need to use crop residues as animal feeds, etc. The survival rate for the trees planted has been only 33 percent. Also, only 18 percent of the farmers planted *Jatropha curcas* as a live fence around their farms. The interest in planting *Jatropha* declined over time.
Rating: Low to moderate

Financial sustainability: Most of the adoption of CA in Zambia is supported by the donor community and international development agencies. This poses a challenge to the project's sustainability, because the government's contribution to the intervention was minimal. Farmers do not have adequate resources to buy required inputs (such as seeds, herbicides, and fertilizer). Without herbicides, there has been pressure for using hired labor for land preparation.
Rating: Low to moderate

Community participation: Farmers were involved in the implementation. A study by Aune, Nyanga, and Johnsen (2012) indicated the project had good rapport with farmers, created incentives for the farmers to adopt CA, and as a result managed to facilitate increased adoption of CA technologies and practices.
Rating: High to very high

Gender consideration: One of the objectives of the project was to increase the number of women involved in CA. Evaluation results showed that this objective was partly achieved, because women faced several limiting factors (for example, lack of labor, lack of land, and lack of access to such inputs as seeds and fertilizers). Women also find it hard to use some of the equipment promoted, such as *chaka* hoes, which require significant strength to use.
Rating: Low to moderate

Complementary investments and partnerships: Various actors have been involved in promoting CA in Zambia, including large-scale private actors, NGOs, the government, and donors. However farmers face challenges to adopt CA due to lack of access to complementary inputs, such as seeds, land, and fertilizers. The government fertilizer and seed subsidy schemes only benefited a few smallholder famers.
Rating: Low to moderate

Capacity building: Training was a major component of the project. The project selected farm coordinators who were trained by CFU field officers. Each farm coordinator trained contact farmers, and each contact farmer trained associate farmers and additional farmers or nonassociated farmers. The project built up an extension system based on regional coordinators (CFU staff), farm coordinators, contact farmers, and associate farmers.
Rating: High to very high

Organized groups: Farmers are organized in various groups that facilitated learning and experience sharing.
Rating: High to very high

Leadership and dedication: CFU has been instrumental in developing CA in Zambia. Formed in 1995, CFU is an independent organization having a collaborative agreement with the Zambian National Farmers Union. Local farmers' organizations promoted CA.
Rating: High to very high

Policies and national-level factors: CA was introduced at the right time, when Zambia needed an intervention to reverse the declining trend in agricultural productivity. The Ministry of Agriculture and Co-operatives has a climate change adaptation and mitigation agenda, and potential adaptation areas have been identified—CA being one. Agricultural policies in Zambia stimulate maize production through input subsidies and purchasing maize at a price higher than given at the regional market. Despite supporting policies, adoption and sustainability of CA in Zambia face a number of constraints, including high prices of inputs, such as herbicides and fertilizers; natural disasters, such as termites and fires; and marketing problems for agricultural produce. These constraints reduce the benefit of the intervention.
Rating: Low to moderate

Source: Authors' evaluation based on Aune, Nyanga, and Johnsen (2012).

(continued)

TABLE 6A.5 (continued)

F: Crop Crisis Control Project (C3P)

Problem definition: The project aimed to address the problem of low productivity of cassava and bananas resulting from cassava mosaic disease (CMD) and banana *Xanthomonas* wilt (BXW), which were priority agricultural constraints for the study countries. Evidence was gathered from various studies, including research by the Association for Strengthening Agricultural Research in Eastern and Central Africa (ASARECA) and the National Agricultural Research System (NARS), in the design of the project and throughout its implementation. For example, food security surveys and analyses were commissioned under C3P to help practitioners better understand the relationship between both CMD and BXW and food security, so as to design the right approaches to reach vulnerable populations.
Rating: High to very high

Choice of the commodity/instruments: The main instruments used were introducing and distributing disease-resistant strains of cassava planting materials to control CMD, and promoting agronomic measures to counter BXW.
Rating: High to very high

Suitability of instruments: C3P aimed at targeting the poorest smallholder farmers who were identified by project staff, opinion leaders, and farmers' group members. Typically, farmers without clean planting materials and vulnerable households (those with members infected with HIV, sick members, elderly, widows, widowers, orphaned children, etc.) were the main beneficiaries. The interventions used by the project were appropriate for addressing productivity constraints identified in the problem definition.
Rating: High to very high

Design and timing of implementation: The project's design included an integrated strategy to intensify and coordinate efforts to combat CMD and BXM. The project was led by an experienced chief of party, with a multidisciplinary team involving agronomists, a social scientist, cassava and banana breeders, a socioeconomist, a plant pathologist, a virologist, and a geographic information system specialist. Technical capacity to implement the project was through a collaborative mechanism via International Institute for Tropical Agriculture, Association for Strengthening Agricultural Research in Eastern and Central Africa, national agricultural research system, nongovernmental organizations, and many other partners, and there were only minor delays in some aspects.
Rating: High to very high

Environmental sustainability: There is no explicit evidence on how environmental sustainability issues were considered.
Rating: Low to moderate

Financial sustainability: The project's solutions relied heavily on project funding from USAID. It became difficult to continue with the production and distribution of clean planting materials after the project's completion. Furthermore, some of the varieties developed have been found to be susceptible to a new threat—cassava brown streak disease.
Rating: Low to moderate

Community participation: Local communities adopted successful measures to control CMD and BXW. They were also involved in various project activities, such as multiplication and distribution of improved varieties and training fellow farmers. *Rating: High to very high*

Gender consideration: There was no specific gender consideration or targeting.
Rating: Low to moderate

Complementary investments and partnerships: C3P involved more than 40 implementing partners across the Great Lakes region, and the roles of the different partners were clear. An evaluation of the project, however, observed that partner communication and coordination could have been improved.
Rating: High to very high

Capacity building: Multiplication and dissemination efforts were accompanied by training and education aimed at achieving better growing techniques and methods of disease prevention. Training activities were conducted through workshops and through on-farm and field visits. Various knowledge products, such as manuals, briefs, and posters, were produced and disseminated. A combination of multicountry learning and lesson sharing and a regional framework fostered by C3P enabled these lessons to be documented and implemented across most of the C3P countries.
Rating: High to very high

Organized groups: The project supported farmers' groups and community-based organizations (CBOs) to participate in managing secondary sites for multiplication of disease-resistant varieties. It also encouraged the formation or strengthening of farmers' groups to facilitate bulking of their cassava and knowledge sharing.
Rating: High to very high

Leadership and dedication: The project was well received within the target countries. National research organizations, farmers' organizations, and individual farmers participated in furthering project objectives. The leadership of ASARECA under the Eastern Africa Root Crops Research Network brought together various partners to implement the intervention.
Rating: High to very high

Policies and national-level factors: There was political will to support the project. The project was implemented at a time when the target countries had committed to revive neglected crops, such as cassava. The project was initiated when national governments were also working to find ways of fighting the two diseases.
Rating: High to very high

G: East Africa Dairy Development Project (EADD)

Problem definition: The project aimed to address issues of low milk productivity, milk perishability, and milk marketing, consistent with constraints identified in national strategy documents in the study countries. The design of EADD was informed by detailed background studies and the incorporation of lessons learned from similar projects by Heifer International in various countries.
Rating: High to very high

Choice of commodity/instruments: Milk is an important commodity for food security and income generation. The project targeted suitable agroecological zones for milk production—namely, dairy-producing areas in Kenya and Rwanda. In Uganda, project sites were predominantly in pastoralist areas.
Rating: High to very high

Suitability of instruments: Although the project aimed at reaching poor smallholder farmers, there were concerns that some of the beneficiaries were not truly poor, because of the approach of focusing on producer organizations. Most poor farming communities may not have qualified to be members of the producer organizations.
Rating: Low to moderate

Design and timing of implementation: The project had a team of highly qualified staff in different technical areas, such as dairy technology, livestock production, and business management. The project had some design problems. For example, implementation started before the problem analysis was comprehensively undertaken. In Kenya, however, there were problems of delayed or slow start-up, mostly caused by postelection violence there. Although baseline studies were conducted, the results were insufficiently used in the design and implementation activities.
Rating: Low to moderate

Environmental sustainability: There is no explicit evidence on how environmental sustainability issues were considered.
Rating: Low to moderate

Financial sustainability: Most of the resources for the project were from the donor (Bill & Melinda Gates Foundation), which can be a threat to sustainability when the donor support is no longer available.
Rating: Low to moderate

Community participation: Local communities were involved heavily during the project's implementation. They participated in the development of dairy hubs and training activities. They were keen to adopt various technologies promoted by the project, including rearing of improved breeds and feed management.
Rating: High to very high

(continued)

TABLE 6A.5 (continued)

Gender consideration: Gender considerations were not adequately taken into account initially. During the second year of the project (in 2009), however, an effort was devoted to identifying emerging internal gender gaps, including the responsiveness of the institutional setup for the promotion of gender equality. Gender focal points were identified at different levels (country and regional office, partners, etc.), and a gender working group was established. A mid-term evaluation noted that EADD had been effective in achieving gender balance among project staff, executive committees in dairy farmer business associations, and dairy service providers.
Rating: High to very high

Complementary investments and partnerships: EADD involved a wide range of partners. It was led by Heifer International in partnership with the International Livestock Research Institute (ILRI), TechnoServe, the World Agroforestry Centre (ICRAF), and the African Breeders Service Total Cattle Management. Several complementary investments were in place through partnerships among governments, the private sector, and milk-processing companies. EADD's business-based approach to development attracted multinationals, such as Nestlé and Tetra Pak, to collaborate with the project. Microfinance associations, village banks, commercial banks, and the chilling plants' check-off system of credit against milk deliveries gave farmers, youth entrepreneurs, and business men and women opportunities to engage in a range of enterprises that extended well beyond the dairy sector.
Rating: High to very high

Capacity building: Capacity building was carried out in various areas, including dairy husbandry, business practices and operation, milk management practices, feeding, and fodder production. A wide range of approaches was used in training and knowledge dissemination, including exchange visits, demonstration plots, and community radio.
Rating: High to very high

Organized groups: Farmers were mobilized into groups to set up and run producer companies. The companies were assisted to set up infrastructure to market milk and deliver inputs to members through the dairy hubs.
Rating: High to very high

Leadership and dedication: This project was implemented at a time when there was a high level of government commitment and attention to agriculture through the Comprehensive Africa Agriculture Development Programme (CAADP) process.
Rating: High to very high

Policies and national-level factors: Although all implementing countries have prioritized promotion of livestock production at the policy level, several factors affected the project. Most milk-producing areas were hit by drought at some point during the project's implementation. This affected the availability of feed and water for livestock. Insecurity in some parts of Uganda (particularly in northern Uganda) was a challenge. In Kenya, the project was affected by the postelection violence in 2007–2008. Transport and marketing challenges, such as poor roads, low access to cooling facilities in milk surplus areas, and lack of appropriate milk transport equipment, negatively affected the project.
Rating: Low to moderate

Source: Authors' evaluation based on TANGO International (2010) and Mutinda (2013).

H: FARM Africa Goat Dairy Improvement Project (FAGDIP)

Problem definition: The project sought to address the problem of the low productivity and growth of the East African indigenous goat as a means of improving milk production and increasing the nutrition, income, and overall livelihoods of the majority of the rural poor with a limited livestock asset base. It was designed to address the problems of small-scale, resource-poor livestock keepers in sustaining a cross-breeding program by themselves, resulting from their small flock sizes and consequent unavailability of good-quality genetic breeding material, as well as their lack of access to government services. The design of the project was informed by the documented experiences of Farm Africa in piloting agricultural interventions that are appropriate for the needs of the poor farmers in East Africa. A participatory planning and project design was undertaken with the community leaders.
Rating: High to very high

Choice of the commodity/instruments: Goats are very important in the highland mixed crop–livestock system, as well as in the pastoral system. They provide meat, milk, manure, skin, asset, security, and sociocultural benefits. Goat rearing is suitable in the production systems where the intervention was promoted. Goats are relatively cheap, and the poor are more likely to be able to afford them than cows.
Rating: High to very high

Suitability of instruments: The project targeted smallholder and resource-poor livestock keepers, especially the vulnerable, such as households affected by HIV/AIDS and headed by women. Instruments used included promotion of improved breeds, training in animal healthcare, and improved feeding technologies. The project imported exotic dairy goat breeds to crossbreed with the indigenous goats.
Rating: High to very high

Design and timing of implementation: The project was implemented through a series of activities that were targeted to specific geographical areas. Implementation of these activities benefited from the long-term experience of the organization in supporting livestock-related activities in mixed and pastoral systems.
Rating: High to very high

Environmental sustainability: The project promoted better goat management strategies to avoid environmental degradation. One example of the strategies was to encourage dairy goat keepers to grow fodder on soil and water conservation structures. Although goats are generally known for environmental destruction, the model promoted by Farm Africa proved that natural resource management can be possible when better goat management techniques are employed.
Rating: High to very high

Financial sustainability: Farm Africa managed and facilitated the operations of the project from the start, and then handed it over in 2004 to the communities through a new umbrella organization, the Meru Goats Breeders Association, which was created as part of the project. Funding to implement project activities was adequate. The poor households have not been able to sustain the technology after the end of the project. Evidence from Kenya and Tanzania suggests that the households that have been able to adopt the technology are those with higher per capita incomes and more effective asset-accumulation strategies than nonadopters.
Rating: Low to moderate

Community participation: Local communities were involved in project activities. Community leaders, extension staff, development workers, and Farm Africa were used to identify resource-poor farmers who were to benefit from the project. They also participated in training other farmers. The willingness to adopt new technologies promoted by the project was evident among the farmers.
Rating: High to very high

Gender consideration: Both men and women were involved in and benefited from the project, although the women were involved more in production but less in the formal marketing of milk, as men had more access to marketing information than women.
Rating: High to very high

Complementary investments and partnerships: Considerable effort was put in place to leverage partnerships among local communities, the private sector, animal health workers, and agrodealers. There was also support for savings and credit funds for small enterprise development.
Rating: High to very high

Capacity building: Capacity building was a key component of the project. It produced and disseminated training materials, and trained beneficiaries on basic animal husbandry, housing, fodder production, management and utilization, group dynamics, record keeping, and conservation. Selected members of the community were nominated to receive further training on basic animal health and breeding techniques. These members became the local providers of animal health and breeding services to the community.
Rating: High to very high

Organized groups: Through the project a local breeder association, MGBA, was established. This association drew membership from registered farmers' groups, whose members have interest in dairy goats. Farmer-managed organizations were established to coordinate and extend services during and after the intervention period.
Rating: High to very high

(continued)

TABLE 6A.5 (continued)

Leadership and dedication: Farmers and group leaders were keen to learn about and adopt the technology. Group leaders participated in leadership training supported by the project. The project was supported by the Ministry of Livestock Development.
Rating: High to very high

Policies and national-level factors: National policies that accommodate community-based livestock improvement initiatives were lacking. Cattle milk was the only officially marketable milk in Kenya during the project's lifetime. Kenya changed its policy only in 2008 to include goat milk in the list of official milk products.
Rating: Low to moderate

Source: Authors' evaluation based on Bradstock (2007) and Davis and Negash (2007).

I: Farm Input Subsidy Program (FISBP), Malawi

Problem definition: To address the problems of recurring food deficits, low maize productivity, high grain prices, and high dependence on food aid, FISBP's main objective is to raise the income of smallholder farmers through improvements in agricultural productivity and food security. The program was informed by past studies that identified low input use as a key constraint in the agriculture sector. The subsidy program has built on and emerged from Malawi's innovative experience in implementing universal starter pack and targeted input programs from 1998 to 2005.
Rating: High to very high

Choice of the commodity/instruments: The target was maize (and smallholder farmers), which is critical to the economy and the livelihoods of most farmers.
Rating: High to very high

Suitability of instruments: Although, the program intended to reach the poor smallholder maize farmers, it faced substantial logistical challenges and systemic development with large-scale registration and targeting of those farmers. Various innovations have been used, but targeting problems are still prevalent. A study by Talip, Whitney, and Paul (2013) found that the poor are not reached and that the failure to target the poor is pervasive at all levels of government (national, district, and community).
Rating: Low to moderate

Design and timing of implementation: The program faced various operational challenges, including shortages of and delays in delivery of inputs, and cumbersome coupon-processing and -redemption systems. The government has limited human and financial capacities to meet the operational demands of the program. There are challenges in input procurement, particularly fertilizer, caused by late completion of the tendering and bid awards. This resulted in some fertilizers being procured at high prices, with large variation in prices and increased input costs.
Rating: Low to moderate

Environmental sustainability: There is no evidence of adequate measures to address environmental impacts.
Rating: Low to moderate

Financial sustainability: This is a very costly program. The government is having challenges in financing program activities, and has limited human and financial capacities to meet the program's operational demands.
Rating: Low to moderate

Community participation: Local community representatives have been used to identify the recipients. Within districts, traditional authorities, local government staff, and Ministry of Agriculture and Food Security staff have had varying roles in coupon allocations, working with village development committees and other local stakeholders to identify recipients. Communities have welcomed the interventions and have applied the inputs.
Rating: High to very high

Gender consideration: The program targets both female- and male-headed households. Some previous programs in Malawi have found that female-headed households are less likely to receive coupons than male-headed households (Chirwa, Matita, and Dorward 2011) and, where female-headed households receive subsidy coupons, they tend to receive fewer compared with their male counterparts (SOAS et al. 2008; Dorward, Chirwa, and Jayne 2010). Recent guidelines issued by the government encourage communities to give priority to female-headed households.
Rating: High to very high

Complementary investments and partnerships: Households received different combinations of maize seed and fertilizer coupons. Some of the operational challenges in input distribution were addressed by involving the private sector. In 2006–2007, for example, the private sector distributed all of the seed and 28 percent of the fertilizer. Because of slow private-sector development, however, there is a shortage of private agrodealers in rural remote areas.
Rating: Low to moderate

Capacity building: There are concerns that information sharing with the beneficiaries of the program is inadequate. There is also need to strengthen the capacity of the private sector to participate in input distribution.
Rating: Low to moderate

Organized groups: There is no evidence of support for building or strengthening the capacity of farmers' groups.
Rating: Low to moderate

Leadership and dedication: The government committed and showed strong political will to implement the subsidy program. The government is also committed to addressing the operational problems facing the program. The traditional leaders and village development committees have demonstrated their commitment and dedication.
Rating: High to very high

Policies and national-level factors: Extreme climatic conditions, such as droughts and floods, and high variability in maize prices have contributed to risks in input use. The high price variability in maize prices, for example, has encouraged the government to intervene in maize markets (for example, setting minimum and maximum prices and banning exports and private trade).
Rating: Low to moderate

J: Farmer Input Support Program (FISPP), Zambia

Problem definition: The program aimed at addressing the low access of smallholder farmers to fertilizers and improved seeds to improve productivity, increase food insecurity, and reduce poverty. The program was identified as being a cornerstone of the country's poverty reduction strategy.
Rating: High to very high

Choice of the commodity/instruments: The program targeted maize, which is an important food and cash crop in Zambia. The crop has received considerable government attention over the years in terms of financial investment to support smallholder farmer access to seed and fertilizer. The need to diversify resulted in the program being expanded to include rice, groundnuts, sorghum, and cotton. The program was implemented in all agroecological zones where these crops are grown within the country.
Rating: High to very high

Suitability of instruments: Although the original objective of the program was to target small-scale maize famers with the capacity to grow 1–5 hectares (ha) of maize and pay 25 percent of the cost of inputs, some studies found that inputs were targeted to the least poor rural households, and that wealthier small- and medium-scale farmers also benefited from the program. Some studies have indicated corruption at the distribution centers to be one of the factors affecting the distribution process (CSPR, Zambia 2011). The low volume of inputs distributed to the farmers has affected the effectiveness of the intervention.
Rating: Low to moderate

(continued)

TABLE 6A.5 (continued)

Design and timing of implementation: Inputs are supplied to districts by private traders selected through a national tender. Local distributors deliver inputs to satellite depots, and issue these inputs to cooperatives and other farmers' organizations. The District Agricultural Committee selects local distributors and farmers' cooperatives. Selected farmers' cooperatives and other farmers' organizations deposit 25 percent of the value of the inputs. The major concerns raised with regard to the implementation of the program have been late delivery of the inputs and inadequate quantities supplied. The problem of poor access to inputs is more serious for small-scale farmers, who have not been adequately reached. The program also supported access to agricultural credit, but this did not do well because of problems of credit defaulting. There were large leakages of project funds, however.
Rating: Low to moderate

Environmental sustainability: There is no articulation of environmental issues or how they may be addressed.
Rating: Low to moderate

Financial sustainability: This program has a heavy financial burden. It consumed the vast majority of the Government of Zambia's agricultural budget. In 2011, for example, 73 percent of the poverty reduction budget was allocated to the program (Burke 2012). Farmers have become dependent on FISPP, which is problematic for sustainability. The program is designed in such a way that every two years beneficiaries should graduate (once they have generated sufficient income); however, this has not been happening.
Rating: Low to moderate

Community participation: Though efforts were made to involve local communities, their level of engagement has been questioned. This is likely because of the low involvement of the private sector in the input supply chain of the program.
Rating: Low to moderate

Gender consideration: Not apparent.
Rating: Low to moderate

Complementary investments and partnerships: The Food Reserve Agency buys maize from farmers at a guaranteed price above market prices. A multidisciplinary approach is promoted. For instance, in addition to enhancing access to inputs, the government encourages extension workers and researchers to support farmers. It is also supportive of the Conservation Agriculture Program.
Rating: High to very high

Capacity building: Awareness has been created by local extension officers explaining at farmers' meetings the rules and modalities governing the program. However, there are some concerns that the capacity enhancement and sensitization activities have not been adequate.
Rating: Low to moderate

Organized groups: Cooperatives and farmers' organizations are the main channels in the distribution of inputs to beneficiary farmers. However, some of the cooperatives have been formulated only for the purpose of accessing inputs from more than one input provider.
Rating: Low to moderate

Leadership and dedication: The Zambian government has demonstrated strong commitment to agriculture and rural development through its allocation of more than 10 percent of the country's total budget to the sector, as per the CAADP targets. There is also support from local leaders, as well as village farmers' committees and farmers' organizations.
Rating: High to very high

Policies and national-level factors: The Fifth and Sixth National Development Plans support agricultural production, and indicate agriculture, livestock, and fisheries to be main priority growth sectors, together with mining, tourism, manufacturing, and commerce and trade. The agriculture budget has been increased in line with CAADP's objectives.
Rating: High to very high

Sources: Authors' evaluation based on CSPR (2011); Jayne et al. (2011); and Burke (2012).

K: Fodder Trees and Shrubs Project (FTSP)

Problem definition: Having identified the problem of limited access to affordable animal feed among smallholder farmers, the intervention aimed to develop fodder technology and promote its adoption across East Africa (Kenya, Uganda, Tanzania, and Rwanda). The intervention was informed by several studies that have established that feed is a major constraint to livestock production in eastern and central Africa.
Rating: High to very high

Choice of the commodity/instruments: The project targeted areas that are most affected by feed problems, in particular under intensive animal production systems where improved animal breeds are kept for dairy production. It promoted fodder trees and shrubs (for example, calliandra, leucena, and mulberry) that provide multiple benefits, such as milk production, animal health, and soil conservation. The shrubs are relatively easy to manage, and fodder trees do not compete with food crops, can be intercropped and, once mature, can be fed to livestock for several years.
Rating: High to very high

Suitability of instruments: The project reached more than 200,000 smallholder dairy farmers in eastern Africa. It considered local demands by promoting fodder shrubs that suited local agroecologies and preferences. Fodder trees and shrubs research and scaling up were motivated mainly by demand for quality dairy feed to increase milk production in the smallholder dairy farming systems of the region, as they offer an affordable alternative source of high-protein supplementary feed for dairy animals.
Rating: High to very high

Design and timing of implementation: Implementation has been taking place through a series of different projects that add value to each other. The World Agroforestry Center has been implementing these projects in partnership with other CGIAR centers, NGOs, national governments, development partners, and other stakeholders.
Rating: High to very high

Environmental sustainability: The project was promoting agroforestry, which has various environmental benefits, including preventing soil erosion by creating soil cover and improving soil fertility by fixing atmospheric nitrogen. Fodder trees and shrubs are also used as fuelwood and, hence, minimize pressure on natural forests.
Rating: High to very high

Financial sustainability: The project's sustainability is very likely, because farmers have taken charge of the intervention, as it does not seem to require significant financial investment and is, hence, easier to maintain.
Rating: High to very high

Community participation: Communities are very involved in dissemination of the technology and training. Farmer-to-farmer dissemination has been a key approach. Communities have embraced the new technologies, and fodder shrubs are now planted by many farmers in eastern Africa.
Rating: High to very high

Gender consideration: There has been considerable gender consideration in promotion of fodder trees and shrubs. Women comprised about 50 percent of farmers planting them in various project sites. Women also were involved in the establishment of demonstration sites and were hired as facilitators.
Rating: High to very high

Complementary investments and partnerships: Collaborative partnerships were developed between research institutions (national and international) and governments. Other organizations promoting fodder trees include farmers' groups, NGOs, CBOs, and private companies. The partners have played different roles based on their comparative advantage. Various studies have been undertaken to monitor and evaluate the interventions. Fodder shrubs have been promoted along with other interventions, such as the East Africa Dairy Development Project.
Rating: High to very high

Capacity building: Training of farmers was a major component of the project. Farmers were trained on feed management. Knowledge products, including brochures and briefs, were produced and disseminated.
Rating: High to very high

(continued)

TABLE 6A.5 (continued)

Organized groups: Farmers were organized in groups to facilitate the dissemination and widespread adoption of the fodder technologies.
Rating: High to very high

Leadership and dedication: There was clear indication of commitment and dedication from all partners (governments, the private sector, and farmers). Farmers were keen to participate in on-farm experiments and learned how to incorporate fodder shrubs into daily feed ratios.
Rating: High to very high

Policies and national-level factors: Growing demand for dairy products is encouraging dairy production. However, the dairy industry faces various policy and institutional constraints that affect milk marketing and, subsequently, the process of fodder adoption, which is dependent on milk marketing trends. Heavy losses of seedlings have resulted from frequent and unpredictable rainfall patterns in the region.
Rating: Low to moderate

Source: Authors' evaluation based on Wambugu et al. (2006).

L: Fuve Panganai Irrigation Scheme (FPIS), Zimbabwe

Problem definition: The scheme aimed to improve the income, food security, and living standards of smallholder rural households by addressing the problem of recurring drought. Technical studies conducted prior to the start of the scheme included soil analysis and physical and socioeconomic analysis. A project feasibility report was also produced.
Rating: High to very high

Choice of commodity/instruments: The project focused on cotton, beans, maize, and groundnuts. It also encouraged farmers to grow vegetables. The interventions under the scheme were suitable for the agroecological zones where the project took place.
Rating: High to very high

Suitability of instruments: Demand for the irrigation system was evident. The scheme was established as a precaution against the inherent variability in rainfall, as well as to ensure year-round cultivation. It targeted small-scale farmers, and reached disadvantaged rural populations, including those living with HIV/AIDS.
Rating: High to very high

Design and timing of implementation: There was no clarity on how the information collected from socioeconomic studies was used in the project's design. There seemed to be a top-down approach in implementation, and there were delays in commencement of the scheme because of financial constraints. The project was planned in the 1970s, but only materialized in the late 1980s, when the German government provided the financial support to the project. However, this financial support was not adequate to sustain the project.
Rating: Low to moderate

Environmental sustainability: Efforts to address the negative impacts that resulted from irrigation, such as the decline in fertility and increase in salinity, were inadequate. The scheme also suffered from high water leakages.
Rating: Low to moderate

Financial sustainability: Farmers lack resources to purchase inputs and maintain the irrigation infrastructure. Also, the benefits have not been sustained. Farmers participating in the scheme have negative cash flow outcomes and find it difficult to cope with rising water charges and disrupted irrigation schedules.
Rating: Low to moderate

Community participation: Though the project seems to have made efforts to involve beneficiaries in its design, it faced resistance from local communities initially. As such, community members were more involved in the project's implementation phase.
Rating: Low to moderate

Gender consideration: Both men and women were key players in the program. For example, women who participated in the scheme reported improvements in their income status as a result of income from crops produced.
Rating: High to very high

Complementary investments and partnerships: There were poor linkages to input supply and no reliable product markets. Previously, farmers accessed markets through the Grain Marketing Board or Cotton Marketing Board. Market deregulations led to the loss of secure market opportunities.
Rating: Low to moderate

Capacity building: Extension provides training on food security crops (maize, beans, groundnuts, and wheat). In the initial years of the project, capacity-building activities were limited. This has changed in recent years.
Rating: Low to moderate

Organized groups: There were no explicit efforts to organize farmers into groups.
Rating: Low to moderate

Leadership and dedication: The project received considerable resistance from the local communities in the initial stages of implementation; however, this resistance subsided with time. The Ministry of Agriculture provided leadership. The government's attention to the development of small-scale irrigation schemes was in a bid to meet its objectives toward decentralizing irrigation schemes, mainly in rural areas for empowerment.
Rating: High to very high

Policies and national-level factors: Economic meltdown led to high water charges and debts.
Rating: Low to moderate

Sources: Authors' evaluation based on Manyame (1998) and Chazovachii (2012).

M: Kaleya Irrigation Project (KIP), Zambia

Problem definition: This intervention was informed by the evidence of inadequate use of irrigation potential in Zambia. The literature indicates that although Zambia has large irrigation potential, less than 30 percent of the land suitable for irrigation has been developed. A needs assessment conducted in the target areas indicated that unemployment was a significant problem. The project aimed at creating jobs for the rural poor.
Rating: High to very high

Choice of commodity/instrument: The target was sugarcane, which is a high-value crop with a ready market at the sugar mill.
Rating: High to very high

Suitability of instruments: The project considered local demand, as the sugarcane it produced was sold to the Zambia Sugar Company (ZSC), which milled the cane into sugar for the local and export markets.
Rating: High to very high

Design and timing of implementation: Kaleya Smallholders Company Limited (KASCOL), as an outgrower, oversaw 1,080 ha of smallholder sugarcane growers (organized under farmers' associations) and 1,100 ha on its own estate farms. Good governance was a key success factor. KASCOL had a board of directors that was elected every three years. KASCOL's approach to business was a combination of its own production and contract farming. A key enabling factor in the initial stages of the KASCOL project was the configuration of expertise and contributions provided by the different shareholders. The Commonwealth Development Corporation (CDC) and ZSC brought production and management expertise, while two banks brought financial resources to the new company.
Rating: High to very high

Environmental sustainability: Not apparent.
Rating: Low to moderate

Financial sustainability: There were adequate financial resources to support the project, through funding by the World Bank, Development Bank of Zambia, ZSC, Barclays Bank, and CDC. But the financial management and administrative capacities of smallholders were limited, because of a strict management agreement between KASCOL and smallholders. Low initial development costs and low debt levels were a plus.
Rating: Low to moderate

Community participation: The KASCOL model involved equity participation and board representation for smallholder outgrower farmers. The Kaleya Smallholder Farmers' Association (KASFA) sits on KASCOL's board of directors.
Rating: High to very high

(continued)

TABLE 6A.5 (continued)

Gender consideration: The KASCOL smallholder scheme, while dominated by men, was deemed accessible by women, albeit with some limitations because of workloads and intrahousehold decisionmaking dynamics. This was illustrated by only 28 percent of the outgrowers being women. Women had a strong voice in decisionmaking within households where women were registered as outgrowers.
Rating: High to very high

Complementary investments and partnerships: Farmers' associations get inputs on credit from KASCOL. KASCOL negotiates fertilizer prices with ZSC. Transport services are outsourced. KASCOL is responsible for infrastructure maintenance and distribution of irrigation water. KASCOL provides social services (health programs on HIV/AIDS, clinics, primary schools, and recreation facilities). The market for farmers' produce was ensured.
Rating: High to very high

Capacity building: A participatory approach was used in agronomic training of farmers in sugarcane production practices. KASCOL provided farmers six months of agronomic training and paid them for managing the land as apprentices. Farmers who were capable are settled on 4 ha of land on a renewable 14-year lease. The scheme has created employment for the community, thus enhancing the financial empowerment of the beneficiaries.
Rating: High to very high

Organized groups: KASFA mediates in all issues pertaining to its farmer members, especially those related to prices.
Rating: High to very high

Leadership and dedication: The project has strong government- and local-level support. The government provided free land for the project. Farmers participated in the project as outgrowers. Their leaders were dedicated and represented the farmers by serving on the KASCOL board.
Rating: High to very high

Policies and national-level factors: The project site was located on an aquifer, which proved to be an invaluable source of irrigation water. Favorable rains and fertile soils in the Mpongwe district also contributed to the project's success. Government support to the agriculture sector is high. There were reforms to encourage private investments in agriculture.
Rating: High to very high

Source: Authors' evaluation based on Mujenja and Wonani (2012).

N: Kenya Dairy Development Programme (KDDP), Kenya

Problem definition: The program aimed to increase livestock productivity, so as to address problems of food insecurity and poverty. It responded to the national priorities of using livestock as a pathway out of poverty. The importance of livestock is emphasized in various government policy documents. The project was informed by past studies that have articulated the constraints facing the dairy industry in Kenya and the huge opportunity for growth through investments in addressing those constraints.
Rating: High to very high

Choice of the commodity/instruments: The program targeted dairy production, an important activity in Kenya for food production and income generation. Dairy products (milk) account for 30 percent of livestock gross domestic product and more than 22 percent of livestock gross marketed products in Kenya. KDDP focused its activities in 16 districts that are highly suitable for dairy production.
Rating: High to very high

Suitability of instruments: Dairy farmers were selected in different geographical regions in the country based on such factors as cattle population and the number of milk market points.
Rating: High to very high

Design and timing of implementation: As the implementer, Land O' Lakes delegated specific functions to such organizations as ILRI, Nairobi Veterinary Centre, Pioneer Technologies, and Kenya Agricultural Research Institute (KARI). All actors played their roles well to implement the project. Resources to implement project activities were adequate. Implementation of the project at the cooperative level faced a number of challenges, such as high turnover of management staff and slow decisionmaking by cooperatives.
Rating: High to very high

Environmental sustainability: Although the program had a component of natural resource management, its implementation was inadequate. There is no explicit information on how the program planned to address any environmental impact resulting from the interventions.
Rating: Low to moderate

Financial sustainability: The beneficiaries have continued with the improved dairy techniques acquired through the program. However, the level of financial resources available to address dairy-related constraints has been limited since the end of the program.
Rating: Low to moderate

Community participation: Use of the learning-by-doing technique to encourage adoption of the promoted technologies and practices and farmer participation in the livestock FFS raised farmers' commitment to the project.
Rating: High to very high

Gender consideration: The rate of program participation by women was about 35 percent.
Rating: High to very high

Complementary investments and partnerships: KDDP developed and strengthened partnerships with several organizations, including the Ministry of Livestock and Fisheries Development, ILRI, the Kenya Dairy Board Dairy Training Institute, Pioneer Technologies, Nairobi Veterinary Centre, KARI, the University of Nairobi, and Kenya Broadcasting Corporation. The partners played different roles in the program based on their comparative advantage.
Rating: High to very high

Capacity building: More than 100,000 farmers were trained on improved dairy management practices and technologies. The program enhanced farmers' accessibility to reliable and efficient artificial insemination (AI) services by training service providers and facilitating establishment of AI service points. The project also disseminated various knowledge products, including bulletins, journals, and education materials.
Rating: High to very high

Organized groups: Farmers were organized into 60 livestock FFS and dairy cooperatives. The program provided technical assistance to the cooperatives and dairy institutions. It established new and strengthened existing cooperatives and dairy institutions. However, the cooperatives have been facing challenges. The high turnover of management staff and slow decisionmaking in cooperatives continue to pose a challenge in turning them into effective service providers to farmers.
Rating: Low to moderate

Leadership and dedication: Farmers were dedicated to adopting the skills they learned. They demanded and used dairy information provided to them.
Rating: High to very high

Policies and national-level factors: Most milk-producing areas were hit by protracted dry spells, especially the droughts in 2004 and 2005. The program was affected by the postelection violence that hit Kenya in 2007–2008. Transport and marketing challenges, such as poor roads, low access to cooling facilities in milk-surplus areas, and lack of appropriate milk transport equipment, negatively affected the project.
Rating: Low to moderate

Sources: Authors' evaluation based on Land O' Lakes (2008).

(continued)

TABLE 6A.5 (continued)

O: National Agricultural Advisory Services (NAADS), Uganda

Problem definition: NAADS aimed to address the challenge of limited access to agriculture extension and advisory services by farmers. The program had inadequate focus on other farmer constraints that must accompany an extension intervention, such as labor, access to inputs, ineffective extension services for crop and animal farmers, and unresolved market access issues (UNFFE 2011). Local communities were involved in the selection of enterprises. NAADS goals and objectives have been relevant, as reflected by the National Development Plan and National Development Strategy.
Rating: High to very high

Choice of the commodity/instruments: Enterprises were selected by community members and extension officers. Interventions were selected according to their agroecological suitability.
Rating: High to very high

Suitability of instruments: Targeting the right beneficiaries was an issue. For example, Okoboi et al. (2011) note that, contrary to NAADS' aim to prioritize support to marginalized households, the proportion of targeted marginalized households was low compared with other households.
Rating: Low to moderate

Design and timing of implementation: Although various implementation guides were developed, they were hardly adhered to, which has been a major challenge to the quality of the program's implementation. Other challenges include late disbursement of funds to districts and subcounties where activities are implemented, embezzlement of funds, distribution of poor-quality inputs, and government disruption of activities (Okoboi et al. 2011).
Rating: Low to moderate

Environment sustainability: NAADS appears to be having more success in promoting adoption of improved varieties of crops and some other yield-enhancing technologies than in promoting improved soil fertility management. This raises concern about the sustainability of productivity increases that may occur, since such increases may lead to more rapid soil nutrient mining, unless comparable success in promoting improved soil fertility management is achieved (Benin et al. 2007).
Rating: Low to moderate

Financial sustainability: Farmers' groups are unable to raise the desired amount of capital from membership contributions to adequately support their activities.
Rating: Low to moderate

Community participation: In general, participation of farmers in group or community activities was considered to be very good or good by most of the groups/communities. Besides attending general meetings, local communities were involved in such activities as enterprise selection, demonstration and training, management of technology development sites, and development of a constitution and/or bylaws.
Rating: High to very high

Gender consideration: According to the NAADS Act, 2001, the program was created to pay more attention to women, people living with disabilities, and youths who were considered marginalized from mainstream economic activity. The program completion report indicated gender imbalances in farmer institutions, with men leading most of the groups. An impact assessment study indicated that the objective of generating gender-responsive services has not been achieved fully.
Rating: Low to moderate

Complementary investments and partnerships: The program involved a public–private extension service delivery approach encouraging farmers to demand and control agricultural advisory services. Despite having diverse partnerships with the National Union of Coffee Agribusinesses and Farm Enterprises, IDA, the Department for International Development, and DANIDA, farmers still faced some constraints. Benin et al. (2011), for example, found that shortage of capital and credit facilities was often cited by farmers as a critical constraint facing them, in addition to scarcity of agricultural inputs, lack of adequate farmland, unfavorable weather patterns, and problems of pests and diseases.
Rating: Low to moderate

Capacity building: Although NAADS supported demonstrations and supply of materials, it has been reported that limited professional and skills competence to guide the capacity development of farmers' institutions is still a challenge. Recent studies indicate that despite investment in capacity-building activities, farmers' understanding of NAADS operations is still limited (UNFFE 2011). There were also concerns about the quality of extension services for crop and animal farmers.
Rating: Low to moderate

Organized groups: NAADS supported formation of farmers' groups to select agricultural activities on which they need information and advice. The groups benefited from NAADS technologies.
Rating: High to very high

Leadership and dedication: NAADS was created in 2001 by an act of parliament. Various government ministries and institutions were involved in its implementation, including the ministries of Agriculture, Finance, Planning and Economic Development; local government; and farmers' institutions.
Rating: High to very high

Policies and national-level factors: The program had substantial support from the government. Agriculture sector reforms implemented by the government culminated in the 25-year NAADS program. The intervention, however, is constrained by unfavorable weather patterns and problems related to pests and diseases.
Rating: High to very high

P: National Agricultural Extension Intervention Program (NAEIP), Ethiopia

---

Problem definition: The program focuses on extension to address low agricultural productivity, with the goal of improving food security and reducing poverty. It is a scale-up of the Participatory Demonstration and Training System (PADETES) approach to boosting cereal yields and output. PADETES was an integrated program of extension, seed, fertilizer, and credit that was piloted by Sasakawa Global 2000 (SG 2000).
Rating: High to very high

Choice of commodity/instrument: The program is mainly focused on cereals, such as maize, wheat, sorghum, teff, and barley, which are identified in government documents to be a priority commodity for food security and poverty reduction.
Rating: High to very high

Suitability of instruments: The program directly reached about half a million farm households over a 10-year period. It targeted high-potential areas and paid inadequate attention to the vast majority of resource-poor farmers. The program has been considered to follow a supply-driven approach, as it did not adequately incorporate the beneficiaries' needs and demands.
Rating: Low to moderate

Design and timing of implementation: Although the program reached many people, efforts to scale up PADETES were less successful than the pilot demonstrated by SG 2000. Various implementation challenges affected the program. For example, an inadequate number of field-level extension officers constrained the effectiveness of the transmission of recommended packages of technology to farmers. A large expansion of the extension program has taken place, increasing the number of extension workers; however, the number of farmers per extension worker is still very high.
Rating: Low to moderate

Environmental sustainability: No clear articulation of how any environmental issues would be addressed.
Rating: Low to moderate

Financial sustainability: The government has made extensive investments in extension. The program distributed massive amounts of production inputs, including improved seeds, fertilizer, and credit. However, because the government was not able to sustain these services, the productivity gains were short lived.
Rating: Low to moderate

Community participation: There have been concerns that local communities were not as adequately involved in the program's planning as they were its implementation, and that the program used a top-down approach.
Rating: Low to moderate

(continued)

TABLE 6A.5 (continued)

Gender consideration: There is no indication as to whether there was any gender targeting.
Rating: Low to moderate

Complementary investments and partnerships: Examples of complementary interventions include improved seeds, fertilizers, and credit. Some of the partners include the International Food Policy Research Institute, the Government of Ethiopia, and the private sector.
Rating: High to very high

Capacity building: Capacity building has been affected by an inadequate number of extension staff. This had led to passive transmission of recommended messages to farmers, with little technology adaptation to local contexts. It has also eroded the credibility of the frontline extension workers among the smallholder farmers.
Rating: Low to moderate

Organized groups: Farmers' groups were not popular. Farmers were trained in the training centers collectively.
Rating: Low to moderate

Leadership and dedication: The Government of Ethiopia spearheaded the implementation of the program, with help from stakeholders (both private and public) in the agriculture sector. Sector and national policies and plans are supportive of the program.
Rating: High to very high

Policies and national-level factors: Unfavorable climatic conditions, such as droughts, negatively affected crop production. Crop production has also been negatively affected by the government's many policy changes and by the shifting roles of the public and private sectors—mainly those related to the marketing of agricultural inputs.
Rating: Low to moderate

Sources: Authors' evaluation based on NEPAD and FAO (2005) and Spielman, Kelemwork, and Alemu (2011).

Q: New Rice for Africa (NERICA) upland rice, Uganda

Problem definition: The project aimed to address low rice productivity in general, and the lack of cash crops in some areas. Rice is a strategic commodity in Uganda. Although rice production is increasing in Uganda, the country is still a net importer of rice. Improving rice productivity has been prioritized to reduce reliance on imports. Various technical and socioeconomic studies were commissioned to inform the project's design and implementation.
Rating: High to very high

Choice of the commodity/instruments: Rice production is a major intervention identified in Uganda's agricultural development strategy and investment plan. The project targeted all areas. However, in areas unsuitable for NERICA rice production (that is, where the profitability of NERICA rice relative to other crops is low), there were massive dropouts from the project—an economic and logical response.
Rating: High to very high

Suitability of instruments: The intervention targets the poor, including the internally displaced population in northern Uganda. Promotional activities target areas that are suitable for upland rice production.
Rating: High to very high

Design and timing of implementation: Various interventions to promote NERICA over the past decade have received financial support from development partners and technical support from research institutions. Several activities have been initiated to cover various aspects, including research, extension, and training of trainers. The high availability of improved seed varieties is stimulating rice production.
Rating: High to very high

Environmental sustainability: The program is not explicit on how it plans to address environmental issues.
Rating: Low to moderate

Financial sustainability: Reliance on external funding sources limits the sustainability of the interventions. So far, JICA and the Food and Agriculture Organization of the United Nations (FAO) have provided a large proportion of the resources. The government has not allocated adequate resources to strengthen human, institutional, and technical capacities. Staff and financial resources have been adequate to continuously disseminate the technologies to the farmers and address the constraints farmers face in adopting them. Some adopters have abandoned the technologies, which is raising a question about the project's sustainability.
Rating: Low to moderate

Community participation: Participatory varietal selection, where farmers select their varieties and evaluate interspecific lines, is used.
Rating: High to very high

Gender consideration: The project targets both men and women. NERICA rice has been beneficial to women in Uganda. Many women seem to think that despite the additional labor burden associated with growing upland rice, they have become more independent and have gained decisionmaking power in their households. Their bargaining power also has been strengthened, and spouses share proceeds through more democratic dialogue. Different studies indicate that female-headed households are experiencing yields per hectare equal to those of male-headed households.
Rating: High to very high

Complementary investments and partnerships: The Government of Uganda, WARDA, JICA, and FAO are working together to promote rice production. The actors with different comparative advantages are targeting different actors along the rice value chain, including farmers, millers, and traders.
Rating: High to very high

Capacity building: Capacity-building efforts include the FFS, research and extension capacity, development of demonstration plots, implementation of various experiments, development of technical manuals and training materials, and activities for farmers, millers, and government leaders. Nonetheless, capacity building, extension services, and awareness among the smallholder farmers are still insufficient, which seems to be limiting desired upland rice adoption rates and levels.
Rating: Low to moderate

Organized groups: Farmers organized themselves into groups, and NERICA seeds were distributed to them.
Rating: High to very high

Leadership and dedication: The project obtained strong support from the Ugandan government, particularly because of its objectives to increase food security and incomes and reduce dependence on food imports.
Rating: High to very high

Policies and national-level factors: Although government policies favor rice production, the rice sector in Uganda faces a number of constraints. These include rainfall variability, which reduces NERICA rice profitability; underdeveloped markets for seeds; inadequate rice-milling services; a weak extension system; credit constraints; and imperfect information about methods of seed production, the quality of seeds, and the rice-milling business.
Rating: Low to moderate

Sources: Authors' evaluation based on Kijima (2008), Kijima et al. (2011), and Lodin (2012).

R: Operation *Mwolyo* Out (OMO), Kenya

Problem definition: The project aimed at addressing the problem of persistent droughts leading to reliance and overdependence on food aid—a problem that ranks high in the government's priorities. A large part of the Mwolyo district is known for its insufficient rains, leading to food shortages. OMO started as an outreach program. Its founder is a retired teacher who lives in the area. A participatory needs assessment was conducted through discussions with the beneficiaries.
Rating: High to very high

(continued)

TABLE 6A.5 (continued)

Choice of commodity/instruments: The project promoted digging dams and using water pans for harvesting water. These efforts were targeted at maize and beans, which are important staple crops. The project also targeted high-value crops, including onions, watermelons, French beans, sweet potatoes, chillies, and various fruits. These commodities grow well in the project areas when water is available. The project also supports livestock production, an important economic activity in the arid agroecological zones.
Rating: High to very high

Suitability of instruments: The project initially targeted or attracted women, but youths and men have gradually joined. Local-level socioeconomic factors and preferences for food and income generation were the main influential factors.
Rating: High to very high

Design and timing of implementation: Project design and implementation are led by the Christian Impact Mission, a local NGO. The design was based on a participatory approach through a seven-point plan (community mobilization, water harvesting, appropriate agricultural technologies, high-value crops, value addition, development of marketing associations, and market linkages). The participants are encouraged to keep records of their agricultural production. Although the project does not have an elaborate M&E system, it has strongly invested in partnerships to support M&E. Several organizations, such as the University of Nairobi, the Regional Strategic Analysis and Knowledge Support System, UNDP, and the World Food Programme, have collaborated with OMO to document lessons from the project. M&E needs further strengthening.
Rating: High to very high

Environmental sustainability: The project promotes use of biogas to reduce dependence on the forest as a source of energy. Local communities are preserving the natural environment for tourism purposes.
Rating: High to very high

Financial sustainability: The project is a good case on how farmers' own resources can be mobilized to minimize dependency on eternal resources, which often challenges sustainability. The local communities themselves dig dams and water pits. Ecotourism and environmental conservation are promoted to attract external funds or sources of nonfarm income for the communities.
Rating: High to very high

Community participation: Members of the community are empowered to participate, as the project builds on indigenous knowledge and practices.
Rating: High to very high

Gender consideration: Gender issues are integrated into the project's design and implementation. During a key informant interview with the beneficiaries, many women indicated how the project has improved their livelihoods, via the training activities and the adoption of technologies and better farming techniques.
Rating: High to very high

Complementary investments and partnerships: Complementary interventions to facilitate marketing, handling of postharvest losses, and access to input are very limited. While some partnerships have been initiated, they are still very few.
Rating: Low to moderate

Capacity building: Farmers were trained on improved agricultural methods. NGOs and CBOs also are being trained on grassroots community participation.
Rating: High to very high

Organized groups: Farmers were organized into groups under the umbrella of the local church, which enhanced group cohesion.
Rating: High to very high

Leadership and dedication: Local communities' interest in and effort to bring about change from their own initiative are high. The strong sense of ownership by the community is attributable to the use of the local population's indigenous knowledge.
Rating: High to very high

Policies and national-level factors: This initiative faces threats from recurrent droughts in the area and from lack of infrastructure and market access.
Rating: Low to moderate

S: Participatory Irrigation Development Project (PIDP), Tanzania

Problem definition: The project aimed at addressing the problem of inadequate access to water for agricultural production. This was in line with the government's overall priority to combat rural poverty by enhancing rural and agricultural development. The project made efforts to undertake technical analysis prior to the interventions. However, the studies did not adequately inform problem identification and project design. For example, the project attempted to undertake the required technical analysis in identifying irrigation schemes for development and/or rehabilitation, but the basis for decisions was in some cases weak because of the lack of data (especially on hydrology). This lack of information led to the selection of some schemes where the available volume of water was insufficient and could not meet the community's needs.
Rating: Low to moderate

Choice of the commodity/instruments: Rice was the targeted commodity, which is an important food and cash crop in Tanzania.
Rating: High to very high

Suitability of instruments: The project targeted marginalized farmers and provided opportunity to the traditionally landless rural population, especially women and youths. However, it was unable to involve the poorest to the extent envisaged, as a key beneficiary-selection criterion is ability to contribute substantial labor, which was not always possible for poorer households and female farmers. There were also some inadequacies with regard to site selection, which led the project to invest in some schemes that did not have sufficient water. The project did not adequately factor in local demand and perceptions.
Rating: Low to moderate

Design and timing of implementation: The project suffered several challenges. In addition to poor site selection, investments in pit latrines were not adequately implemented, because the community did not demand them. Other challenges included low institutional capacity at the district level, a limited range of water-harvesting technologies used, underestimation of construction costs, a lengthy tendering process, low capacity of contractors, and unclear land rights of the "new" landowners.
Rating: Low to moderate

Environment sustainability: Efforts to address environmental problems resulting from irrigation were inadequate.
Rating: Low to moderate

Financial sustainability: Although policy statements support irrigation development, they have yet to be translated into concrete commitments in budgets to ensure that extension agents can continue to provide advice to the community on a wide range of issues, after the phasing out of the project.
Rating: Low to moderate

Community participation: The project's design appears to have been participatory. All irrigation schemes were designed after consultation and planning involving beneficiary communities, development committees, and local government authorities. A tripartite agreement was drawn up among the implementing agencies on their respective roles in the development and maintenance of the irrigation structures, feeder roads, buildings, and wells. A public–private approach was applied in the development and management of infrastructure.
Rating: High to very high

Gender consideration: The project was successful in involving women in water use associations (WUAs). In some cases, the 70:30 participation target ratio of men to women was surpassed.
Rating: High to very high

Complementary investments and partnerships: Partnerships with universities, NGOs, consulting companies, and private contractors worked well to bring together different knowledge skills and capacities. The roles and responsibilities of the partners were well defined.
Rating: High to very high

(continued)

TABLE 6A.5 (continued)

Capacity building: The project supported institutional and personal capacity development. It invested in providing extension services and training farmers and their leaders. Areas of training included management of irrigation infrastructure, establishment and management of WUAs, governance, leadership, management of an O&M fund, savings and credit cooperatives, group dynamics, and a logical framework approach in planning and monitoring results. Capacity building was carried out by the project through FFS, on-farm trials, and study tours, all of which proved to be effective and efficient. Through the extension services subcomponent, training on best agronomic practices was conducted in all schemes.
Rating: High to very high

Organized groups: The project promoted collective action in agriculture. It supported the formation and strengthening of self-help organizations (such as WUAs) that contributed to the establishment, operation, and maintenance of the irrigation schemes.
Rating: High to very high

Leadership and dedication: Building on experiences from similar projects in the past, the government promoted irrigation and water-harvesting systems, which were reflected in its National Irrigation Policy. The Ministry of Agriculture, Food Security and Cooperatives took the lead, and district councils were responsible for actual program implementation. The project had a high level of government dedication and leadership.
Rating: High to very high

Policies and national-level factors: There was no irrigation policy to guide irrigation activities at the time of the project's implementation. Two severe droughts followed by extensive rain affected the project's performance. Planned agronomic trials of high-yielding varieties and training on agro-nursery management were hampered by the severe droughts.
Rating: Low to moderate

T: Push–Pull Technology (PPT)

Problem definition: The PPT project was developed to address a combination of productivity constraints, such as insect pests, the parasitic weed *Striga*, land degradation, and poor soil fertility. PPT was informed by technical, feasibility, and socioeconomic studies, which together identified insect pests and *Striga* as key challenges to cereal production in East Africa.
Rating: High to very high

Choice of commodity/instruments: PPT targeted cereal crops (maize, millet, and sorghum). PPT involves intercropping maize with an insect-repellent plant (such as *Desmodium*) and an attractive-trap plant (such as Napier grass) as a border crop, which fits well with traditional mixed-cropping systems in East Africa.
Rating: High to very high

Suitability of instruments: PPT targeted resource-poor smallholder farmers. The technology has been very appealing to farmers, because it addresses multiple challenges they face concurrently (Khan et al. 2006). The insect-repellent or insect-trap plants are also used as animal feeds, thereby solving the problem of fodder availability for mixed crop–livestock farmers.
Rating: High to very high

Design and timing of implementation: PPT is implemented by the International Centre of Insect Physiology and Ecology (ICIPE) and key partners. ICIPE has good technical and project implementation capacity. Because the project was well funded, it was able to implement activities in line with its work plan.
Rating: High to very high

Environmental sustainability: PPT contributes to soil fertility management through nitrogen fixation, natural mulching, improved biomass, and control of erosion. It also supports biodiversity through the variety of plant and animal species on the farm. Adopters of the technology have benefited from reduced runoff and soil erosion, enhanced soil fertility, and minimized use of agrochemicals.
Rating: High to very high

Financial sustainability: PPT is a low-cost technology. It uses minimal inputs, is based on locally available plants, and requires minimal farmer management capacity. Thus, its benefits are likely to continue.
Rating: High to very high

Community participation: There was high level of involvement among local communities. The government, represented by KARI, was involved in the design stage. Farmers were consulted, especially during the research stages of the project's design. Farmers were also consulted and trained during the initial stages of the project's implementation. Farmers showed strong commitment and have adopted the technology.
Rating: High to very high

Gender consideration: PPT integrated both men and women. PPT manages ecological weeds, which reduces the workload for women and youths, who typically do most of the weeding on the farm.
Rating: High to very high

Complementary investments and partnerships: Public- and private-sector partnerships in East Africa include ICIPE, Rothamsted Research, Heifer International Project–Kenya, KARI, Kenyan Ministry of Agriculture, the Ministry of Livestock and Fisheries Development through the National Agriculture and Livestock Extension Programme, the Lake Zone Agricultural Research and Developing Institute in Tanzania, NGOs, and farmers' groups.
Rating: High to very high

Capacity building: The project used trial and demonstration plots, media (print and audio), existing field-extension backstopping, FFS, and strengthening of farmer-to-farmer extension.
Rating: High to very high

Organized groups: Farmers were organized into groups via the FFS. These groups have been instrumental for disseminating the technology.
Rating: High to very high

Leadership and dedication: ICIPE, KARI, and the Ministry of Agriculture are supporting PPT. Farmers are dedicated to developing PPT, and agreed to be involved in promoting its adoption through the learning-by-doing approach, as well as through participatory ecological field studies.
Rating: High to very high

Policies and national-level factors: Adoption of PPT is constrained by small land sizes among smallholder farmers. It is also constrained by lack of strong national extension support; lack of information; and shortage of inputs, particularly *Desmodium* seed.
Rating: Low to moderate

U: Regional Land Management Unit (RELMA)

Problem definition: RELMA addressed various land and water management issues, including land degradation and water scarcity as a result of poor spatial distribution and timing of rainfall. Although the project was informed by various studies, there were shortcomings in the problem definition stage, which contributed to including components that did not address local priorities. The participatory needs assessment was inadequate. Four subcomponents were discontinued by the project management following the mid-term review of RELMA in 2005.
Rating: Low to moderate

Choice of the commodity/instruments: RELMA targeted food crops (such as maize and beans), high-value tree crops (such as those for fodder, fruits, and wood), and livestock rearing. It promoted different soil and water management practices to suit different agroecological zones.
Rating: High to very high

Suitability of instruments: RELMA targeted smallholder farmers in rural areas. RELMA's activities generated a noticeable impact on its clients during the implementation period. Most significantly, its subcomponents on soil fertility, conservation agriculture, dryland/livestock management, and rainwater harvesting continue to offer benefits to smallholder farmers.
Rating: High to very high

(continued)

TABLE 6A.5 (continued)

Design and timing of implementation: The project's shortcomings included lack of oversight during the planning process, which led to overestimating the interest of some clients and designing some project components that were irrelevant. The initial implementation faced a number of challenges (with staffing and financial management), which led to a slow start. The project's management was successful in accelerating the implementation momentum by effecting productive project staff reallocations within the ICRAF structure and promoting stricter budgetary and fiscal discipline among the RELMA staff. However, it did not fully compensate for weaknesses in subcomponent design or client support strategies resulting from the compressed planning period, or for the shortage of RELMA in-house expertise outside the core land, livestock, and water management and publication competencies.
Rating: Low to moderate

Environmental sustainability: RELMA promoted improved methods of land and water management, conservation farming technology, and rainwater-harvesting techniques.
Rating: High to very high

Financial sustainability: Although RELMA promoted simple and low-cost water supply and environmental management techniques, the continued implementation of some of the RELMA-promoted interventions relies on external support. RELMA was funded by the Swedish International Development Agency. When the project ended, maintaining its sustainability became a challenge.
Rating: Low to moderate

Community participation: Local communities were not adequately involved during the project's formulation phase.
Rating: Low to moderate

Gender consideration: RELMA was sensitive to gender issues. Training was provided to both men and women.
Rating: High to very high

Complementary investments and partnerships: RELMA worked and coordinated with various partners, including the Southern and Eastern Africa Rainwater Network, the International Rainwater Harvesting Alliance, the Centre for Science and Environment, ICRAF, the United Nations Environment Programme, UN Habitat, and national rainwater associations.
Rating: High to very high

Capacity building: RELMA created awareness through training (with training materials in local languages), media coverage in print and audio, and extension services to enhance knowledge in land management.
Rating: High to very high

Organized groups: RELMA promoted the formation of common interest groups and FFS, which provided technical support. The beneficiaries had capacity to engage effectively in the project.
Rating: High to very high

Leadership and dedication: Local communities were dedicated to implementing the interventions promoted by the project. Governments were positive about the program's support of environmental management.
Rating: High to very high

Policies and national-level factors: Timing was right because of the widespread awareness about environmental challenges. National governments in the target countries were generating policies and strategies to support environmental management.
Rating: High to very high

V: Sasakawa-Global 2000 Agricultural Program (SG2000-AP)

Problem definition: The NGO SG 2000 worked to address the problem of low agricultural productivity and food insecurity by introducing yield-enhancing agricultural technologies. The project was designed based on the documented information and data on causes of famine in various parts of Africa, which indicated that there was unexploited potential to increase food production through crop and livestock intensification.
Rating: High to very high

Choice of commodity/instruments: The program supported staple food crops (for example, maize, wheat, rice, legumes, and roots and tubers) and common livestock (cattle) kept by the smallholder farmers in the project areas.
Rating: High to very high

Suitability of instruments: SG 2000 promoted different commodities in different areas based on the suitability of local conditions. It also promoted agricultural intensification with appropriate, financially viable technology.
Rating: High to very high

Design and timing of implementation: SG 2000 was formulated through a partnership between the Sasakawa Africa Association and Global 2000 of the Carter Center. The NGO was mainly financed by the Sasakawa Foundation (now called the Nippon Foundation). It also worked closely with many other partners. The roles and responsibilities of different partners were clearly articulated. For instance, the Sasakawa Africa Association was responsible for program management, while Global 2000 specialized in policy-related interventions. SG 2000 has adequate qualified staff to implement project activities. Six expatriate field directors managed and supervised the 12 SG 2000 country projects with the assistance of local professionals and support staff. Two expatriate staff supervised multicountry programs to strengthen university-level extension education and agroprocessing microenterprise development.
Rating: High to very high

Environmental sustainability: SG 2000 promoted various strategies for better soil and water management. For example, in Mali, the strategy included efforts to combat wind and water erosion and use natural phosphates and legumes.
Rating: High to very high

Financial Sustainability: Sustaining the same level of support to the farmers after the end of the project has presented some challenges. The project worked closely with ministries of agriculture and national extension systems as a way of enhancing sustainability. However, various constraints within government extension systems have affected the project's sustainability.
Rating: Low to moderate

Community participation: SG 2000 worked with farmers and ministries of agriculture to test and promote adoption of appropriate, profitable technologies that increase yields and improve soil fertility. It involved government systems and farmers in technology transfer.
Rating: High to very high

Gender consideration: Not apparent.
Rating: Low to moderate

Complementary investments and partnerships: SG 2000 supported various complementary interventions along the crop and livestock value chains. It supported access to inputs (fertilizer, seed); contributed to value addition through agroprocessing, so as to reduce postharvest losses; promoted improved storage techniques and technologies; invested in promoting public–private partnerships, so as to leverage contributions from other partners in implementing these activities; and worked with ministries of agriculture and national extension services, as well as with national and international agricultural research systems and other development organizations.
Rating: High to very high

Capacity building: SG 2000 trained farmers on improved farming techniques through the use of high-yielding technologies. It supported the national extension system in the study countries, so as to enhance access to agricultural extension. Various knowledge transfer approaches were used, including experimental plots and farmer-owned demonstration plots.
Rating: High to very high

Organized groups: The project helped farmers to organize into groups and cooperatives. Farmers also were encouraged to create rural savings and loan associations.
Rating: High to very high

(continued)

TABLE 6A.5 (continued)

Leadership and dedication: Beneficiaries were dedicated in participating in the project. Extension efforts were centered on the production test plot, which is a half-hectare parcel owned or managed by a participant farmer who agrees to test the new technology on his or her own field and share experiences with others.
Rating: High to very high

Policies and national-level factors: Efforts to introduce new technologies to farmers in semiarid areas, such as Burkina Faso and Mali, were confronted by a more fragile ecosystem (for example, nutrient-poor and badly drained soils, and insufficient and erratic rainfall). The project's effectiveness was also negatively affected by other factors, such as highly variable producer prices, weak marketing infrastructure, and poor input-responsive millet varieties.
Rating: Low to moderate

W: Speciality Coffee Program (SCP), Rwanda

Problem definition: The intervention aimed at addressing the priority problem of farmers producing a low-quality coffee crop that was not attractive in the international market. The project was informed by past studies on the trends and performance of the coffee industry in Rwanda. The importance of coffee is articulated in the national policy and strategy documents. The project focused on building capacity in the coffee sector to produce specialty coffee of high value in international markets.
Rating: High to very high

Choice of commodity/instrument: Rwanda has a long history of coffee production. Coffee is an important cash crop in Rwanda, which has suitable agroclimatic conditions for growing the crop.
Rating: High to very high

Suitability of instrument: SCP targets smallholder farmers and rural communities.
Rating: High to very high

Design and timing of implementation: The Rwanda Coffee Development Agency is promoting speciality coffee, in collaboration with other stakeholders in the country. A number of development partners have been supporting the Rwandan government on the intervention, and the numbers of coffee-processing factories and coffee-washing stations have increased as a result of the program.
Rating: *High to very high*

Environmental sustainability: Coffee processors are making efforts to address environmental issues, but these efforts are not adequate, as some other players in the coffee value chain are also required to invest in addressing environmental problems.
Rating: *Low to moderate*

Financial sustainability: The coffee sector is mainly supported by development partners. Government financial commitment is still low. Underutilization of the increased number of washing stations is threatened by low coffee production.
Rating: Low to moderate

Community participation: SCP brings together stakeholders in the coffee industry, including representatives of local communities. The Rwanda Coffee Development Authority involves stakeholders in the industry to agree on a minimum weekly price for coffee.
Rating: High to very high

Gender consideration: The project has made deliberate efforts to empower female coffee growers.
Rating: High to very high

Complementary investments and partnerships: Several complementary investments are in place, including marketing infrastructure and coffee-washing stations. There is collaboration among various actors, including local communities, the government, development partners, the private sector, NGOs, and research institutions.
Rating: High to very high

Capacity building: Technical support is provided through training in improved coffee production, coffee processing, washing station management, and coffee marketing. Access to credit has been promoted. Coffee-washing stations have been constructed. The capacity to fully wash coffee beans has significantly increased in the recent past.
Rating: High to very high

Organized groups: Farmers are organized in groups through cooperatives. However, the management problems in some cooperatives require further investments in strengthening planning, administration, and financial management skills.
Rating: Low to moderate

Leadership and dedication: The government has been dedicated to the promotion of coffee as a cash crop and to the development of the coffee value chain. Local leaders and other agriculture stakeholders are supportive of the initiative.
Rating: High to very high

Policies and national-level factors: The government has made a number of reforms in the coffee industry in the past two decades. In late 1990s, the government opened up the market for coffee export to increase competition, and began to focus on improving the coffee value chain. In 2002, the government unveiled a national coffee strategy for capturing a larger share of the specialty coffee sector. Despite such policy reforms, several challenges remain, including old coffee trees, low-yielding varieties, high transportation costs, and the high costs and weak management skills of the washing stations.
Rating: Low to moderate

X: System of Rice Intensification (SRI), Rwanda

Problem definition: The project aimed at addressing low rice yields. It was informed by various research activities undertaken by the Institut des Sciences Agronomiques du Burundi. The project supports the national priorities. The Support Project for the Strategic Plan for the Transformation of Agriculture (PAPSTA II) identified rice as one of the high-value crops in the country and one of the cereal commodity chains that will serve as a major source of internal agricultural markets in Rwanda.
Rating: High to very high

Choice of commodity/instruments: Rice consumption is on the rise as a result of urbanization and population growth. The government aims to meet the growing demand through domestic production in the marshlands. SRI involves changing the management of plants, soil, water, and nutrients, including early, quick, and healthy plant establishment; reduced plant density; improved soil conditions through enrichment with organic matter; and reduced and controlled water application.
Rating: High to very high

Suitability of instruments: Interventions targeted resource-poor smallholders, who are members of targeted cooperatives located in the Kibaza, Bugesera, and Rwabutazi/Kihere districts, which are suitable for rice production.
Rating: High to very high

Design and timing of implementation: The technology is being spread through pilot projects funded by development partners, including IFAD and JICA. Through research, technical support and other promotional activities have been carried out. SRI projects have been well staffed and implemented as planned. M&E activities have been in place.
Rating: High to very high

Environmental sustainability: SRI is a way of producing more with less by using fewer inputs, particularly less water, seed, and chemical fertilizer. With SRI technology, the soil is kept alternately dry and wet, allowing the plants' roots to take oxygen from the ground surface.
Rating: High to very high

(continued)

TABLE 6A.5 (continued)

Financial sustainability: Most of SRI promotion activities are supported by projects funded by development partners, which can be a problem when the project ends.
Rating: Low to moderate

Community participation: Farmers' groups are involved in the design and implementation of SRI activities. Their adoption of the technology has resulted in increased rice yields among the beneficiaries.
Rating: High to very high

Gender consideration: Not apparent.
Rating: Low to moderate

Complementary investments and partnerships: Various partners are involved, including IFAD, the National Agriculture Research Institute, the Ministry of Agriculture, extension workers, and union of rice cooperatives. Farmers face several challenges that limit their willingness or ability to adopt the technology, including insufficient storage infrastructure for surplus produce, scarce access to mineral fertilizers, and lack of regular follow-up by SRI technicians at every stage of SRI implementation.
Rating: Low to moderate

Capacity building: Although there is training of rice producers, demonstration farmers, and extension officers (who were trained by Malagasy experts), there is inadequate regular follow-up by SRI technicians at every stage of SRI implementation.
Rating: Low to moderate

Organized groups: Farmer cooperative schemes were set up by the government. These were built on local networks. The groups have been beneficial in promoting technology and sharing experiences.
Rating: High to very high

Leadership and dedication: The program is supported by the government. It was introduced in Rwanda under the Projet d'Appui au Plan Stratégique pour la Transformation de l'Agriculture (Support Project for the Strategic Plan for the Transformation of Agriculture—PAPSTA) and co-financed by IFAD. However, there is resistance by some smallholder farmers to adopt SRI technologies on their plots.
Rating: High to very high

Policies and national-level factors: The government has shown its support for the project through the national rice policy. It has set rice production as a priority, especially in the valley bottom marshlands. However, the rice sector faces a number of challenges, such as marketing problems and lack of access to both inputs and storage facilities.
Rating: Low to moderate

Y: Wei Wei Integrated Development Project (WWIDP), Kenya

Problem definition: The project aimed at addressing the problem of scarcity of water for crop production, with the objective of mitigating food security problems, as stipulated in national strategy documents. A cost–benefit analysis was conducted, and results were highly in favor of the project.
Rating: High to very high

Choice of commodity/instruments: The project introduced new technology for gravity-fed, overhead irrigation in the area. Farmers were encouraged to grow relevant food crops suitable for the area, including maize, sorghum, green gram, and cow peas.
Rating: High to very high

Suitability of instruments: The project targeted districts (such as West Pokot) that were vulnerable because of unfavorable dry climatic conditions. It used irrigation interventions that have increased crop productivity in other places with similar biophysical conditions. The project was built on local indigenous knowledge and farming practices. Pokot people living in the project area were already accustomed to growing crops using the traditional furrow irrigation system. The project recognized this and invested in improving the irrigation system by constructing a modern gravity-fed pipeline system.
Rating: Low to moderate

Design and timing of implementation: The project seemed to be well designed and implemented in a timely manner. Resources were adequate, and an M&E system was put in place to track project performance. There are several documented case studies regarding this intervention.
Rating: High to very high

Environmental sustainability: Measures to mitigate potential erosion problems that could result from irrigation were put in place. The project introduced an external vegetation windbreak between every four plots. In addition, at the perimeters of the plots natural vegetation was maintained. The project distributed Vetiver grass for planting across most water channels as a means to reduce the speed of water and its erosive capacity. The project was successful in the rejuvenation of vegetation.
Rating: High to very high

Financial sustainability: The gravity-fed irrigation system does not require energy for its operation. Maintenance costs for the system are minimal, and there is relatively less waste from seepage and evaporation, as was the case with the traditional furrows. Farmers have continued to realize consistently improved yields of the crops promoted by the project over the years.
Rating: High to very high

Community participation: Local people were consulted and involved in the implementation of the project right from the beginning. A plot allocation committee was created to ensure fairness in the allocation of plots. The committee consists of project staff, the executive committee of the Wei Wei Farmers Association, the local councilor, and local traditional leaders.
Rating: High to very high

Gender consideration: Not apparent.
Rating: Low to moderate

Complementary investments and partnerships: Partnerships were formed among various actors, including the governments of Kenya and Italy, private companies (seed companies), and local communities (farmers' associations). An integrated approach was used to address multiple constraints.
Rating: High to very high

Capacity building: Farmers were trained on improved soil and water management practices. Training was carried out through farm-level activities, workshops, and seminars. Training materials were also developed and distributed to the beneficiaries.
Rating: High to very high

Organized groups: Farmers formed groups and supported each other by sharing both knowledge on agriculture technologies and marketing information.
Rating: High to very high

Leadership and dedication: The government demonstrated its commitment to make the project sustainable by creating a joint management structure to continue with the project's implementation after the end of the donor support. Local leaders and farmers were very keen to implement the project, and there was strong sense of ownership of the project.
Rating: High to very high

Policies and national-level factors: The project is located in a remote area with recurrent droughts and frequent intertribal and interclan clashes.
Rating: Low to moderate

Source: Authors' evaluation based on Mugova and Mavunga (2000).

TABLE 6A.6 Performance in meeting the overall productivity objective or target

Project name and productivity performance indicators and achievements

Project: Animal Health Services Rehabilitation Programme (AHSRP)—Kenya

Target vaccination rate: 75 percent of the herd

Achieved vaccination rate: 37.5 percent

% of target achieved: 50 percent

Overall performance rating: Very poor

Project: Agriculture Productivity Enhancement Programme (APEP)—Uganda

	Yield (tons/ha)		
Target crops	**Target**	**Achieved**	**% of target achieved**
Coffee	1.1	1.0	91
Cotton	1.3	0.9	69
Sunflowers	1.8	1.4	78
Rice	1.8	2.8	156
Maize	4.5	3.6	80
Flowers	33.7	28.7	85
Bananas	20.2	24.1	119
Green vanilla beans	0.9	0.72	80

Overall performance rating: Moderate

Project: Agricultural Sector Development Programme (ASDP-irrigation)—Tanzania

Performance indicators	**Target**	**Achieved**	**% of target achieved**
Irrigation schemes to rehabilitate	1,520	1,325	87.2
Irrigated area	380,000 ha	363,514 ha	95.7
Rice yield	6 tons/ha	5 tons/ha	83.3

Overall performance rating: Moderate

Project: Cassava Enterprise Development Project (CEDP)—Nigeria

Target crop: Cassava

Target yield: 25.0 tons/ha

Achieved yield: 27.2 tons/ha

% of target achieved: 109 percent

Overall performance rating: Good

Project: Conservation Agriculture Project 1 (CAP1)— Zambia

Target crop: Maize

Target yield: 5.0 tons/ha (target is based on Conservation Farming Unit, 2007)

Achieved yield (average for all conservation agriculture approaches): 3.3 tons/ha

% of target achieved: 57 percent

Overall performance rating: Poor

Project: Crop Crisis Control Programme (C3P)—Kenya, Uganda, Tanzania, Rwanda, Burundi, and DRC.
Target crops: Cassava and banana

Performance indicators	Target	Achieved	% of target achieved
Area planted with disease-resistant crops (ha)	542	697	128.6
Number of farmers trained	6,000	47,631	793.9
Number of extensionists participating in training	310	1,000	322.6

Overall performance rating: Good

Project: East Africa Dairy Development Project (EADD)—Kenya, Uganda, and Rwanda
Target (objective): Increased milk production

	Production (liters/day)		
Project countries	Target*	Achieved	% of target achieved
Kenya	7.2	5.5	77
Rwanda	8.2	6.3	77
Uganda	6.3	4.9	77

*Target is based on EADD phase II target of doubling baseline milk production.

Overall performance rating: Moderate

Project: FARM Africa Dairy Goat Improvement Project (FADGIP)—Meru, Kenya

Baseline milk yield: 0.2–1.0 liters/day (local goat breeds)

Target yield: 3.0 liters/animal/day

Achieved yield: 2.9 liters/animal/day

% of target achieved: 96 percent

Overall performance rating: Moderate

Project: Farm Input Subsidy Program (FISBP)—Malawi

Target crop: Maize

Target yield: 3.0 tons/ha (computed as baseline yield of 1.3 tons/ha * 2.3)

Achieved yield: 2.7 tons/ha

% of target achieved: 90 percent

Overall performance rating: Moderate

(continued)

TABLE 6A.6 (continued)

Project: Farmer Input Support Program (FISPP)—Zambia
Target crop: Maize
Target yield: 3.3 tons/ha (computed as baseline yield of 1.44 tons/ha × 2.3)
Achieved yield: 2.24 tons/ha
% of target achieved: 68 percent
Overall performance rating: Poor

Project: Fodder Trees and Shrubs Project (FTSP)—Kenya, Rwanda, Tanzania, and Uganda
Objective: Increase milk yield
Target yield: 4 liters/animal/day (computed by doubling the milk baseline yield of 2 liters/animal/day)
Achieved yield: 3.5 liters/animal/day
% of target achieved: 88 percent
Overall performance rating: Moderate

Project: Fuve Panganai Irrigation Scheme (FPIS)—Zimbabwe
Target crop: Rice
Target yield: 6.0 tons/ha
Achieved yield: 1.6 tons/ha
% of target achieved: 27 percent
Overall performance rating: Very poor

Project: Kaleya Irrigation Project (KIP)—Zambia
Target crop: Sugarcane
Target yield: 100 tons/ha
Achieved yield: 110–115 tons/ha
% of target achieved: 110–115 percent
Overall performance rating: Good

Project: Kenya Dairy Development Programme (KDDP)—Kenya
Target product: Milk
Target: 40 percent increase in milk productivity
Achieved yield: 19 percent increase in milk productivity on average for all participants
% of target achieved: 48 percent
Overall performance rating: Very poor

Project: National Agricultural Advisory Services (NAADS) Phase 1—Uganda
Target (objectives): Increase access to agricultural advisory services, adoption of agricultural technologies, and yields

	NAADS		Non-NAADS	
Target crops	**Yield in 2004 (kg)**	**% change 2000–2004**	**Yield in 2004 (kg)**	**% change 2000–2004**
Groundnuts	426	**57**	433	**−0.6**
Maize	669	**64**	835	**27.3**
Bananas	5,942	**−5**	3,323	**55.3**
Sorghum	449	**77**	389	**34.8**
Sweet potatoes	1,761	**18**	1,392	**7.3**
Cassava	1,244	**46**	4,340	**−9.4**
Beans	572	**62**	721	**17.2**
Coffee	516	**−28**	2,090	**81.3**
Irish potatoes	1,003	**260**	1,369	**285.4**

Based on Benin et al. (2007). Difference-in-differences—that is, using columns in bold—was applied to calculate differences in percentage change.

Overall performance rating: Moderate

Project: National Agricultural Extension Intervention Program (NAEIP)—Ethiopia
Target (objective): Increased yield of maize, teff, wheat, and sorghum

	Yield (tons/ha)		
Target crops	**Non-NAEIP**	**NAEIP farmers**	**% difference**
Maize	1.9	2.9	52.6
Teff	0.9	1.1	22.2
Wheat	1.4	1.9	35.7
Sorghum	1.5	1.9	26.7

Based on World Bank (2007).

Overall performance rating: Very poor

Project: New Rice for Africa (NERICA)—Uganda

Target crop: Upland rice

Target yield: 3.30 tons/ha (based on Uganda rice strategy)

Achieved yield: 2.85 tons/ha

% of target achieved: 86 percent

Overall performance rating: Moderate

(continued)

TABLE 6A.6 (continued)

Project: Operation *Mwolyo* Out (OMO)—Kenya

Target crop: Maize

Target yield: 3.0 tons/ha (based on baseline yield of 1.3 tons/ha × 2.3)

Achieved yield: 5.4 tons/ha under irrigated conditions

% of target achieved: 180 percent

Overall performance rating: Very good

Project: Participatory Irrigation Development Programme (PIDP)—Tanzania

Target yield: 4 tons/ha

Achieved yield: 2 tons/ha

% of target achieved: 50 percent

Overall performance rating: Poor

Project: Push–Pull Technology (PPT)—Kenya, Tanzania, Uganda, and Ethiopia
Target crops: Maize and sorghum

	Yield (tons/ha)			
Target crops	Baseline	Target	Achieved	% of target achieved
Maize	1.5	3.0	6.0	200
Sorghum	1.0	2.5	3.0	120

Overall performance rating: Very good

Project: Regional Land Management Unit (RELMA), water harvesting component—Eritrea, Ethiopia, Kenya, Tanzania, Uganda, and Zambia

Target yield: 2.30 tons/ha (based on baseline yield of 1.0 ton/ha × 2.3)

Achieved yield: 1.97 tons/ha

% of target achieved: 86 percent

Overall performance rating: Moderate

Project: Sasakawa Global 2000 Agricultural Program (SG2000-AP)—Ghana, Sudan, Tanzania, Benin, Togo, Mozambique, Eritrea, Guinea, Burkina Faso, Malawi, Mali, Nigeria, Ethiopia, and Uganda

Objective: Increased cereals yields

Target yield: 2.3 tons/ha (based on baseline yield of 1.0 ton/ha × 2.3 for Uganda)

Achieved yield: 4.0 tons/ha

% of target achieved: 174 percent

Overall performance rating: Very good

Project: Specialty Coffee Program (SCP)—Rwanda

Target: 57.6 percent fully washed coffee by 2012

Achieved: 30.0 percent fully washed in 2012

% of target achieved: 52 percent

Overall performance rating: Poor

Project: System of Rice Intensification (SRI)—Rwanda

Target yield: 7 tons/ha (based on the target of the Government of Rwanda's National Rice Development Strategy)

Achieved yield: 7.5 tons/ha

% of target achieved: 107 percent

Overall performance rating: Good

Project: Wei Wei Integrated Development Project (WWIDP)—Kenya
Target crops: Maize and sorghum as the main crops

	Yield (tons/ha)			
Target crops	**Baseline**	**Target**	**Achieved**	**% of target achieved**
Maize	1.5	3.00	4.93	164.3
Sorghum	0.5	1.25	4.40	352.0

Overall performance rating: Very good

Source: Authors' compilation from literature review, project documents, and expert consultations. (Table 6A.1 contains details on each project.)

Notes: Overall productivity performance rating: very poor = less than 50% of target achieved; poor = 50.0–74.9% of target achieved; moderate = 75.0–99.9% of target achieved; good = 100.0–149.9% of target achieved; and very good = 150% or more of target achieved.

References

Abdul, I., R. Alhadu, L. M. Sidek, M. Nor, M. Desa, N. Elzin, and A. Basri, 2011. "Multi Criteria Analysis in Environmental Management: Selecting the Best Stormwater Erosion and Sediment Control Measure in Malaysian Construction Sites." *International Journal of Energy and Environment* 2 (5): 853–862.

Abramovich, J., and D. Zook. 2015. *How Good Coffee Becomes Good Business for African Farmers*. Report to the Bill & Melinda Gates Foundation. Accessed June 2015. http://www.impatientoptimists.org/Posts/2015/03/How-Good-Coffee-Becomes-Good-Business-for-African-Farmers#.VkvNJ00w8qQ.

AfDB (African Development Bank, African Development Fund). 2010. *Appraisal Report: Small-Scale Irrigation Project (SIP), Republic of Zambia*. http://www.afdb.org/fileadmin/uploads/afdb/Documents/Project-and-Operations/Zambia_-_Smallscale_Irrigation_Project_-_Appraisal_Report.pdf.

Andersen, E. S., D. Birchall, S. A. Jessen, and A. H. Money. 2006. "Exploring Project Success." *Baltic Journal of Management* 1 (2): 127–147.

Aune, J. B., P. Nyanga, and F. H. Johnsen. 2012. *A Monitoring and Evaluation Report of the Conservation Agriculture Project 1 (CAP_1) in Zambia.* Noragric Report 68. Aas, Norway: Norwegian University of Life Sciences, Department of International Environment and Development Studies. http://www.umb.no/statisk/noragric/publications/reports/2012/noragricrep68.pdf.

Ayele, Z., and C. Peacock. 2003. "Improving Access to and Consumption of Animal Source Foods in Rural Households: The Experiences of a Women-Focused Goat Development Program in the Highlands of Ethiopia." *The Journal of Nutrition* 133 (11): 3981S–3986S.

Baccarini, D., and A. Collins. 2003. "Critical Success Factors for Projects." In *Surfing the Waves: Management Challenges; Management Solutions,* edited by A. Brown. Proceedings of the 17th ANZAM Conference, Fremantle, Western Australia, December 2–5, Australian and New Zealand Academy of Management.

Baltenweck, I., and G. Mutinda. 2013. *Dairy Hubs in East Africa: Lessons from the East Africa Dairy Development Project.* A "livestock live" talk held at the International Livestock Research Institute (ILRI) Nairobi campus, June 26. http://www.slideshare.net/ILRI/dairy-hubs-in-east-africa-lessons-from-the-east-africa-dairy-development-project–23612057.

Bangwe, L., and B. van Koppen. 2012. *Smallholder Outgrowers in Irrigated Agriculture in Zambia.* AgWater Solutions Project Case Study. Colombo, Sri Lanka: International Water Management Institute; Stockholm Environment Institute; Rome: Food and Agriculture Organization of the United Nations; Washington, DC: International Food Policy Research Institute; Denver, CO: International Development Enterprises; and Englewood, CO: CH2MHill. http://awm-solutions.iwmi.org/Data/Sites/3/Documents/PDF/Country_Docs/Zambia/zambia-outgrowers.pdf.

Benin, S., E. Nkonya, G. Okecho, J. Pender, S. Nahdy, S. Mugarura, E. Kato, and G. Kayobyo. 2007. *Assessing the Impact of the National Agricultural Advisory Services (NAADS) in the Uganda Rural Livelihoods.* IFPRI Discussion Paper 00724. Washington, DC: International Food Policy Research Institute.

Benin, S., E. Nkonya, G. Okecho, J. Randriamamonjy, E. Kato, G. Lubadde, M. Kyotalimye, and F. Byekwaso. 2011. *The Impact of Uganda's National Agricultural Advisory Services Program.* Research Monograph. Washington, DC: International Food Policy Research Institute.

Boudreaux, K. C. 2010. "A Better Brew for Success: Economic Liberalization in Rwanda's Coffee Sector." In *Yes Africa Can: Success Stories from a Dynamic Continent/World Bank.* Washington, DC: World Bank. http://siteresources.worldbank.org/AFRICAEXT/Resources/258643–1271798012256/rwanda_Coffee.pdf.

Boussard, J. M., B. Daviron, F. Gérard, and T. Voituriez. 2005. *Food Security and Agricultural Development in Sub-Saharan Africa: Building a Case for More Support.* Background Document. Paris: CIRAD for the Food and Agriculture Organization of the United Nations. Accessed November 29, 2011. http://www.fao.org/tc/tca/work05/cirad.pdf.

Bradstock, A. 2007. *Kenya Dairy Goat and Capacity Building Programme: A Socioeconomic Baseline.* Report. Working Paper No. 7. London: Farm Africa.

Burke, W. J. 2012. "Determinants of Maize Yield Response to Fertilizer Application in Zambia: Implications for Strategies to Promote Smallholder Productivity." PhD dissertation, Michigan State University, East Lansing, MI.

Byerlee, D., D. J. Spielman, D. Alemu, and M. Gautam. 2007. *Policies to Promote Cereal Intensification in Ethiopia: A Review of Evidence and Experience.* IFPRI Discussion Paper No. 707. Washington, DC: International Food Policy Research Institute.

Catley, A., J. Burns, D. Abebe, and O. Suji. 2008. *Participatory Impact Assessment: A Guide for Practitioners.* Boston: Tufts University, Feinstein International Center.

Chazovachii, B. 2012. "The Impact of Small Scale Irrigation Schemes on Rural Livelihoods: The Case of Panganai Irrigation Scheme Bikita District, Zimbabwe." *Journal of Sustainable Development in Africa* 14 (4).

Chemonics International, Inc. 2006. *Assessing USAID'S Investments in Rwanda's Coffee Sector—Best Practices and Lessons Learned to Consolidate Results and Expand Impact.* Kigali, Rwanda: United States Agency for International Development.

Chirwa, E. W., M. Matita, and A. Dorward. 2011. *Factors Influencing Access to Agricultural Input Subsidy Coupons in Malawi.* FAC Working Paper 027. Brighton, UK: Future Agricultures Consortium.

Copestake, J., S. Johnson, and K. Wright. 2004. "Impact Assessment of Microfinance: Towards a New Protocol for Collection and Analysis of Qualitative Data" In *Methods, Knowledge and Power: Combining Quantitative and Qualitative Development Research,* edited by J. Holland and J. Campbell. Swansea, UK: IT Publications and University of Swansea Centre for Development Studies.

Croppenstedt, A., M. Goldstein, and N. Rosas. 2013. *Gender and Agriculture: Inefficiencies, Segregation, and Low Productivity Traps.* Policy Research Working Paper 6370. Washington, DC: World Bank.

CSPR (Civil Society for Poverty Reduction), Zambia. 2011. *An Assessment of the Implementation and Viability of the Farmer Input Support Program in Zambia: With Case Studies of Mazabuka and Monze Districts.* Technical Report. Lusaka, Zambia.

Davis, K. and M. Negash. 2007. *Gender, Wealth, and Participation in Community Groups in Meru Central District, Kenya.* CAPRi Working Papers 65. Washington, DC: International Food Policy Research Institute.

De Brucker, K., C. Macharis, and A. Verbeke. 2011. "Multi-criteria Analysis in Transport Project Evaluation: An Institutional Approach." *European TransportTrasporti Europei* 47 (2011): 3.

Denning, G., P. Kabambe, P. Sanchez, A. Malik, R. Flor, R. Harawa, P. Nkhoma, C. Zamba, C. Banda, C. Magombo, M. Keating, J. Wangila, and J. Sachs. 2009. "Input Subsidies to Improve Smallholder Maize Productivity in Malawi: Toward an African Green Revolution." *PLoS Biol* 7(1): e1000023. doi:10.1371/journal.pbio.1000023.

Diagne, A., S. G. Midingoyi, M. Wopereis, and I. Akintayo. 2010. *The NERICA Success Story: Development, Achievements and Lessons Learned.* Draft Report. Cotonou, Benin: The Africa Rice Center (AfricaRice).

Dorward, A., E. Chirwa, and T. Jayne. 2010. *The Malawi Agricultural Inputs Subsidy Programme, 2005/6 to 2008/9.* Final Report. Lilongue: Government of Malawi, Ministry of Agriculture and Food Security.

Dorward, A., R. Sabates Wheeler, I. MacAuslan, C. P. Buckley, J. Kydd, and E. Chirwa. 2006. *Promoting Agriculture for Social Protection or Social Protection for Agriculture: Strategic Policy and Research Issues.* Discussion Paper. Pretoria: Food, Agriculture and Natural Resources Policy Analysis Network. http://www.fanrpan.org/documents/d00128/.

Dowswell, C. 2011. "Sasakawa-Africa Association/Global 2000 SG 2000 Agricultural Program in Africa: Our Future Directions." Presentation made at the World Agroforestry Centre (ICRAF), Nairobi, Kenya, June 30. http://www.slideshare.net/agroforestry/saa-our-future-directionsicraf-presentation.

Druilhe, Z., and J. Barreiro-Hurlé. 2012. *Fertilizer Subsidies in sub-Saharan Africa.* ESA Working Paper No. 12–04. Rome: Food and Agriculture Organization of the United Nations, Agricultural Development Economics Division. http://www.fao.org/3/a-ap077e.pdf.

EC (European Commission). 2010. *Strategic Environmental Assessment (SEA) of the Sugar Sector in Zambia.* Paris.

EPRC (Economic Policy Research Centre). 2011. "An Insight into the Performance of the National Agricultural Advisory Services Programme." Briefing paper presented to the Sessional Committee on Ministry of Agriculture, Animal Industry and Fisheries (MAAIF), Kampala, Uganda, August.

Erikson J. 2008. *Improved Land Management for Sustainable Development (RELMA-in ICRAF).* Sida Evaluation 2008:56. Stockholm: Swedish International Development Cooperation Agency.

Farm Africa. 2007. *A Goat Model: A Proven Approach to Reducing Poverty among the Small-holder Farmers in Africa by Developing Profitable Goat Enterprises and Sustainable Support Services*. Working Paper No. 9. London: Farm Africa.

Feder, G., J. R. Anderson, R. Birner, and K. Deininger. 2010. *Promises and Realities of Community-Based Agricultural Extension*. IFPRI Discussion Paper 00959. Washington, DC: International Food Policy Research Institute.

Fischler, M. 2010. *Impact Assessment of Push Pull Technology in Eastern Africa*. Nairobi, Kenya: International Centre of Insect Physiology and Ecology (ICIPE) Science Press.

Fonchingong, C., and L. N. Fonjong. 2003. "The Concept of Self-reliance in Community Development Initiatives in the Cameroon Grassfields." *Nordic Journal of African Studies* 12 (2): 196–219.

Franzel, S., and C. Wambugu. 2007. "The Uptake of Fodder Shrubs among Smallholders in East Africa: Key Elements That Facilitate Widespread Adoption." In *Forages: A Pathway to Prosperity for Smallholder Farmers*, edited by M. D. Hare and K. Wongpichet, 203–222. Proceedings of an International Symposium. Ubon Ratchathani, Thailand: Ubon Ratchathani University, Faculty of Agriculture.

Gaitano, S. 2011. "The Design of M&E Systems: A Case of East Africa Dairy Development Project." Paper Presented at the International NGO and Research Training Center (INTRAC) 7th Monitoring and Evaluation Conference. Nairobi, Kenya, June 14–16. http://www.intrac.org/data/files/ME_conference_papers_2011/Working_groups_papers/Working_group_8/East_Africa_Dairy_Development_Project_Simiyu.pdf.

Gathiru, K., and C. Ong. 2006. *Evolution of Land Management Approaches. Two Decades of Research and Development in Eastern and Southern Africa*. Technical report. Nairobi, Kenya: World Agroforestry Center.

Gawler, M. 2005. *Project Design in the Context of Project Cycle Management: Sourcebook*. World Wildlife Fund Introductory Course. Prevessin-Moens, France: Artemis Services for Nature Conservation and Human Development.

Gebreselassie, S. 2006. "Intensification of Smallholder Agriculture in Ethiopia: Options and Scenarios." Paper prepared for the Future Agricultures Consortium Workshop, Institute of Development Studies, University of Nairobi, Kenya. 20-22 March 2006; http://r4d.dfid.gov.uk/Output/178530/Default.aspx.

Haggblade, S., and P. Hazell. 2010. *Successes in African Agriculture: Lessons for the Future*. IFPRI Issue Brief 63. Washington, DC: International Food Policy Research Institute.

Haggblade, S., and G. Tembo. 2003. "Conservation Farming in Zambia." Conference paper number 11, presented at the InWEnt, IFPRI, CTA conference "Success in Agriculture," Pretoria, South Africa, December 1–3.

Hyvari, I. 2006. "Success of Projects in Different Organizational Conditions." *Project Management Journal* 37 (4): 31–41.

ICIPE (International Centre of Insect Physiology and Ecology). 2003. *Habitat Management Strategies for Control of Stemborers and Striga Weed in Cereal Based Farming Systems in Eastern Africa*. Annual Scientific Report 2002–2003. Nairobi, Kenya.

IFAD (International Fund for Agricultural Development). 1993. *Animal Health Services Rehabilitation Programme—Kenya*. Project evaluation report. Nairobi, Kenya. http://www.ifad.org/evaluation/public_html/eksyst/doc/prj/r188kece.html.

———. 2007. *Participatory Irrigation Development Programme, Completion Evaluation*. Report No. 1831-TZ - Rev. 1. Dar es Salaam: United Republic of Tanzania.

———. 2009. *Making Rice a Cash Cow in Rwanda*. Kigali, Rwanda: IFAD, Rwanda Country Programme. http://ciifad.cornell.edu/sri/countries/rwanda/RwandaIFADSRIflyerJan09.pdf.

Illovo Sugar. 2014. "Zambia Socio-Economic Impact Assessment: Internal Management Report." Lusaka, Zambia: Illovo Sugar.

ITAD. 2008. *Performance Evaluation of National Agricultural Advisory Services (NAADS)*. Final report. Hove, East Sussex, UK.

Jayne, T. S., N. M. Mason, W. Burke, A. Shipekesa, A. Chapoto, and C. Kabaghe. 2011. *Mountains of Maize, Persistent Poverty*. Policy Synthesis No. 48. Lusaka, Zambia: Food Security Research Project.

Jayne, T. S., D. Mather, N. M. Mason, and J. Ricker-Gilbert. 2013. "How Do Fertilizer Subsidy Programs Affect Total Fertilizer Use in Sub-Saharan Africa? Crowding Out, Diversion, and Benefit/Cost Assessments." *Agricultural Economics* 44: 687–703.

Jayne, T. S., and S. Rashid. 2013. "Input Subsidy Programs in Sub-Saharan Africa: A Synthesis of Recent Evidence." *Agricultural Economics* 44: 547–562.

Karamouz, M., R. Kerachian, B. Zahraie, and S. Araghi-Nejhad. 2002. "Monitoring and Evaluation Scheme Using the Multiple-Criteria-Decision-Making Technique: Application to Irrigation Projects." *Journal of Irrigation and Drainage Engineering* 128 (6): 341–350.

Khan, Z. R., J. A. Pickett, L. J. Wadhams, A. Hassanali, and C. A. O. Midega. 2006. "Combined Control of *Striga hermonthica* and Stemborers by Maize—*Desmodium* spp. Intercrops." *Crop Protection* 25 (9): 989–995.

Khandker, S. R., G. B. Koolwal, and H. A. Samad. 2010. *Handbook on Impact Evaluation: Quantitative Methods and Practices*. Washington, DC: World Bank. https://openknowledge.worldbank.org/bitstream/handle/10986/2693/520990PUB0EPI1101Official0Use0Only1.pdf?sequence=1.

Khang, D. B., and T. L. Moe. 2008. "Success Criteria and Factors for International Development Projects: A Lifecycle-based Framework." *Project Management Journal* 39 (1): 72–84.

Kherallah, M., C. Delgado, E. Gabre-Madlin, N. Minot, and M. Johnson. 2002. *Reforming Agricultural Markets in Africa.* Washington, DC: International Food Policy Research Institute; Baltimore: Johns Hopkins University Press.

Kijima, Y. 2008. *New Technology and Emergence of Markets: Evidence from NERICA Rice in Uganda.* Discussion Paper No. 165. Nagoya, Japan: Nagoya University, Graduate School of International Development.

Kijima, Y., K. Otsuka, and D. Sserunkuuma. 2011. "An Inquiry into Constraints on a Green Revolution in Sub-Saharan Africa: The Case of NERICA Rice in Uganda." *World Development* 39: 77–86.

Kijima, Y., D. Sserunkuuma, and K. Otsuka. 2006. "How Revolutionary Is the 'NERICA Revolution'? Evidence from Uganda." *Developing Economies* 44 (2): 252–267.

Kimenye, L., and A. Bombom. 2009. *Best-Bet Research Outputs for Enhancing Agricultural Productivity in Eastern and Central Africa: Abstracts.* Entebbe, Uganda: Association for Strengthening Agricultural Research in Eastern and Central Africa.

Kristjanson, P., A. Waters-Bayer, N. Johnson, A. Tipilda, J. Njuki, I. Baltenweck, D. Grace, and S. MacMillan. 2010. *Livestock and Women's Livelihoods: A Review of the Recent Evidence.* Discussion Paper No. 20. Nairobi, Kenya: International Livestock Research Institute.

Land O' Lakes. 2008. *Kenya Dairy Development Program (KDDP).* Final report. Arden Hills, MN. Accessed June 2013. http://pdf.usaid.gov/pdf_docs/Pdacl801.pdf.

Lewis, S. E. 2011. *Failure Report 2011.* Toronto: Engineers without Borders Canada. http://legacy.ewb.ca/mainsite/pages/whoweare/accountable/FailureReport2012.pdf.

Lin, H., Q. Wan, X. Li, J. Chen, and Y. Kong. 1997. "GIS-Based Multicriteria Evaluation for Investment Environment." *Environment and Planning B: Planning and Design* 24 (3): 403–414.

Linton, J. D., S. T. Walshand, and J. Morabito. 2002. "Analysis, Ranking and Selection of R&D Projects in a Portfolio." *R&D Management* 32 (2): 139–148.

Lodin, J. B. 2012. "Intra-household Bargaining and Distributional Outcomes Regarding NERICA Upland Rice Proceeds in Hoima District, Uganda." *Gender, Technology and Development* 16 (3): 253–278.

MAFC (Ministry of Agriculture, Food Security and Cooperatives). 2010. *Agricultural Sector Development Program (ASDP) Performance Review (2006/07–2009/10).* Dar es Salaam, Tanzania.

Mangwezi, K. 2011. "The Role of 'the Canalero' in the Management of Irrigation Systems around the World." Thesis submitted in partial fulfillment of the requirements of Master of Science in International Land and Water Management at Wageningen University, the Netherlands.

Mansuri, G., and V. Rao. 2013. *Localizing Development: Does Participation Work?* Washington, DC: World Bank. DOI: 10.1596/978–0–8213–8256–1.

Manyame, C. 1998. "A Review of the Production Performances of Small Scale Irrigation Schemes and Their Role in Rural Development: Focus on the Fuve-Panganai Irrigation Schemes in Zaka District." Unpublished dissertation submitted in partial fulfillment of the requirements of Bachelor of Arts Honours Degree in Geography, Department of Geography, University of Zimbabwe, Harare.

Mapiye, C., M. Mwale, J. F. Mupangwa, M. Chimonyo, R. Foti, and M. J. Mutenje. 2008. "A Research Review of Village Chicken Production Constraints and Opportunities in Zimbabwe." *Asian-Australasian Journal of Animal Science* 21 (11).

Masika, T. 2011. "Community Impact Mission (CIM)." Unpublished report. Nairobi, Kenya.

Mazungu, E. 1999. "Strategies of Smallholder Irrigation Management in Zimbabwe." Unpublished dissertation submitted in partial fulfillment of the requirements of Doctor of Philosophy at Wageningen University, the Netherlands.

Mkonya, E., L. K. Gezehegn, M. Kilasara, and F. Kahimba. 2013. *Impact Evaluation of the Irrigation Investment of the Agricultural Sector Development Programme (ASDP).* Technical report. Dar es Salaam, Tanzania: Ministry of Agriculture and Cooperatives.

Mugova, A., and J. Mavunga. 2000. *Wei Integrated Development Project: A Success Story in Desertification Control.* Desertification Control Bulletin No. 36. UNEP Success Stories in Desertification Control—Award 1999. New York: United Nations Environment Programme. Accessed November 30, 2011. http://www.iapad.org/wei_wei_project/docs/wei_wei_project_unep_evaluation_report.pdf.

Mujenja, F., and C. Wonani. 2012. *Long-Term Outcomes of Agricultural Investments: Lessons from Zambia.* London: International Institute for Environment and Development.

Muller, R., and R. Turner. 2007. "The Influence of Project Managers on Project Success Criteria and Project Success by Type of Project." *European Management Journal* 25 (4): 298–307.

Munda, G., P. Nijkamp, and P. Rietveld. 1994. "Qualitative Multicriteria Evaluation for Environmental Management." *Ecological Economics* 10: 97–112.

Murgo, B. 2015. "Raising Millionaires in Yatta." *Smart Farmer Magazine.* Accessed April 8. http://smartfarmerkenya.com/page/article/133-raising-millionaires-in-yatta.

Mutinda, G. 2011. "Stepping Out in the Right Direction: Integrating Gender in EADD." Presentation for the Workshop on Gender and Market-Oriented Agriculture, Addis Ababa, Ethiopia, January 31– February 2.

———. 2013. "Gender in the East Africa Dairy Development Project." Presentation for the East Africa Dairy Development Project, Nairobi, Kenya.

Nelson, C., B. MkNelly, C. Garber, E. Edgcomb, N. Horn, G. Gaile, and K. Lippold. 2000. *Learning from Clients: Assessment Tools for Microfinance Practitioners.* Washington, DC: The Small Enterprise Education and Promotion Network.

NEPAD and FAO (New Partnership for Africa's Development and Food and Agriculture Organization of the United Nations). 2005. *Bankable Investment Profile, Human Resource Development for Extension.* Rome: FAO. ftp://ftp.fao.org/docrep/fao/008/ae667e/ae667e00.pdf.

NFSD (Novartis Foundation for Sustainable Development). 2005. *Project Management Handbook: A Working Tool for Project Managers.* Basel, Switzerland.

Noble, A. D., J. Prety, F. W. T. Penning de Vreis, and D. Bossio. 2005. "Development of Bright Spots in Africa: Cause for Optimism?" In *Bright Spots Demonstrate Community Successes in African Agriculture,* edited by F. W. T. Penning de Vries. Working Paper 102. Colombo, Sri Lanka: International Water Management Institute.

Nubukpo, K., and M. Galiba. 1999. "Agricultural Intensification in West Africa: Insights from Sasakawa Global 2000's Experience." Paper presented at the Workshop on Agricultural Transformation, sponsored by Tegemeo Institute, Eastern and Central Africa Programme for Agricultural Policy Analysis, Michigan State University, and United States Agency for International Development, Nairobi, Kenya, June 27–30.

OECD (Organisation for Economic Co-operation and Development). 2010. "Glossary of Key Terms in Evaluation and Results Based Management." www.oecd.org/dac/evaluation.

Ojango, J. M. K., C. Ahuya, A. M. Okeyo, and J. E. O. Rege. 2010. *The FARM-Africa Dairy Goat Improvement Project in Kenya: A Case Study.* Nairobi, Kenya: International Livestock Research Institute. http://agtr.ilri.cgiar.org/index.php?option=com_content&view=article&id=202&Itemid=239.

Okoboi, G., F. R. Muwanika, X. Mugisha, and M. Nyende. 2011. *Economic and Institutional Efficiency of the National Agricultural Advisory Services' Programme: The Case of Iganga District.* Economic and Social Research Foundation Research Series Number 76. Kampala, Uganda: National Institute of Economic and Social Research.

Ouma, R., L. Njoroge, D. Romney, P. Ochungo, S. Staal, and I. Baltenweck. 2007. *Targeting Dairy Interventions in Kenya: A Guide for Development Planners, Researchers and Extension Workers.* Nairobi, Kenya: Smallholder Dairy Project/Kenya Dairy Development Programme.

Patton, M. 1986. *Utilization-Focused Evaluation,* 2nd ed. Newbury Park, CA: SAGE Publications.

PCU (Projects Coordinating Unit). 2007. *Production and Area Cultivated by Smallholder Farmers.* Abuja, Nigeria: Federal Ministry of Agriculture and Rural Development.

Peacock, C. 2005. "Goats: Unlocking Their Potential for Africa's Farmers." Farm Africa Working Papers Series No. 2. Paper presented at the Seventh Conference of Ministers Responsible for Animal Resources, Kigali, Rwanda, October 31–November 4.

Pender, J., F. Place, and S. Ehui, eds. 2006. *Strategies for Sustainable Land Management in the East African Highlands.* Washington, DC: International Food Policy Research Institute.

Penning de Vries, F. W. T., ed. 2005. *Bright Spots Demonstrate Community Successes in African Agriculture.* Working Paper 102. Colombo, Sri Lanka: International Water Management Institute.

Peterman, A., A. Quisumbing, J. Behrman, and E. Nkonya. 2010. *Understanding Gender Differences in Agricultural Productivity in Uganda and Nigeria.* IFPRI Discussion Paper 01003. Washington, DC International Food Policy Research Institute. http://www.ifpri.cgiar.org/sites/default/files/publications/ifpridp01003.pdf.

Petticrew, M., and H. Roberts. 2006. *Systematic Reviews in the Social Sciences: A Practical Guide.* Nairobi, Kenya: Wiley-Blackwell Publishing Ltd.

Place, F., R. Roothaert, L. Maina, S. Franzel, J. Sinja, and J. Wanjiku. 2009. *The Impact of Fodder Trees on Milk Production and Income among Smallholder Dairy Farmers in East Africa and the Role of Research.* ICRAF Occasional Paper No. 12. Nairobi, Kenya: World Agroforestry Centre.

Poole, N., and A. de Frece. 2010. *A Review of Existing Organisational Forms of Smallholder Farmers' Associations and Their Contractual Relationships with Other Market Participants in the East and Southern African ACP Region.* AAACP Paper Series—No. 11. Rome: Food and Agriculture Organization of the United Nations. Accessed June 22, 2014. http://eprints.soas.ac.uk/8623/1/FAO_AAACP_Paper_Series_No_11[1].pdf.

Poulton, C., and A. Dorward. 2008. "Getting Agriculture Moving: Role of the State in Increasing Staple Food Crop Productivity with Special Reference to Coordination, Input Subsidies, and Credit and Price Stabilisation." Paper prepared for AGRA Policy Workshop, Nairobi, Kenya, June 23–25.

Pretty, J. 1995. *Regenerating Agriculture: Policies and Practice for Sustainability and Self-Reliance.* Washington, DC: National Academies Press; London: Joseph Henry Press.

Raju, K. S., and D. N. Kumar. 2005. "Fuzzy Multicriterion Decision Making in Irrigation Planning." *Irrigation and Drainage* 54 (4): 455–465.

RELMA (Regional Land Management Unit). 2005. *Potential for Rainwater Harvesting in Africa: A GIS Overview.* Nairobi, Kenya: World Agroforestry Centre and United Nations Environment Programme.

Rondinelli, D. A. 1986. *Development Management in Africa: Experience with Implementing Agricultural Development Projects*. Document Number PN-AAL–081. AID Evaluation Special Study No. 44. Washington, DC: United States Agency for International Development. http://pdf.usaid.gov/pdf_docs/PNAAL081.pdf.

Saaty, R. W. 1987. "The Analytical Hierarchy Process: What It Is and How It Is Used." *Mathematical Modelling* 9 (3–5): 161–176.

Sahee Foundation. 2008. *Sustainability of Rural Development Projects in Swaziland: Why Projects Fail or Succeed*. Davos, Switzerland. Accessed November 30, 2011. http://www.sahee.org/pdfs/Rural-Development-Swaziland08.pdf.

Seeley, J., R. Grellier, and T. Barnett. 2004. "Gender and HIV/AIDS Impact Mitigation in Sub-Saharan Africa—Recognising the Constraints." *SAHARA-J: Journal of Social Aspects of HIV/AIDS* 1 (2): 87–98.

SOAS (School of Oriental and African Studies), Wadonda Consult, Overseas Development Institute, and Michigan State University. 2008. *Evaluation of the 2006/7 Agricultural Input Supply Programme, Malawi: Final Report*. London: SOAS.

SOFA (State of Food and Agriculture) Team. 2011. *Gender Differences in Assets*. ESA Working Paper No. 11–12. Rome: Food and Agriculture Organization of the United Nations, Agricultural Development Economics Division. http://www.fao.org/publications/sofa/en/.

SOFA (State of Food and Agriculture) Team and C. Doss. 2011. *The Role of Women in Agriculture*. ESA Working Paper No. 11–02. Rome: Food and Agriculture Organization of the United Nations, Agricultural Development Economics Division. http://www.fao.org/publications/sofa/en/.

Spielman, D. J., D. Kelemwork, and D. Alemu. 2011. *Seed, Fertilizer, and Agricultural Extension in Ethiopia*. Ethiopian Strategy Support Program (ESSP II) Working Paper 20. Washington, DC: International Food Policy Research Institute.

Spielman, D., and R. Pandya-Lorch. 2009. *Million Fed: Proven Successes in Agricultural Development*. Washington, DC: International Food Policy Research Institute.

Tadelle, D., and B. Ogle. 2001. "Village Poultry Production Systems in the Central Highlands of Ethiopia." *Tropical Animal Health and Production* 33 (6): 521–537.

Talip, K., E. Whitney, and P. Winters. 2013. *Decentralized Beneficiary Targeting in Large-Scale Development Programs: Insights from the Malawi Farm Input Subsidy Program*. World Bank Policy Working Paper 6713. Washington, DC. http://papers.ssrn.com/sol3/papers.cfm?abstract_id=2360879.

TANGO International. 2009. *Sustainability of Rural Development Projects: Best Practices and Lessons Learned by IFAD in Asia*. The eighth in a series of discussion papers produced by the Asia and the Pacific Division, IFAD—Occasional Papers: Knowledge for Development Effectiveness. Rome: International Fund for Agricultural Development. http://www.ifad.org/operations/projects/regions/pi/paper/8.pdf.

———. 2010. East Africa Dairy Development Project—Mid-Term Evaluation. Nairobi, Kenya.

———. 2011. "East African Dairy Development (EADD) Project, Mid-Term Evaluation, Regional Synthesis Report (Draft)." Unpublished. Nairobi, Kenya.

Touwen, A. 2001. *Handbook for Projects: Development Management and Fundraising*. Geneva: International Federation of University Women. http://www.ifuw.org/training/pdf/project-handbook.pdf.

UNFFE (Uganda National Farmers Federation). 2011. *Farmers' Fireplace Conversations. Action Research to Audit the Effectiveness of NAADS and Proposals for Reform: The NAADS That Farmers Want*. Progress Report. Kampala, Uganda.

Uphoff, N. 1986. *Getting the Process Right: Improving Irrigation Water Management with Farmer Organization and Participation*. Working paper prepared for the Water Management Synthesis II Project, Consortium for International Development. Ithaca, NY: Cornell University.

URP (United Republic of Tanzania). 2011. *Controller and Auditor General 2011*. Dar es Salaam, Tanzania.

USAID (United States Agency for International Development). 2007. *Uganda Agricultural Productivity Enhancement Program (APEP), Fourth Annual Progress Report October 2006 to September 2007*. Kampala, Uganda. Accessed June 2014. http://pdf.usaid.gov/pdf_docs/Pdack800.pdf.

———. 2008. *Kenya Dairy Development Program (KDDP)*. Final Report. Nairobi, Kenya. http://pdf.usaid.gov/pdf_docs/PDACL801.pdf.

Voogd, H. 1982. "Multicriteria Evaluation with Mixed Qualitative and Quantitative Data." *Environment and Planning B: Planning and Design* 9 (2): 221–236.

Wambugu, C., S. Franzel, J. Cordero, and J. Stewart. 2006. *Fodder Shrubs for Dairy Farmers in East Africa: Making Extension Decisions and Putting Them into Practice*. Nairobi, Kenya: World Agroforestry Centre; Oxford, UK: Oxford Forestry Institute.

Wang, J., Y. Jing, and C. Zhang. 2009. "Weighting Methodologies in Multi-criteria Evaluations of Combined Heat and Power Systems." *International Journal of Energy Research* 33 (12): 1023–1039.

WARDA (West Africa Rice Development Association). 2001. *NERICA Rice for Life.* Technical report. Washington, DC: CGIAR. http://www.warda.cgiar.org/publications/NERICA8.pdf.

White, L. G. 1986. *Managing Development Programs: Management Strategies and Project Interventions in Six African Agricultural Projects.* AID Evaluation Special Study No. 38. Washington, DC: United States Agency for International Development. Accessed June 2014. http://pdf.usaid.gov/pdf_docs/PNAAL071.PDF.

World Bank. 1990. *Bura Irrigation Settlement Project.* Project Completion Report. Washington, DC.

———. 1994. *Kenya Animal Health Services Project* (IDA CREDIT 1758-KE; IFAD LOAN 188-KE; OPEC LOAN 407). Project Completion Report No. 13836. December 29. Washington, DC.

———. 2007. *Agricultural Water Development for Poverty Reduction in Eastern & Southern Africa: Main Report.* Washington, DC.

Zavadskas, E. K., Z. Turskis, J. Tamošaitiene, and V. Marina. 2008. "Multicriteria Selection of Project Managers by Applying Grey Criteria." *Ukio Technologinis ir Ekonominis Vystymas* 14 (4): 462–477.

Chapter 7

CONCLUSIONS AND IMPLICATIONS FOR RAISING AND SUSTAINING HIGH AGRICULTURAL PRODUCTIVITY IN AFRICA

Samuel Benin

After more than a decade since the Comprehensive Africa Agriculture Development Programme (CAADP) was launched in 2003, many African countries have begun to articulate an agricultural transformation or Green Revolution agenda. These two approaches—like previous agriculture-led development frameworks, priorities, and strategies—hinge on a fundamental issue: how to raise and maintain high agricultural productivity. With the majority of the population living in rural areas and depending on agriculture for their livelihoods, and with typical household sizes of five to eight family members that together farm only 1–3 hectares (ha) of land characterized by low agricultural productivity, it is easy to understand why rural poverty is so prevalent and persistent—and why raising agricultural productivity in a sustainable manner remains a fundamental development goal for Africa.

The recent high global food prices of 2007–2008 and later periods, which caused food crises in many African countries, have heightened the urgency of increasing food production and agricultural productivity. Because raising agricultural productivity can catalyze a wide range of different direct and indirect outcomes, the purpose for raising productivity matters. For example, it may seem that if raising rural incomes to reduce poverty is the issue, then raising labor productivity may be what matters most. But if providing cheap food for the urban poor is the issue, then raising yields may be what matters most. However, because many rural households in Africa are also net buyers of food, then yield increases that reduce the price of food could also raise the purchasing power of rural households and reduce rural poverty.

To help address these issues, the research studies presented in this book aimed at improving our understanding of the status of and trends in African agricultural productivity and its determinants. This chapter summarizes the main findings, with their implications for options for raising and maintaining high agricultural productivity across different parts of Africa.

Main Findings and Implications

Agricultural productivity in Africa has increased at a moderate rate over time, but there is variation in the rate of growth in land productivity, labor productivity, and total factor productivity (TFP), which vary in different parts of Africa.

Between 1961 and 2012, land productivity increased the fastest, at an annual average rate of 3.3 percent, followed by labor productivity at 2.0 percent, and then TFP at 0.7 percent (Figure 7.1). However, the growth paths of the three indicators over time are quite different. Whereas the rate of growth in land productivity followed an inverted U-shaped path, those of labor productivity and TFP followed U-shaped paths (Figure 7.1). Since the mid-1980s, growth in land productivity has slowed from an annual average high of 3.9 percent in the 1980s to 2.2 percent in the 2000s, which is similar to the average rate achieved in the 1960s. Growth in labor productivity and TFP, however, has been increasing since the mid-1980s, reaching 3.2 percent and 2.0 percent, respectively, in the last decade. The stagnation or decline in TFP observed prior to the mid-1980s was the result of loss of efficiency in agricultural production. From the mid-1980s onward, growth in both efficiency and technical change contributed positively and equally to TFP growth.

With the majority of Africa's poor population living in rural areas and depending on agriculture for their livelihoods, the rapid growth in agricultural productivity in the last decade (2001–2012), particularly labor productivity and TFP, is consistent with the World Bank (2012) finding of a decline in the poverty rate in the continent from the long-standing average of 50 percent to 47 percent in 2008. Therefore, sustaining the high growth in labor productivity and TFP will be critical for deepening the gains achieved in poverty reduction.

Because land expansion has accounted for the bulk of growth in food and agricultural output in the past, the decline in growth in land productivity (yields) is indicative of the slowdown in available fertile farmland. Therefore, intensifying the production process (that is, obtaining more output from the same amount of land) will be critical for reversing the declining growth in land productivity to help avert future food crises.

Policy improvements and complementary investments that accelerate the expansion of Africa's technical frontier and continue to improve efficiency in

FIGURE 7.1 Land, labor, and total factor productivity (TFP) growth and TFP growth decomposition in Africa (%, annual average 1961–2012)

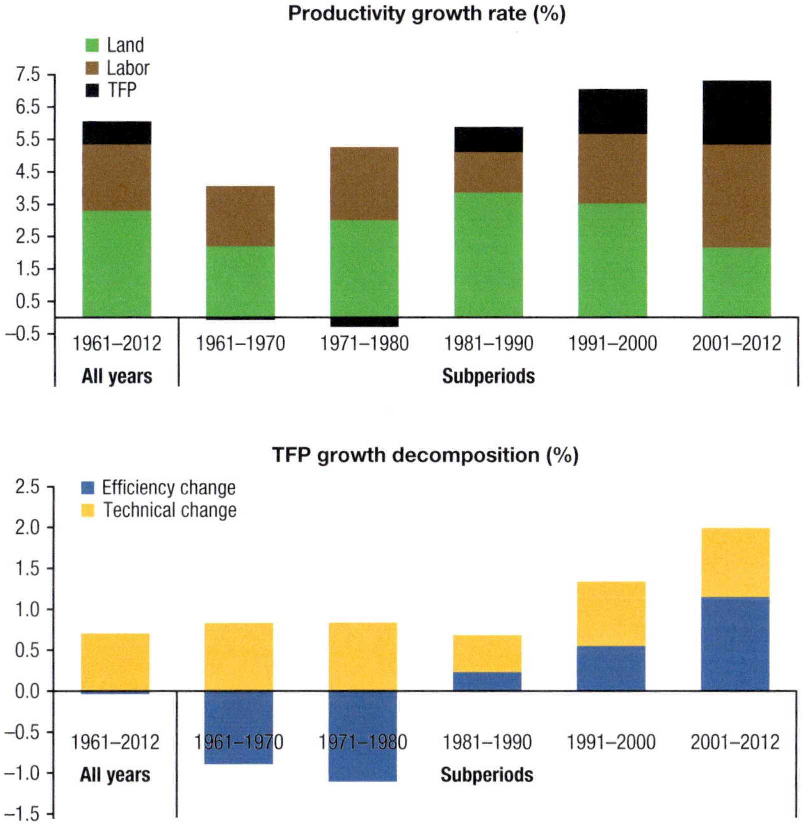

Source: Authors' illustration based on productivity model results in Chapter 2.

the production systems will be critical for increasing and sustaining high agricultural productivity growth. Although technical change has accounted for half of the growth in TFP in Africa, annual average technical change has not surpassed 0.8 percent in any decade (Figure 7.1), which lags far behind technical change rates achieved in other developing regions of the world.

Agricultural productivity growth trends vary in different parts of Africa, which is the result of differences in input use and capital intensities in agricultural production.

Although overall land productivity in Africa as a whole and in the majority of the subregions and countries increased faster than labor productivity, in leading agricultural economies, such as Egypt and South Africa, labor productivity increased faster than land productivity. This increase in labor productivity influenced the overall relative growth trends in land and labor productivity in southern Africa and in the Southern African Development Community (SADC) and Union du Maghreb Arabe (UMA) Regional Economic Communities (RECs) in 1961–2012.

Regarding TFP, growth was fastest in northern Africa and in the Common Market for Eastern and Southern Africa (COMESA) REC, at more than the 1.0 percent average for Africa as whole. TFP growth was slowest in western Africa, at 0.1 percent. Prior to the mid-1980s, the negative efficiency change observed in Africa as a whole was severest in central and western Africa and in the Economic Community of West African States and UMA RECs, averaging more than –1.4 percent per year.

During the periods of recovery and turnaround that started in the mid-1980s, growth in technical change contributed to more than 70 percent of TFP growth in northern Africa, both in the low-income mineral-rich countries economic group and in the EAC and COMESA RECs. However, during the same period, growth in technical change contributed to only 7 percent of TFP growth in the low-income countries with less favorable agricultural conditions. In the last decade of 2001–2012, when TFP grew the fastest, 6 countries (Benin, Sudan, Sierra Leone, Cameroon, Republic of Congo, and Kenya) had annual average TFP growth rates of at least 3 percent, 5 countries had 2–3 percent growth, 11 had 1–2 percent growth, 11 had less than 1 percent but positive growth rates, and the remaining 12 of the 45 countries studied had negative TFP growth rates.

The technological frontier for low-input, low-capital-intensive countries has been moving very slowly or not at all, compared with the frontier for the high-input, high-capital-intensive countries, where it has been moving steadily and faster. In particular, countries with a larger endowment of crop capital and using more fertilizer and feed per worker (likely also those with more commercial-oriented agriculture) are those that experienced rapid technical change.

There is substantial spatial variation in agricultural productivity growth trends, which emphasizes that blanket interventions for raising productivity are unnecessary and inefficient.

From the spatial analysis presented in Chapter 3, land productivity ranged from a low average of $240–$290[1]/ha in the agropastoral-millet/sorghum system of eastern Africa and the pastoral system of western Africa to a high average of $1,125/ha in the humid coastal systems of western Africa, where cash crops are widespread. The spatial variation in labor productivity was much greater: from $206/worker in the highland temperate mixed system of eastern Africa to $3,620/worker in the large commercial and smallholder systems in southern Africa, where large commercial operators are highly mechanized in comparison with the rest of Africa south of the Sahara. The results of the spatial analysis suggest that there is value in applying a consistent production systems framework across the continent for revealing and assessing the scope for potential regionwide technology generation and transfer.

Therefore, there is no escaping the significant role that the geographic context plays in conditioning both the baselines and the likely trajectory of productivity growth possibilities in different parts of the continent. Whereas this finding is not new, it emphasizes the need for targeting and formulating location-specific agricultural policies, investments, and interventions. It is no longer justifiable, it is unnecessary, and it is inefficient to formulate and implement blanket agricultural interventions that do not adequately account for important spatial variation in the production environments and constraints faced by different farmers.

Economic, market, sociodemographic, and environmental factors are important determinants of agricultural productivity.

The large spatial variation observed in African agricultural productivity performance, including its evolution over time, is the result of several factors related to population density, infrastructure and market access, policies, culture, social factors, environment, etc. To help improve targeting and formulation of location-specific, productivity-enhancing interventions, the typology analysis in Chapter 4 identified spatially identical production areas in terms of similarity in farming system, resource bases, and economic, market, and sociodemographic characteristics—called agricultural productivity zones (APZs).

A total of 543 APZs are identified across the 43 countries analyzed, with several APZs being common to several countries. The distribution of APZs varies substantially across the 43 countries analyzed, with Tanzania, Kenya, Nigeria, and Sudan exhibiting the most diversity (Figure 7.2). Each of these four countries has more than 24 APZs. In contrast, Djibouti, Equatorial

1 All currency is in US dollars, unless specifically noted as "international dollars."

FIGURE 7.2 **Distribution of agricultural productivity zones in Africa**

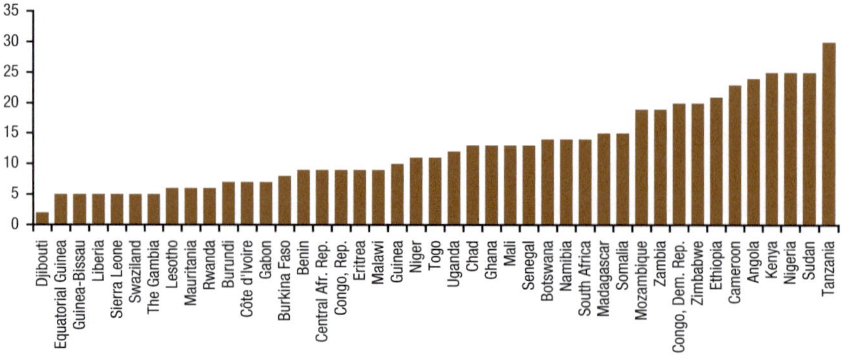

Source: Authors' illustration based on typology model results in Chapter 4.

Guinea, Guinea-Bissau, Liberia, Sierra Leone, Swaziland, and The Gambia exhibited the least diversity, with each having 5 or fewer APZs. Countries that have a common APZ can benefit also from each other's experience in overcoming binding constraints in the agriculture sector through cross-country learning and shared lessons.

Focusing on fertilizer within a specific APZ, Chapter 5 assessed the current patterns of agricultural intensification in Africa to address the broader question of identifying the best strategy for different APZs. Because fertilizer is a labor-intensive technology in Africa, there is a positive relationship between fertilizer use per hectare and population density, as expected. However, the statistical significance of the relationship is low, and there is high variability in the use of fertilizer at different levels of population density. At low population densities, for example, population pressure on land increases fertilizer use, but not necessarily fertilizer intensity (that is, the amount per hectare of land). As such, fertilizer is used as an instrument for land expansion, and not for yield increases. It is only at high-density levels that fertilizer use is correlated with increased yields. Given the high cost of fertilizer and other factors (for example, water) required to make its use profitable, wide use of the technology that brings about substantial yield increases may be limited, unless farmers are complemented by other investments that increase labor productivity.

For an African Green Revolution, some best-bet possibilities of success for using a fertilizer-focused technology package are in Ethiopia, Kenya, Uganda, and Malawi. These countries have more than 60 percent of potential agricultural land in favorable agroecologies and high population densities in those areas. For the bulk of the land under root crop, tree crop, perennial highlands, and forest-based systems, substantial investment in agricultural research and development (R&D) will be required to increase the output and productivity of root crops and tree crops in the most productive agroecologies, as international spillovers for these crops and ecologies are expected to be lower than those for cereals. Increasing output and productivity will also require opening new markets, especially for staple crops, such as cassava, which are nontradables and constrained to small domestic markets.

This finding raises an important question as to which agricultural productivity interventions have worked well and what the lessons are for scaling them up or for replicating them in other places. Aspects of this topic are addressed in Chapter 6, which identified determinants of successful project implementation based on 25 productivity-enhancing interventions in various APZs. The results show that successful implementation of interventions takes local conditions into account, internalizes the views of potential beneficiaries and involves them in all phases of implementation, and addresses environmental issues.

Unfortunately, many of the projects initially intended for evaluation in the study had to be dropped because of lack of adequate information on project formulation, implementation, and performance. Furthermore, the evaluation conducted offers insights about implementation—an important measure along the impact pathway—rather than inferences of outcomes and impact, which require more rigorous qualitative or quantitative project evaluation measures. As a result, there are few projects (successful or failed) to draw concrete lessons on implementation from for all the many different local production environments. Achieving widespread productivity growth in Africa, however, requires investments at larger scales than the small projects that seem to dominate require. There is also need for more commitments and actions by national governments and national stakeholders to ensure that good interventions are sustained. Most of the seemingly successful interventions lasted only for the duration of the project.

Implications for Raising and Sustaining High Agricultural Productivity

Raising and sustaining high levels and growth rates of agricultural productivity in Africa above those achieved in recent times, or at levels and rates consistent with the CAADP agricultural growth rate target of 6 percent year, face fundamental challenges of rapid population growth and slowdown in land availability on one hand, and reversing the underinvestment in the agriculture sector and rural areas (among others) on the other hand.

The analysis in Chapter 5 suggests that the rapid growth in labor productivity achieved in many countries in recent years has depended on their ability to incorporate more land into agricultural production. As such, more rapid increases in labor productivity are essential to compensate for rapid population growth and to improve rural incomes. This will require accelerating the expansion of Africa's technical frontier through a combination of policy improvements and significant investments in agricultural R&D, together with complementary investments in such areas as irrigation, market infrastructure, and institutions (Diao, Headey, and Johnson 2008; von Braun et al. 2008; Diao et al. 2012; Mogues and Benin 2012).

Large incremental expenditure and investment in agriculture will be required to raise and maintain a high level of agricultural productivity and growth in Africa.

Agricultural research infrastructure and capacities in Africa have been eroded through years of neglect, primarily because of lack of public funding for agricultural R&D. Only recently has growth in spending on agricultural R&D and the number of researchers picked up (Chapter 1).

Considering agricultural spending and investments in general, for example, the 2003 Maputo Declaration set a target for agricultural financing by governments at 10 percent of total national expenditures. Whereas several countries have increased the share of total spending allocated to the agriculture sector, only 13 countries (Burundi, Burkina Faso, Republic of the Congo, Ethiopia, Ghana, Guinea, Madagascar, Malawi, Mali, Niger, Senegal, Zambia, and Zimbabwe) have surpassed the target in any single year since 2003 when the declaration was made (Benin and Yu 2013). Most of the large African agricultural economies spent less than 5 percent of their total national budgets on agriculture, resulting in the low performance for Africa as a whole.

Regarding agricultural R&D, the New Partnership for Africa's Development has set a national agricultural R&D investment target of at least 1.0 percent of agricultural gross domestic product (GDP). Most countries have spent far less than this level. In 2008, for example, the amount spent on agricultural R&D as a percentage of agricultural GDP is estimated at about 0.6 percent, with only 8 countries (out of 31 countries studied) meeting the 1.0 percent target (Beintema and Stads 2011).

What is the magnitude of investment required to raise and maintain a high level of agricultural productivity and growth—for example, to attain the CAADP target of 6 percent annual average growth in agricultural GDP? Whereas the response to this question depends on the efficiency and effectiveness of investments, as well as on the desired development objective, the results of the economic and economywide modeling used in CAADP planning indicate that although it is possible for many African countries to reach the 6 percent target, they will require substantial additional growth across different key subsectors and commodities to do so. This in turn will require substantial additional investments to stimulate the necessary acceleration in growth in the key subsectors (Diao et al. 2012). In the majority of the cases, the additional investments required are in excess of the 10 percent of total expenditures commitment agreed to under the Maputo Declaration or in the agriculture budgets that are articulated in national agricultural investment plans (NAIPs).

The types of agricultural investments and policies are important because they have different effects; those that deliver location-specific technologies and that account for the diversity of farmers will be critical.

Because different policies and types of investments have different effects on growth and other development outcomes, the right focus has to be found for different contexts. The studies by Fan (2008) and Mogues and Benin (2012), as well as several earlier studies, show that different types of spending across different geographic areas deliver substantially different returns and impacts on different development objectives. Moreover, the returns and impacts vary over time, suggesting that prioritization and proper sequencing of policies and investments are essential if the policies and investments are to be effective.

With substantial heterogeneity in the production environment (Chapters 3 and 4), investments and policy interventions need to deliver location-specific technologies that are tailored to the relevant agroecological characteristics and production systems, and that account for the considerable diversity of potentials and constraints faced by farmers (Chapters 3, 4, and 5).

The case studies of several agricultural productivity investment projects examined in Chapter 6 suggest that there are successful projects that are short lived (three to five years), as well as thinly scattered across the continent. These have not been successfully scaled up and out. Tackling the issue of sustaining success is an aspect that cannot be overemphasized. There is a need for more commitments and actions by national governments and other stakeholders to ensure that any intervention is well documented and that good interventions are sustained.

Because many African countries have small economies and limited capacities, regional agricultural strategies will be helpful, emphasizing complementary policies and extension systems that maximize the spillovers of technologies.

Many countries in Africa have small economies and thus limited capacities and resources for developing effective agricultural R&D systems. Focusing on regional agricultural R&D strategies can help fill these gaps and facilitate economies of scale.

Studies, such as those carried out by Omamo et al. (2006), Nin-Pratt et al. (2011), and Johnson et al. (2014), shed light on the potential gains from implementing such regional agricultural R&D strategies. Regarding the SADC REC, for example, Johnson et al. (2014) show that the returns to agricultural R&D in the region differ by the country of origin of the technologies, as well as by commodities. The assumptions of the study by Johnson et al. (2014), particularly those underlying the probabilities of successful spill-outs and spill-ins of agricultural R&D, highlight areas of the policy front that are important to enhance and maximize the benefits of cross-border cooperation in agricultural R&D.

These ideas—cross-border collaboration, and enhancement of regional knowledge and technology spillover—are not new. Indeed, they constitute the fundamental rationale for regional economic institutions and agricultural research organizations. Nevertheless, they deserve re-emphasis to ensure that the core roles and responsibilities of cross-border institutions are persistently reaffirmed and acted upon. Cross-border institutions are more than platforms for the statement of national interests; rather, they present real opportunities to add value that national entities otherwise could not—opportunities that could serve to further enhance and accelerate national ambitions for productivity growth.

Of course, a regional strategy must overcome many institutional and administrative barriers to management and coordination across national boundaries, which can lead to high transaction costs, especially given the large spatial variation in the production environment in the continent, as well as different levels of development of national R&D systems and political economies. Inevitably, any cross-country collaboration will be affected by each country's own program needs, as well as the desire to maintain a bargaining position for domestic resources. Looking for ways to minimize the transaction costs will be critical. That is why the African centers of excellence initiatives are laudable.

Notable recent efforts are two large subregional programs, the Eastern Africa Agricultural Productivity Program and the West Africa Agricultural Productivity Program.[2] In turn, these two programs are funding subregional centers of excellence for particular crops and commodities—maize and wheat in Ethiopia, dairy in Kenya, cassava in Uganda, roots and tubers in Ghana, and rice in Mali and Tanzania, to mention a few.[3] To be successful, these initiatives will require complementary policies and agricultural extension systems that enhance and maximize the spillovers of the targeted technologies to different parts of Africa.

The potential impacts of climate change should be taken into account in the design and implementation of policies and strategies for raising and maintaining high agricultural productivity.

One of the key challenges faced by CAADP is how to deal with emerging issues related to climate change. The studies reviewed in Chapter 3 (Seo et al. 2008; Nelson et al. 2010) provide strong evidence that climate change could impose serious costs for agricultural growth, and that the change is likely to have different effects in different locations.

The studies show that farms in the savanna areas appear to be most vulnerable to higher temperatures and reduced precipitation, whereas those in the subhumid or humid forest could gain even from a severe climate change. Similarly, households in the cereal-root crop mixed, dryland mixed, agropastoral, and pastoral farming systems (common to the savanna AEZs) are likely to be the most vulnerable to climate change. However, because global warming is likely to increase livestock income while reducing crop income, climate change may have a zero net effect on the total agricultural income of households engaging in both crop and livestock production in these systems, depending on the relative importance of the two subsectors in their livelihoods.

Such analyses have yet to be internalized in the CAADP-country NAIPs, which may have budgetary implications for achieving the stated development objectives. Therefore, the strategies for raising and maintaining high

---

2 The Eastern Africa Agricultural Productivity Program is implemented by the Association for Strengthening Agricultural Research in eastern and central Africa. The West Africa Agricultural Productivity Program is implemented by Conseil Ouest et Centre Africain pour la Recherche et le Développement Agricoles/West and Central African Council for Agricultural Research for Development. Both programs were developed with assistance from the World Bank.

3 See http://waapp.org.gh/ and http://www.eaapp.org/ for details.

agricultural productivity should also be based on impact assessments of climate change, to identify the most attractive adaptation options and to develop location-specific implementation approaches.

Overall Policy Implications

For most countries in Africa, especially those with large rural populations and rapid urbanization, there is no more pressing development objective than raising the level and rate of growth of agricultural productivity. Because the majority of Africa's poor lives in rural areas and depends on agriculture for their livelihoods, raising labor productivity—which in turn raises rural incomes—is a key strategy to reduce rural poverty.

To avert a future food crisis of the nature associated with the recent global high food prices, raising land productivity (yields) is equally important. To reverse the declining growth in land productivity, to sustain or strengthen the recent rapid growth in labor productivity, and to expand the technological frontier continuously, the core of a sustainable development strategy for Africa must be to make full use of regional and subregional alliances capable of promoting and disseminating well-designed and appropriately targeted technological innovations in agriculture.

References

Beintema, N. M., and G. J. Stads. 2011. *African Agricultural R&D in the New Millennium: Progress for Some, Challenges for Many.* Washington, DC: International Food Policy Research Institute.

Benin, S., and B. Yu. 2013. *Complying with the Maputo Declaration Target: Trends in Public Agricultural Expenditures and Implications for Pursuit of Optimal Allocation of Public Agricultural Spending.* ReSAKSS Annual Trends and Outlook Report 2012. Washington, DC: International Food Policy Research Institute.

Diao, X., D. Headey, and M. Johnson. 2008. "Toward a Green Revolution in Africa: What Would It Achieve, and What Would It Require?" *Agricultural Economics* 39 (1): 539–550.

Diao, X., J. Thurlow, S. Benin, and S. Fan, eds. 2012. *Strategies and Priorities for African Agriculture: Economywide Perspectives from Country Studies.* IFPRI Issue Brief 73. Washington, DC: International Food Policy Research Institute.

Fan, S., ed. 2008. *Public Expenditures, Growth, and Poverty: Lessons from Developing Countries.* Baltimore: Johns Hopkins University Press.

Johnson, M., S. Benin, L. You, X. Diao, P. Chilonda, and A. Kennedy. 2014. *Exploring Strategic Priorities for Regional Agricultural Research and Development Investments in Southern Africa*. IFPRI Discussion Paper 1318. Washington, DC: International Food Policy Research Institute.

Mogues, T., and S. Benin, eds. 2012. *Public Expenditures for Agricultural and Rural Development in Africa*. London: Routledge.

Nelson, G. C., M. W. Rosegrant, A. Palazzo, I. Gray, C. Ingersoll, R. Robertson, S. Tokgoz, et al. 2010. *Food Security, Farming, and Climate Change to 2050: Scenarios, Results, and Policy Options*. Washington, DC: International Food Policy Research Institute.

Nin-Pratt, A., M. Johnson, E. Magalhaes, L. You, X. Diao, and J. Chamberlin. 2011. *Yield Gaps and Potential Agricultural Growth in West and Central Africa*. Washington, DC: International Food Policy Research Institute.

Omamo, S. W., X. Diao, S. Wood, J. Chamberlain, L. You, S. Benin, U. Wood-Sichra, and A. Tatwangire. 2006. *Strategic Priorities for Agricultural Development in Eastern and Central Africa*. IFPRI Research Report 150. Washington, DC: International Food Policy Research Institute.

Seo, S. N., R. Mendelsohn, A. Dinar, R. Hassan, and P. Kurukulasuriya. 2008. *A Ricardian Analysis of the Distribution of Climate Change Impacts on Agriculture across Agro-Ecological Zones in Africa*. Policy Research Working Paper 4599. Washington, DC: World Bank.

von Braun, J., S. Fan, R. Meinzen-Dick, M. W. Rosegrant, and A. Nin-Pratt. 2008. *International Agricultural Research for Food Security, Poverty Reduction, and the Environment: What to Expect from Scaling Up CGIAR Investments and "Best Bet" Programs*. Washington, DC: International Food Policy Research Institute.

World Bank. 2012. *The State of the Poor: Where Are the Poor and Where Are They Poorest?* Poverty Reduction and Equity Department Note. Washington, DC.

Authors

Samuel Benin (s.benin@cgiar.org) is a research fellow in the Development Strategy and Governance Division at the International Food Policy Research Institute (IFPRI). He received a bachelor's degree from the University of Ghana, Legon; a master's from the University of Massachusetts, Amherst; and a PhD with specialty in econometrics, natural resource economics, and development economics from the University of California, Davis. Benin's research focuses on the impact of public spending and on strategies that can better connect public investments and policies with development goals. His recent publications include "Public Investments in and for Agriculture," coauthored with T. Mogues and S. Fan (*European Journal of Development Research*, 2015); *Strategies and Priorities for African Agriculture: Economywide Perspectives from Country Studies*, coedited with X. Diao, J. Thurlow, and S. Fan (IFPRI, 2012); and *Public Expenditures for Agricultural and Rural Development in Africa*, coedited with T. Mogues (Routledge, 2011).

Pius Chilonda is an independent consultant based in Lusaka, Zambia. At the time of this publication, he was head of the southern Africa office of the International Water Management Institute (IWMI) in Pretoria, South Africa. Chilonda joined IWMI in 2006 as a researcher to start the Regional Strategic Analysis and Knowledge Support System (ReSAKSS) project for southern Africa, and was promoted as head of office in 2010. Prior to joining IWMI, Chilonda worked as a livestock information analyst at the Food and Agriculture Organization of the United Nations (FAO). He holds a PhD in agricultural economics from the University of Ghent, Belgium. His publications include *Exploring Strategic Priorities for Regional Agricultural Research* *and Development Investments in Southern Africa*, coauthored with M. Johnson, S. Benin, L. You, X. Diao, and A. Kennedy (IFPRI, 2014); and *Agricultural Growth Trends and Outlook for Southern Africa: Inter-temporal Trends and Patterns in Agricultural Investment Spending in Southern Africa*, coauthored with G. Matchaya and S. Nhlengethwa (IFPRI and IWMI, 2012).

Zhe Guo (z.guo@cgiar.org) is a senior geographic information system (GIS) coordinator with IFPRI's Environment and Production Technology Division. He holds a bachelor's degree in plant nutrition and soil sciences from China Agriculture University, a master's in natural resource science from the Chinese Academy of Sciences, and a master's in geography from the University of Maryland, College Park. His research areas of interest include spatial modeling, spatial statistics, data mining, and remote sensing and land classification. Guo has worked on multiple projects funded by the Bill & Melinda Gates Foundation (BMGF), the United States Agency for International Development (USAID), and the World Bank. His publications include "What Is the Irrigation Potential for Africa? A Combined Biophysical and Socioeconomic Approach," coauthored with L. You, C. Ringler, U. Wood-Sichra, R. Robertson, S. Wood, T. Zhu, G. Nelson, and Y. Sun (*Food Policy*, 2011); and "Fertilizer Profitability in East Africa: A Spatially Explicit Policy Analysis," coauthored with J. Koo and S. Wood (International Association of Agricultural Economists, 2009).

Paul Guthiga (p.guthiga@cgiar.org) is a senior policy analyst at the International Livestock Research Institute (ILRI) in Nairobi. Before joining ILRI in 2010, he was a lecturer at the University of Nairobi and a postdoctoral fellow at the Kenya Institute for Public Policy Research and Analysis (KIPPRA). Guthiga holds a PhD in agricultural economics from the University of Bonn, Germany. His recent publications include "Assessment of Adoption and Impact of Rainwater Harvesting Technologies on Rural Farm Household Income: The Case of Rainwater Harvesting Ponds in Rwanda," coauthored with J. J. Okello and A. Zingiro (*Environment, Development and Sustainability,* 2014); "REDD+ and Community-Controlled Forests in Low-Income Countries: Any Hope for a Linkage?" coauthored with R. Bluffstone and E. Robinson (*Ecological Economics*, 2013); and "Averting Biodiversity Collapse in Tropical Protected Areas," coauthored with W. F. Laurance, D. C. Useche, J. Rendeiro, et al. (*Nature*, 2012).

Joseph Karugia (j.karugia@cgiar.org) is the coordinator of ReSAKSS for eastern and central Africa (ReSAKSS-ECA), based at ILRI. He holds bachelor's and master's degrees from the University of Nairobi, and a PhD in agricultural economics from the University of Alberta, Canada. Karugia has taught at the University of Nairobi for more than 20 years and served as chairman of the Department of Agricultural Economics in 2002–2003. Before joining ILRI, he served as the research manager at the African Economic Research Consortium. His publications include *The Role of Livestock in the Kenyan Economy: Policy Analysis Using a Dynamic Computable General Equilibrium Model for Kenya,* coauthored with E. Engida, P. Guthiga, and H. Nyota (IFPRI, 2015); *Agricultural Productivity in the COMESA, EAC and IGAD: Status, Trends and Drivers,* coauthored with S. Massawe, P. Guthiga, M. Ogada, and J. Wambua (IFPRI, 2011); and *The Impact of Non-tariff Barriers on Maize and Beef Trade in East Africa,* coauthored with J. Wanjiku, J. Nzuma, S. Gbegbelegbe, E. Macharia, S. Massawe, A. Freeman, M. Waithaka, S. Kaitibie, and A. Gulan (Alliance for a Green Revolution in Africa and ILRI, 2011).

Stella Clara Massawe (s.massawe@cgiar.org) is a monitoring and evaluation (M&E) analyst with ReSAKSS-ECA. Her work involves strategic analysis, agriculture M&E, knowledge management, and provision of technical support to the implementation of the Comprehensive Africa Agriculture Development Programme (CAADP) to ECA countries. Before joining ReSAKSS, Massawe worked as a GIS analyst in ILRI's Targeting Theme. Her research areas of interest include M&E in international development; use of spatial information systems in informing agricultural policy, programming, and M&E; strategic planning and management in international development; and capacity building in various areas related to development M&E. Her recent publications include *Investment Opportunities for Livestock in the North Eastern Province of Kenya: A Synthesis of Existing Knowledge*, coauthored with M. Rakotoarisoa, A. Mude, R. Ouma, A. Freeman, G. Bahiigwa, and J. Karugia (IFPRI, 2015); and "Impact of Cross-Border Trade in Food Staples on Child Nutrition in East Africa," coauthored with P. Guthiga, M. Ogada, J. Ogutu, and J. Karugia (29th Triennial Conference of the International Conference of Agricultural Economists, 2015).

Emmanuel Chibanda Musaba (emusaba51@gmail.com) is a senior lecturer in agricultural economics and agribusiness at Mulungushi University in Kabwe, Zambia. Formerly, he was a coordinator of ReSAKSS-Southern Africa, and an agricultural economics researcher at IWMI. He also served as lecturer in agricultural economics and chaired the departments of Agricultural Economics at the University of Zambia and University of Namibia. Musaba holds a bachelor's degree from the University of Zambia; a master's degree in agricultural economics and business from the University of Guelph, Canada; and a PhD in agricultural economics from the University of Saskatchewan, Canada. He has coauthored several *Agricultural Sector Performance Annual Trends and Outlook Reports* for ReSAKSS-Southern Africa (IFPRI and IWMI).

Manson Nwafor (m.nwafor@cgiar.org) is a policy analyst with ReSAKSS-West Africa, based at the International Institute of Tropical Agriculture (IITA). He holds a PhD in economics from the University of Nigeria, Nsukka. Prior to joining IITA, Nwafor worked as an economist with the USAID Budget Support Project in Nigeria, where he engaged in capacity building of public-sector officials in the area of budget formulation. In IITA, he has supported West African countries with modeling, helping to formulate national agriculture investment plans. Nwafor currently supports West African countries in building Strategic Analysis and Knowledge Support Systems. His research interests include analysis of the micro impacts of macroeconomic policy, agricultural growth modeling, trade liberalization, and Computable General Equilibrium Modeling. His publications include *The Impacts of Trade Liberalization on Poverty in Nigeria: Dynamic Simulations in a CGE Model,* coauthored with A. Adenikinju and K. Ogujiuba (Poverty and Economic Policy Research Network, 2005); *Does Subsidy Removal Hurt the Poor: A CGE–Microsimulation Analysis,* coauthored with K. Ogujiuba and R. Asogwa (Secretariat for Institutional Support for Economic Research in Africa, 2006); and *A 2006 Social Accounting Matrix for Nigeria: Methodology and Results,* coauthored with X. Diao and V. Alpuerto (IFPRI, 2010).

Alejandro Nin-Pratt (a.ninpratt@cgiar.org) is a senior research fellow in IFPRI's Environment and Production Technology Division. He joined IFPRI in 2005 as a research fellow in the Development Strategy and Governance Division. He holds a bachelor's in agronomy, a master's in international economics from the Universidad de la República in Uruguay, and a PhD in agricultural economics from Purdue University. After working as a postdoctoral fellow in Purdue's Agricultural Economic Department, Nin-Pratt worked for ILRI in Ethiopia and Kenya. His publications include "Agricultural Intensification in Ghana: Evaluating the Optimist's Case for a Green Revolution," coauthored with L. McBride (*Food*

Policy, 2014); "Reducing the Environmental Efficiency Gap in Global Livestock Production" (*American Journal of Agricultural Economics*, 2013); and *Policy Changes and the Recovery of Agricultural TFP in Sub-Saharan Africa* (in K. O. Fuglie, S. L. Wang, and V. E. Ball, eds., *Productivity Growth in Agriculture: An International Perspective*, CABI, 2012).

Maurice Juma Ogada (ogadajuma@yahoo.co.uk) holds a bachelor's in economics and geography from Egerton University, Kenya; a master's in economic policy management from Makerere University, Uganda; and a PhD in agricultural economics from Kenyatta University, Kenya. An experienced impact evaluator and capacity builder, Ogada is currently an independent research consultant in the areas of agriculture, natural resource management, trade, and climate change. Previously, he worked as a policy researcher at the ILRI ReSAKSS-ECA platform, and as a policy analyst with KIPPRA's Productive Sector Division. His publications include "Forest Management Decentralization in Kenya: Effects on Household Farm Forestry in Kakamega," coauthored with W. Nyangena and G. Sikei (in S. T. Holden, K. Otsuka, and K. Deininger, eds., *Land Tenure Reforms in Asia and Africa: Assessing Impacts on Poverty and Natural Resource Management*, Palgrave Macmillan, 2013); "Technical Efficiency of Kenya's Smallholder Food Crop Farmers: Do Environmental Factors Matter?" coauthored with M. J. Ogada, D. Muchai, G. Mwabu, and M. Mathenge (*Environment, Development and Sustainability*, 2014); and "Farm Technology Adoption in Kenya: A Simultaneous Estimation of Inorganic Fertilizer and Improved Maize Variety Adoption Decisions," coauthored with G. Mwabu and D. Muchai (*Agricultural and Food Economics*, 2014).

Stanley Wood (stanley.wood@gatesfoundation.org) is a senior program officer in the BMGF Agricultural Development program. He received a master's in water resources technology from Birmingham University, and a master's in agricultural development and PhD in agricultural economics from the University of London. Wood began his career as a hydrologist and a water resource modeler in the United Kingdom, and then progressed to data, information systems, and model development in agriculture and natural resources as a consultant based in Indonesia, Italy, Colombia, and Libya. Before joining BMGF, he was a senior research fellow at IFPRI, where he led efforts to improve the spatially explicit dimensions of productivity and policy analysis research. His publications include "Land Quality, Agricultural Productivity and Food Security: A Spatial Perspective," coauthored with K. Sebastian and J. Chamberlin (in K. D. Wiebe, ed., *Land Quality, Agricultural Productivity and Food Security: Biophysical Processes and Economic Choices at Local, Regional, and Global Levels,* Edward Elgar, 2003); "Drivers of Change in Global Agriculture," coauthored with P. Hazell (*Philosophical Transactions of the Royal Society B [Biological Sciences]*, 2008); and *Pilot Analysis of Global Ecosystems: Agroecosystems,* coauthored with S. Scherr, K. Sebastian, and N. Batjes (World Resources Institute and IFPRI, 2000).

Ulrike Wood-Sichra (u.wood-sichra@cgiar.org) is a senior research analyst in IFPRI's Environment and Production Technology Division. Her work includes updating and running the Spatial Production Allocation Model, assembling crop production data globally, maintaining and operating the DREAM (Dynamic Research Evaluation for Management) software, feeding the HarvestChoice website, and linking model results with GIS applications. Before joining IFPRI, Wood-Sichra was a freelance consultant with FAO and with a number of projects funded by USAID, the Asian Development Bank, and the World Bank. She holds a master's degree from the Technical University in Vienna, Austria. Her publications include "Mapping Global Cropland and Field Size," coauthored with S. Fritz, L. See, I. McCallum, L. You, et al. (*Global Change Biology*, 2015); several chapters in *Atlas of African Agriculture Research and Development* (K. Sebastian, ed., IFPRI, 2014); and "Generating Global Crop Distribution Maps," coauthored with L. You, Z. Guo, and L. Wang (*Agricultural Systems*, 2014).

Bingxin Yu (byuifpri@gmail.com) was a research fellow in IFPRI's Development Strategy and Governance Division. She holds a bachelor's in management science from the University of Science and Technology of China in Hefei, China, and a master's in biometry/statistics and PhD in agricultural economics from the University of Nebraska-Lincoln. Yu has worked on productivity, technology adoption, public expenditure, and other development-related issues. Her publications include "Trends and Composition of Public Expenditures: A Global and Regional Perspective," coauthored with S. Fan and E. Magalhaes (*European Journal of Development Research*, 2015); "Parametric Decomposition of the Malmquist Index in Output-Oriented Distance Function: Productivity in Chinese Agriculture," coauthored with X. Liao and H. Sheng (*Modern Economy*, 2014); and "A Typology of Food Security in Developing Countries," coauthored with Y. Liang (*China Agricultural Economics Review*, 2008).

Index

Made in the USA
Las Vegas, NV
03 November 2026

57423306R00231